Lecture Notes in Networks and Systems

Volume 804

The series "Lecture Notes in Networks and Systems" publishes the latest developments in Networks and Systems—quickly, informally and with high quality. Original research reported in proceedings and post-proceedings represents the core of LNNS.

Volumes published in LNNS embrace all aspects and subfields of, as well as new challenges in, Networks and Systems.

The series contains proceedings and edited volumes in systems and networks, spanning the areas of Cyber-Physical Systems, Autonomous Systems, Sensor Networks, Control Systems, Energy Systems, Automotive Systems, Biological Systems, Vehicular Networking and Connected Vehicles, Aerospace Systems, Automation, Manufacturing, Smart Grids, Nonlinear Systems, Power Systems, Robotics, Social Systems, Economic Systems and other. Of particular value to both the contributors and the readership are the short publication timeframe and the world-wide distribution and exposure which enable both a wide and rapid dissemination of research output.

The series covers the theory, applications, and perspectives on the state of the art and future developments relevant to systems and networks, decision making, control, complex processes and related areas, as embedded in the fields of interdisciplinary and applied sciences, engineering, computer science, physics, economics, social, and life sciences, as well as the paradigms and methodologies behind them.

Indexed by SCOPUS, INSPEC, WTI Frankfurt eG, zbMATH, SCImago.

All books published in the series are submitted for consideration in Web of Science.

For proposals from Asia please contact Aninda Bose (aninda.bose@springer.com).

Maja Arslanagić-Kalajdžić · Naida Ademović ·
Tijana Tufek-Memišević
Editors

Interdisciplinary Advances in Sustainable Development II

Proceedings of the BHAAAS International Symposium on Sustainable Urban Development 2023

Springer

Editors
Maja Arslanagić-Kalajdžić
School of Economics and Business
University of Sarajevo
Sarajevo, Bosnia and Herzegovina

Naida Ademović
Faculty of Civil Engineering
University of Sarajevo
Sarajevo, Bosnia and Herzegovina

Tijana Tufek-Memišević
CANDARC LLC
Arlington Heights, IL, USA

ISSN 2367-3370 ISSN 2367-3389 (electronic)
Lecture Notes in Networks and Systems
ISBN 978-3-031-46691-5 ISBN 978-3-031-46692-2 (eBook)
https://doi.org/10.1007/978-3-031-46692-2

This Springer imprint is published by the registered company Springer Nature Switzerland AG
The registered company address is: Gewerbestrasse 11, 6330 Cham, Switzerland

Paper in this product is recyclable.

Preface

The book you are about to read is a result of the International Symposium on Sustainable Urban Development 2023, which took place in Tuzla, Bosnia and Herzegovina from 1 to 3 June 2023, as part of the 14th Days of BHAAAS Conference. Throughout this symposium, professionals in various fields, researchers, academics, scientists, students and engineers gathered to share their findings and contribute to the advancement of sustainable development over 3 inspiring days. The symposium with its interdisciplinary approach was supported by the Federal Ministry of Spatial Planning of Bosnia and Herzegovina, the Regional Cooperation Council for Southeast Europe (RCC), The University of Sarajevo and other partners.

The symposium commenced with a special session on disaster management and risk reduction, organized in collaboration with the RCC for South East Europe. Keynote speakers and panelists included representatives from Japan International Cooperation Agency (JICA), UN Office for Disaster Risk Reduction (UNDRR), European Civil Protection and Humanitarian Aid Operations (ECHO) and others. A keynote lecture was delivered by Mr. Satoru Nishikawa, Professor at Nagoya University, presenting good practices and strategies for disaster preparedness. His lecture provided valuable insights that the Southeast European region can consider implementing based on Japan's example, particularly in terms of earthquakes. The keynote was followed by a presentation by Doc. dr Burchan Sakarya, from Turkiye Presidency Strategy and Budget Office, discussing available methodologies for calculating the ex-post cost of disasters, with a focus on the recent devastating earthquakes in Turkiye and Syria. Additionally, an overview of earthquake risk assessment in Bosnia and Herzegovina was presented by Prof. dr Naida Ademovic, Faculty of Civil Engineering at the University of Sarajevo. Her presentation provided an understanding of the specific challenges and approaches to assessing earthquake risks in the Western Balkans region.

The session then transitioned into a panel discussion that brought together several experts in the field coming from Europa RE, UNDRR and DPPI SEE. The panelists addressed various topics, including regional financial risk sharing in South East Europe (SEE), principles of humanitarian aid implementation in disaster-stricken regions, the implementation of the Sendai Framework and Sustainable Development Goals in disaster management, and regional cooperation among civil protection agencies in South East Europe. This panel discussion offered diverse perspectives and insights into the regional cooperation efforts and challenges related to disaster management and risk reduction. Engaging discussions and a deep exploration of regional cooperation efforts that shape disaster management practices were key highlights of this session.

The next chapters of this book cover a wide range of topics that illuminate various facets of sustainable urban development. Researchers delve into diverse subjects, such as improving spatial adaptability in small inflexible apartments, conducting critical assessments of industrial heritage recording processes or understanding the ethical implications of neuromarketing in sustainable development. This collection showcases the rich diversity of research and findings that contribute to our collective pursuit of sustainable urban environments.

Two additional special sessions were held during the symposium. First, on topic of sustainability of socialist-era hotel buildings, where invited expert panelists explored challenges and opportunities in preserving and repurposing these iconic structures. Valuable insights into architectural approaches that breathe new life into such buildings are shared, while also discussing incentives and opportunities for their owners to keep up with contemporary hospitality requirements. The second one was held on topic of sustainability standards for green and open spaces in collaboration with the Association of Landscape architects in Bosnia and Herzegovina, with distinguished keynotes and panelists from the University of Belgrade. The current state of planning in terms of green infrastructure was presented and good practice standards for implementation were discussed, particularly for countries in the Western Balkans.

As the symposium progressed, we had the privilege of hosting two distinguished keynote speakers. Professor Haris Alibašić, Ph.D. from the University of West Florida shared insights on adaptive governance of sustainability under political and security uncertainty, using the case of Bosnia and Herzegovina as an illustrative example. Afterwards, Prof. Admir Mašić, Ph.D. from the Massachusetts Institute of Technology (MIT) captivated the audience with a lecture on the latest research on ancient materials and what they can teach us about designing sustainable materials for the future.

This book captures the essence of the International Symposium on Sustainable Urban Development 2023, presenting a collection of research, insights and discussions that emerged from the symposium's sessions. It is a testament to the dedication and expertise of the participants who contributed to the advancement of sustainable development. We hope that the chapters within this book inspire further exploration, foster meaningful dialogues and contribute to the ongoing global efforts in creating a sustainable and thriving urban future.

In total, 19 peer-reviewed papers have been accepted for publication and 3 invited opinion papers. This allowed presentation of a comprehensive exploration of sustainable urban development across seven distinct parts. The first part focuses on disaster management, where the three invited experts shared their perspectives, covering methodologies for calculating the cost of disasters, earthquake risk assessment and regional cooperation. The second part delves into energy efficiency and the environment, examining energy audits, efficiency in public buildings and soil remediation. The third part centres around cities, discussing topics such as urban mobility, the 15-minute city concept and the linear city concept. The fourth part focusses of heritage preservation, including industrial heritage recording processes and neglected urban cultural heritage. The fifth part addresses preservation and conservation, exploring global approaches, ancient monument revival and the challenges of preserving socialist-era hotel buildings. The sixth part concerns sustainable space and design, considering aspects like the persistence of socialist apartment buildings and spatial adaptability in small apartments. Finally, the seventh part explores social aspects, including adaptive governance, ethical implications of neuromarketing and corporate social responsibility during the COVID-19 pandemic. This book provides a comprehensive and multidisciplinary approach to sustainable urban development, offering valuable insights for researchers, practitioners and policymakers alike.

Disaster Management and Risk Reduction

In this part, three key topics related to disaster management and risk reduction are discussed. The explored topics include methodologies for calculating the ex-post cost of disasters, earthquake risk assessment in Bosnia and Herzegovina, and the role of regional cooperation in disaster management and risk reduction policies in South East Europe.

Energy Efficiency and Environment

This part delves into topics related to energy efficiency and the environment. The discussed subjects include the accuracy of energy audits in estimating required investments in energy efficiency measures, energy efficiency in public buildings in Bosnia and Herzegovina, and the potential of using St. John's wort for remediating soil contaminated with heavy metals.

Cities

The topics covered in this part revolve around urban environments. The subjects discussed include the concept of the linear city, urban mobility under the context of the COVID-19 pandemic, the role of cultural landscapes of post-industrial heritage as a driver for placemaking, and the principles of the 15-minute city concept and its repercussions on the city's health.

Heritage, Preservation and Conservation

This part explores various aspects of heritage and preservation of the built environment. The discussed subjects include industrial heritage recording processes in Bosnia and Herzegovina, neglect of urban cultural heritage, implementation of global approaches in the local context, reviving an ancient monument, and the perspectives and challenges of sustainable preservation of hotels from the socialist period.

Sustainable Spaces and Designs

This part addresses the topic of sustainable space and design. The discussed subjects include the persistence of socialist apartment buildings and the functionalist design approach over time and usage, the use of atriums in high-performance buildings, and strategies to improve spatial adaptability in small and inflexible apartments.

Social Aspects

This part explores the social aspects of sustainable development. The discussed subjects include the adaptive governance of sustainability under political and security uncertainty, the ethical implications of neuromarketing in the context of sustainable development goals, the role of corporate social responsibility in enhancing organizational resilience, and a bibliometric analysis of literature related to sustainable cities and communities in South East Europe.

In closing, this book encapsulates the knowledge and insights shared during the International Symposium on Sustainable Urban Development 2023, providing a comprehensive exploration of diverse topics within the field. From disaster management and risk reduction, through energy efficiency and cities, to heritage preservation, sustainable space and design, and social aspects, each part contributes to our understanding of sustainable urban development. The chapters shed light on challenges, strategies and innovative approaches, offering valuable resources for researchers, practitioners and policymakers. May this book inspire further collaboration, dialogue and action as we strive to create sustainable and thriving urban environments for generations to come.

Sarajevo, Bosnia and Herzegovina
Sarajevo, Bosnia and Herzegovina
Arlington Heights, USA

Maja Arslanagić-Kalajdžić
Naida Ademović
Tijana Tufek-Memišević
Chairs of this Session

Contents

Disaster Management and Risk Reduction

Available Methodologies in Calculating the Ex-post Cost of Disaster: Case of Türkiye

Burçhan Sakarya

Abstract This paper explores the methodologies used in estimating the costs and losses of natural disasters, focusing on the Turkish experience following a series of unprecedented earthquakes in February 2023. The analysis emphasizes the need for a variety of methodologies to obtain a comprehensive understanding of the economic impacts of disasters. In the Turkish context, the Strategy and Budget Office (SBB) played a central role in coordinating post-disaster assessment efforts. By utilizing its expertise and coordinating data collection through local administrations and line ministries, the SBB ensured a swift and comprehensive analysis. Various methodologies, such as replacement cost, market valuation, and input–output analysis, were employed to estimate damages and losses in different sectors, and depending on data availability. The study highlights the importance of institutional capacity and expertise in conducting post-disaster assessments. It also emphasizes the need for a multi-method approach to capture the complexity of economic impacts.

Keywords Disaster · Earthquakes · Impact assessment · Macro modelling

1 Introduction

On February 6th, Türkiye was struck by a series of devastating earthquakes with epicenters in Pazarcık (Mw 7.7; focal depth: 8.6 km) and Elbistan (Mw 7.6; focal depth: 7 km) districts of Kahramanmaraş. The first earthquake occurred at 04:17 local time, followed by another at 13:24 local time. Adding to the impact, on February 20, 2023, a third earthquake with a magnitude of Mw 6.4 struck Yayladağı, Hatay, at 20:04 local time. These earthquakes were unprecedented in their scope and the magnitude of destruction they caused, even for a region prone to earthquakes.

B. Sakarya (✉)
TR Presidency Strategy and Budget Office, Ankara, Turkey
e-mail: bsakarya@sbb.gov.tr

M. Arslanagić-Kalajdžić et al. (eds.), *Interdisciplinary Advances in Sustainable Development II*, Lecture Notes in Networks and Systems 804,
https://doi.org/10.1007/978-3-031-46692-2_1

In response to these massive disasters, the Turkish Government immediately declared a state of emergency, and the international community demonstrated remarkable solidarity. Search and rescue teams from 90 countries, comprising over 10,000 personnel, were deployed to support the national efforts in Türkiye.

Simultaneously, a coordinated national initiative was launched to assess the cost of these consecutive disasters. This article provides a review of post-disaster damage and loss estimation methodologies, which form the basis of the techniques employed in this rapid post-disaster assessment. In the following section, a brief description of available methodologies for calculating the ex-post cost of disasters will be provided, followed by an explanation of the recent Turkish experience.

2 Methodology

From an economic perspective, a natural disaster represents a systemic risk that disrupts economic activity [1]. To estimate the costs of a disaster, several methodologies are available, which should be considered as complementary rather than substitutes. These methodologies can be classified into two distinct approaches: direct damage assessment and indirect loss assessment.

Direct damage assessment involves quantifying the physical damage caused by the disaster, such as the destruction of infrastructure, buildings, and other assets. The costs associated with repairing or replacing these damaged assets are calculated to estimate the direct economic losses.

On the other hand, indirect loss assessment considers the additional economic impacts that result from a disaster. These include business interruptions, loss of productivity, reduced output, and supply chain disruptions. This methodology attempts to capture these effects by analyzing changes in economic activity before and after the disaster. Hallegatte et al. [2] offer a practical and robust framework for implementing a loss analysis that can be adjusted according to the typology of the disaster. The indirect loss assessment is highly relevant, especially in the context of the COVID-19 lockdown period, where no harm was caused to physical assets, but economic activity was severely interrupted. Crompton et al. [3] emphasizes the importance of accurate and comprehensive economic impact assessments for policymakers and decision-makers in disaster management, as they guide effective mitigation and recovery efforts.

Calculating the damages and losses presents a challenge. For disasters such as earthquakes, where the replacement of damaged construction is a priority, the Replacement Cost Method stands out as the initial approach. This method estimates the cost of replacing damaged or destroyed assets with new ones at current market prices or unit costs provided by public authorities. It considers factors such as depreciation and inflation to accurately calculate the replacement cost. Gunasekera et al. [4] provide a concise review of the categorization of physical assets and the use of economic replacement value.

Another methodology is the Market Valuation Method, which relies on market prices and transactions to determine the value of the losses. For example, this may involve analyzing property sales data to estimate the value of destroyed real estate or assessing the decline in stock market values.

Economists also utilize the Shadow Pricing Method, which assigns a monetary value to non-market goods and services affected by the disaster but lacking explicit market prices. This approach requires assigning shadow prices to the affected assets and services based on their pre-disaster values. The loss is then calculated as the difference between the shadow prices and the actual value of the assets and services after the disaster.

Input–Output (IO) Analysis is another method that uses input–output tables to track the interdependencies among various sectors of the economy. By quantifying the flow of goods and services between sectors, it helps estimate the ripple effects of a disaster throughout the economy. Okuyama and Santos [5] offer a range of economic modeling frameworks, highlighting the practicality and use of IO models. They also mention the benefits of this model, such as capturing intricate economic interconnections within a regional or national economy. However, its simplicity has drawbacks, such as its linear nature, inflexible structure regarding input and import substitutions, the absence of explicit resource limitations, and the inability to respond to price fluctuations. Outdated IO tables exacerbate these drawbacks. Additionally, other methodologies such as CGE models and Social Accounting Matrices (SAM) can be employed to estimate and analyze the impact of a disaster.

3 Turkish Experience

When discussing the specific methodologies that have been used to calculate our post-disaster assessment, we need to briefly describe the recent earthquake disaster in Türkiye. The typology of the disaster as Hallegatte and Przyluski [6] coins, defines the scope of your analysis. Türkiye was hit by a series of debilitating earthquakes in February 2023. The disaster was unprecedented in its scope and the magnitude of the destruction it caused. The earthquakes covered a territory about of 110,000 km^2, causing major devastation in a total of 11 provinces. The earthquakes claimed the lives of more than 50,000 people, wreaked damage on over half a million buildings as well as communication and energy structures and led to significant financial losses.

Similar to all post-disaster assessment reports, the main objective was to conduct a rapid assessment of the impacts and offer a preliminary overview of post-earthquake needs across sectors, with the aim of informing future reconstruction and recovery planning.

The initial challenge was data availability and data collection. The scope of the disaster and the quality of the data imposes the choice of methodologies. With the declaration of state of emergency, Strategy and Budget Office (SBB) took the lead in post disaster assessment. This has several reasons behind it. First of SBB is the intuition which prepares long-term development plans for Türkiye. Thus apart from technical capacity, this institution has a long history of analysing Turkish economy both sectoral and regional perspective. Hence, a task force was formed under SBB.

Simultaneously a data collection and verification system was set to compile and filter data collected from the field. To achieve this, SBB have utilized local administrations and line ministries to fill out a damage spreadsheet on daily basis. These spreadsheets are compiled and crosschecked via relevant experts. This data collection feeds in to the process of direct damage assessment. The data classification or the dimension of this database have distinctions between physical assets, asset type, damage status, such as heavy, moderate or limited with sectoral code, and ownership.

The damage/costs caused by the earthquake, on the other hand, were addressed by sector in terms of convergence with international standards. In this context, the damage were examined by social sectors (housing, education, healthcare, employment, environment); infrastructural sectors (water and sanitation, municipal services, energy, transportation, communication), and economic sectors (agriculture, mining, manufacturing industry, tourism).

Hence, this helps the task force to choose an appropriate sub-methodology in calculating the cost. For some sectors like health, hospitals, or drinking water network, sewage network etc. replacement cost method is used. As for many of these infrastructure projects the feasibility assessment and investment monitoring were done by the SBB. For other damages such as vehicles, moveable assets of the private entities, market valuation method, utilizing insurance sector data was chosen.

For estimating the loss, during the earthquake phase the task force had some advantages. Turkstat had recently provided GDP figures by provinces. Thus, painting an economic picture of the earthquake region was accurate. Moreover, high frequency data collected from regional ports about foreign trade activity offered a highly valuable information regarding the economic activity.

To estimate the economic loss, the data collected via line ministries were aggregated. This aggregated data then classified in to two. One being infrastructure the other being non-infrastructure. This aggregated data coming from bottom up, then fed in to a production function. This is a similar work to Hallegatte et al. [2] with innovations specific to the scale of Turkish economy and the regional weight of the disaster. Briefly describing the SBB model, we utilize a standard Cobb–Douglas production function. To estimate the loss, we impose a deterioration of the capital stock due to earthquake. The magnitude of the deterioration comes from the compiled, aggregated data. Following the classification, the new calculated capital stock, enters to the model with different capacity utilization rates. This model yields estimates under different scenarios are then checked for robustness.

The next phase involves estimating the magnitude of the demand-side losses. While the supply-side estimates offer a figure derived from a perceived deviation from the de facto production possibilities frontier, it is essential to ensure their robustness from the aggregate demand perspective. To achieve this, we have leveraged our short-term economic modeling capabilities, which include high-frequency data-fed models and single equation models, among others. Essentially, our aim was to examine the multiplier effect resulting from increased government spending on reconstruction efforts. The outcomes and forecasts generated by these models were subsequently checked for robustness and consistency with the supply-side model. Hence, given the limited period, we have been able to provide a damage assessment with sectoral breakdown and a board loss assessment, which has internal consistency, and coherent with the observation. During this period, our main challenge was to get an understanding of the compiled data as early as possible, since there were a constant flow of data at the early stages. Setting a data cut off time and getting the models to work is also a critical stage. This is important to check for reliability of alternative estimates.

4 Concluding Remarks

The recent experience in employing available methodologies to calculate the ex-post cost and loss of disasters in Türkiye, underlines the stark reality that there is no single methodology. Usually this analysis needs to be done swiftly, as policy makers and over all public demands urgent action. Turkish experience once again proved that it is important to use a variety of methodologies to get a complete picture of the costs. The choice of methodology depends on the availability of data, the specific objectives of the analysis, and the nature of the disaster being studied. Moreover, our experience show that institutional structure is also key in initiating such studies with sufficient expertise and assessment capacity. SBB, a central governmental organization overseeing the economic and social development of Türkiye, proved to be essential in coordinating all of the central and local governmental bodies to feed into data collection process, while implementing appropriate methodologies to draw a broad picture of the ex-post cost.

This type of analysis should offer the post-disaster estimation for recovery and reconstruction plans and finance. In addition, they should underline the pre-disaster estimation in order to assess the mitigation strategies. This should construct a foundation for short term and long-term policies. There is no question that immediate rescue efforts needs to be followed by meeting the needs of all vulnerable groups, including providing health, shelter, food, education, and psychosocial support. Furthermore, the medium term needs, such as meeting labor demands, promoting regional employment, enhancing the capacity of relief organizations, and long-term needs, such as building back better, smarter, promoting collaboration among stakeholders at local, national, and international levels would be drawn from these ex-post disaster assessment frameworks.

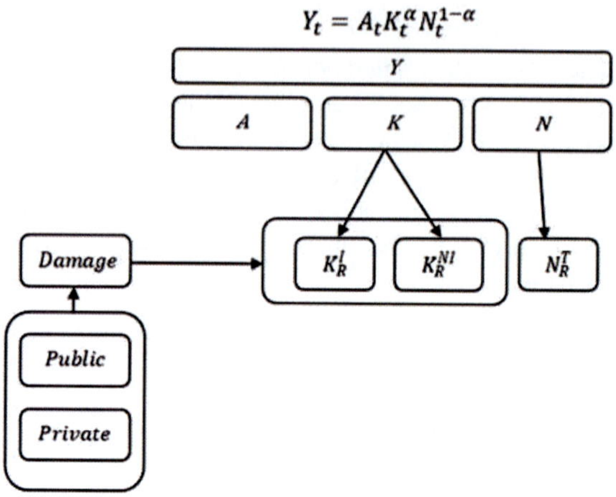

Fig. 1 A stylized framework of the data collection, classification process feeding into the production function, where Y_t is output, A_t is the total factor productivity, K_t is end of period capital stock and N_t is employment. The capital stock is grouped under "infrastructure (I)" and "non-infrastructure (NI)" as K^I_R and K^{NI}_R, while subscript R stands for the disaster region

References

1. SBB. Türkiye earthquakes recovery and reconstruction assessment. https://www.sbb.gov.tr/wp-content/uploads/2023/03/Turkiye-Recovery-and-Reconstruction-Assessment.pdf. Last accessed 06 June 2023
2. Hallegatte S, Jooste C, McIsaac F (2021) Modeling the macroeconomic consequences of natural disasters: capital stock, recovery dynamics, and monetary policy. World Bank Policy Research Working Papers no. 9943
3. Crompton RP, McAneney J, Chen K, Leigh R, Hunter L (2008) Natural hazards and property loss. In: Newton PW (ed) Transitions: pathways towards sustainable urban development in Australia. CSIRO Publishing, pp 281–293
4. Gunasekera R, Ishizawa O, Aubrecht C, Blankespoor B, Murray S, Pomonis A, Daniell J (2015) Developing an adaptive global exposure model to support the generation of country disaster risk profiles. Earth Sci Rev 150:594–608
5. Okuyama Y, Santos JR (2014) Disaster impact and input–output analysis. Econ Syst Res 26(1):1–12
6. Hallegatte S, Przyluski V (2010) The economics of natural disasters: concepts and methods. World Bank Policy Research Working Papers no. 5507. https://doi.org/10.1596/1813-9450-5507

Evaluating the Earthquake Risk Assessment in Bosnia and Herzegovina: A Call for Improved Preparedness

Naida Ademović

Abstract Bosnia and Herzegovina is located in a seismically active region, making it susceptible to earthquakes. As a result, it is crucial to prioritize the seismic assessment of existing buildings and implement effective mitigation strategies. This opinion paper argues that a comprehensive approach, involving rigorous assessments and proactive mitigation measures, is necessary to ensure the safety and resilience of the built environment. As earthquakes can strike without warning, it becomes crucial for nations to undertake robust earthquake risk assessments to mitigate potential damages and safeguard lives. The earthquake risk assessment in BIH requires substantial improvement to ensure the safety and well-being of its citizens. The current limitations pose significant challenges to effective preparedness. The government needs to demonstrate leadership and allocate sufficient resources to earthquake risk assessment and management. Financial investment, policy integration, and long-term commitment are imperative for the sustainability and effectiveness of the framework. By implementing these recommendations, BIH can significantly improve its earthquake risk assessment capabilities, minimize the potential impact of earthquakes, and protect lives, infrastructure, and the economy.

Keywords Earthquake · Risk assessment · Improvement preparness

1 Introduction

Bosnia and Herzegovina is located in a seismically active region, making it susceptible to earthquakes. As a result, it is crucial to prioritize the seismic assessment of existing buildings and implement effective mitigation strategies. This opinion paper argues that a comprehensive approach, involving rigorous assessments and proactive mitigation measures, is necessary to ensure the safety and resilience of

N. Ademović (✉)
Faculty of Civil Engineering, University of Sarajevo, Patriotske lige 30, 71 000 Sarajevo, Bosnia and Herzegovina
e-mail: naida.ademovic@gf.unsa.ba; naidadem@gmail.com

M. Arslanagić-Kalajdžić et al. (eds.), *Interdisciplinary Advances in Sustainable Development II*, Lecture Notes in Networks and Systems 804,
https://doi.org/10.1007/978-3-031-46692-2_2

the built environment. Bosnia and Herzegovina is in the midst of South-Eastern Europe (SEE), which is home to one of Europe's most intricate tectonic formations. According to the Euro Mediterranean Seismic Hazard Map, the area of Bosnia and Herzegovina is subject to a moderate seismic risk with a Peak Ground Acceleration (PGA) between 0.08 and 0.24 g, although the southwest of the nation is subject to a high risk (PGA > 0.24 g). As earthquakes can strike without warning, it becomes crucial for nations to undertake robust earthquake risk assessments to mitigate potential damages and safeguard lives. This opinion paper aims to critically examine the current state of earthquake risk assessment in Bosnia and Herzegovina, highlighting the strengths and weaknesses of existing approaches and emphasizing the need for improved preparedness.

2 Assessing the Current Situation

The earthquake risk assessment in Bosnia and Herzegovina (BIH) is marred by several limitations. Firstly, the seismic hazard maps that has been produced in 2018 and became a part of the national annex to the Eurocode 8 is not obligatory as still the old standards from the ex-Yugoslavia (1981) and enforced. In that respect, it is very difficult to accurately depict the vulnerable regions within the country. An additional problem is that BIH is within six countries that lack building inventory data. The application of several existing approaches created for the vulnerability assessment of existing buildings is significantly limited by this. Ademović et al. [1] conducted a rapid earthquake risk assessment providing a map of the relative risk seismic risk of Bosnia and Herzegovina. Three components were taken into account: hazard seismic map as per the national annex [2], exposure of the population, and type of buildings as per Census 2013 [3]. Three elements of buildings' vulnerability were considered the construction age, the type of material, and the number of floors. The calculation revealed the most endangered areas as shown in Fig. 1.

It was determined that the three most populous cities-Sarajevo, Banja Luka, and Tuzla-have a significant seismic risk to earthquake ground motion. The authors [4–6] have since done more research. Tuzla has the biggest susceptibility among these three cities in terms of people and buildings. Building vulnerability in Banja Luka is extremely close to 1 (zero to one level of vulnerability; zero minimum and one maximum degree of risk), however, population vulnerability is somewhat greater than 0.8. The scenario in Banja Luka is the opposite of that in Sarajevo, the capital of BIH. Without reliable data on historical seismic activity and ground motion characteristics, it becomes challenging to develop effective strategies for disaster management. BIH faces challenges regarding the assessment of structural vulnerabilities. A significant portion of the country's buildings, particularly in older urban areas, were constructed without considering seismic codes, resilience, and lack of maintenance. This oversight increases the risk of collapse and widespread damage during earthquakes. Furthermore, the current state of infrastructure monitoring and evaluation is

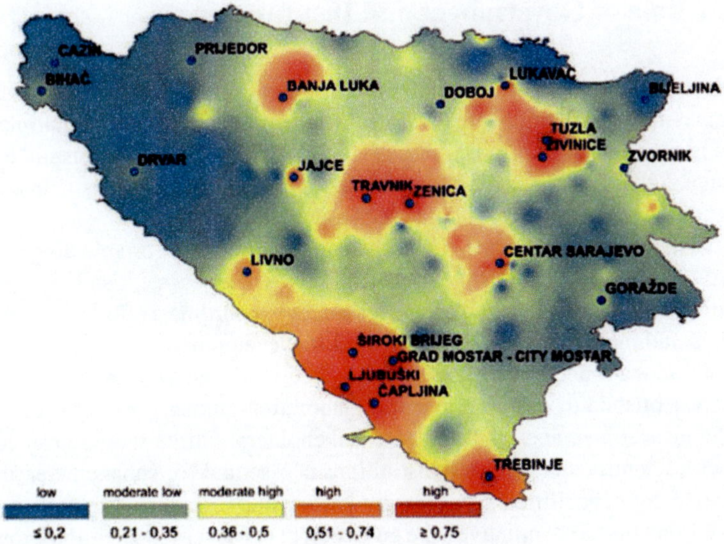

Fig. 1 Distribution of relative seismic risk in Bosnia and Herzegovina [1]

inadequate. However, the resources allocated to these activities are limited, leading to delays in identifying and addressing potential risks.

The absence of a standardized framework for assessing the socioeconomic impact of earthquakes is another critical concern. Understanding the potential consequences of seismic events in terms of economic losses, displacement of populations, and disruption of essential services is crucial for effective risk management. Without such assessments, it becomes challenging to allocate resources, develop contingency plans, and provide timely assistance to affected areas.

It has been noted that the dissemination of information related to earthquake risks to the general public is inadequate. Public awareness campaigns, early warning systems, and evacuation plans are essential components of any robust risk assessment framework, yet they receive insufficient attention. This lack of awareness and preparedness among the population exacerbates the potential impact of earthquakes, putting lives at risk. Additionally, the involvement of local communities in the risk assessment process is often limited. It is worth noting that while some progress has been made in recent years, including initiatives by national and international organizations, the overall earthquake risk assessment landscape in BIH remains inadequate and fragmented.

Addressing the identified challenges requires a multi-faceted approach that involves collaboration between government bodies, research institutions, NGOs, and local communities. Moreover, it is crucial to recognize that earthquake risk assessment should not be seen as a one-time endeavor but rather as an ongoing process. In conclusion, Bosnia and Herzegovina faces significant challenges in its current earthquake risk assessment framework.

3 The Role of Government and Institutions

Government agencies and institutions have a crucial role to play in earthquake risk assessment in BIH. A coordinated and integrated approach involving multiple stakeholders is essential to effectively address the challenges posed by seismic hazards. Unfortunately, the current state of coordination among these entities is inadequate, leading to fragmented efforts and a lack of a unified approach.

Firstly, it is imperative for the government to take the lead in establishing a centralized authority or body responsible for earthquake risk assessment and management. This entity should have the mandate and resources to coordinate efforts across various sectors, including urban planning, infrastructure development, emergency response, and public education. Clear lines of communication and decision-making structures must be established to ensure efficient collaboration among government agencies, scientific institutions, and other relevant stakeholders. Furthermore, close collaboration with scientific and research institutions is essential for enhancing earthquake risk assessment capabilities. By fostering partnerships, sharing data and expertise, and supporting research initiatives, the government can leverage scientific knowledge and advancements to develop more accurate and reliable seismic hazard assessments. Additionally, collaborations can facilitate the development of standardized methodologies and guidelines for assessing structural vulnerabilities, retrofitting techniques, and land-use planning strategies.

Government support is also crucial in terms of financial investment. Allocating sufficient resources to earthquake risk assessment and management initiatives is necessary to fund research projects, data collection and analysis, public awareness campaigns, and infrastructure upgrades. Adequate funding can help strengthen the capacity of institutions and agencies involved in earthquake risk assessment, ensuring they have the necessary tools and expertise to carry out their responsibilities effectively. Moreover, the government should prioritize the integration of earthquake risk assessment into national policies and regulations.

Lastly, the government must actively engage with local communities and empower them to participate in the risk assessment process. This can be achieved through community workshops, training programs, and the establishment of local committees or task forces. Involving communities in decision-making processes, raising awareness about earthquake risks, and providing them with the necessary tools and information can strengthen their resilience and foster a culture of preparedness at the grassroots level. The government and institutions in BIH have a critical role to play in earthquake risk assessment.

4 Recommendations for Improvement

To address the limitations and challenges identified in the earthquake risk assessment in BIH, several recommendations can be put forth to strengthen the preparedness and resilience of the country:

- Develop Comprehensive Seismic Hazard Maps: The government should invest in the development of up-to-date and accurate seismic hazard maps that consider historical data, fault lines, and ground motion characteristics. These maps should serve as a foundation for designing effective disaster management strategies, land-use planning, and infrastructure development.
- Formation of a comprehensive database of buildings: The data should include all necessary information for the researchers to perform a seismic vulnerability assessment. The physical characteristics of each building are essential (building type, number of stories, construction materials, occupancy type, etc.). The precise geographical coordinates of each building. Details about the structural system of the building and construction year. Information about compliance with local building codes and standards during construction is valuable. Records of any maintenance activities or structural upgrades performed on the buildings should be included. Data about previous seismic assessments or retrofitting studies, post-event damage data on buildings, etc.
- Strengthen Infrastructure Monitoring and Evaluation: Allocate resources and establish regular inspection and evaluation protocols for critical infrastructure such as bridges, dams, and hospitals. Identify vulnerabilities and prioritize retrofitting efforts to reduce the potential impact of earthquakes on essential services.
- Enhance Structural Resilience: Prioritize the assessment of structural vulnerabilities and retrofitting of critical buildings and infrastructure, particularly in older urban areas.
- Conduct Socioeconomic Impact Assessments: Develop a standardized framework to assess the potential socioeconomic consequences of earthquakes, including economic losses, displacement of populations, and disruption of essential services. This assessment will facilitate informed decision-making, resource allocation, and the development of contingency plans.
- Improve Coordination and Collaboration: Foster collaboration between government bodies, research institutions, NGOs, and local communities. Establish a unified platform for knowledge sharing, coordination, and decision-making to enhance the effectiveness of earthquake risk assessment efforts.
- Promote Public Awareness and Education: Implement public awareness campaigns that educate citizens about earthquake risks, preparedness measures, and response strategies. Conduct regular drills and training programs to enhance public readiness and resilience.
- Increase Research, Development and Data Collection: Invest in research and development initiatives related to earthquake engineering, including studies on building vulnerability, structural resilience, and retrofitting techniques. Support

scientific institutions and research projects that contribute to advancing earthquake risk assessment methodologies and technologies. This includes conducting field studies, collecting and analyzing geological and geophysical data, and collaborating with international institutions and experts to improve the understanding of seismic hazards specific to BIH.
- Involve Local Communities: Empower local communities through education, training, and participatory decision-making processes. Engage community leaders, stakeholders, and residents in the risk assessment process to gather valuable insights into local vulnerabilities, cultural considerations, and community-specific needs.
- Ensure Long-Term Commitment: Recognize earthquake risk assessment as an ongoing process that requires regular updates, continuous monitoring, and incorporation of new scientific findings. Maintain a long-term commitment to improving earthquake resilience through sustained funding, institutional support, and policy integration.
- Public–Private Partnerships: Encouraging collaboration between government entities, private sector organizations, and academia can lead to innovative solutions, technological advancements, and resource-sharing that strengthen the overall preparedness and resilience of the country.
- Cross-Border Cooperation: Address the need for cross-border cooperation and information exchange regarding earthquake risk assessment. Collaborative efforts with neighboring countries can facilitate data sharing, harmonize standards, and enable joint research and response initiatives.
- Integration with Climate Change Adaptation: Acknowledge the interplay between earthquake risk assessment and climate change adaptation. Climate change can potentially influence seismic activity, ground conditions, and the vulnerability of infrastructure. Integrating earthquake risk assessment with climate change adaptation strategies and energy efficiency can lead to more comprehensive and resilient planning and preparedness measures.
- Consideration of Cultural Heritage: Highlight the importance of preserving and protecting cultural heritage sites during earthquake risk assessment. Many historic buildings and structures in BIH hold significant cultural value and should be preserved for future generations.
- International Best Practices: Discuss the relevance of studying and adopting international best practices in earthquake risk assessment.
- Allocation of Adequate Resources: Emphasize the need for sustained financial resources and institutional support for earthquake risk assessment activities.

By implementing these recommendations, BIH can significantly enhance its earthquake risk assessment framework, mitigate the potential impact of seismic events, and safeguard the lives and well-being of its population. Stakeholders at all levels must prioritize these actions and work collaboratively to build a resilient and earthquake-ready nation.

5 Conclusion

In conclusion, the earthquake risk assessment in BIH requires substantial improvement to ensure the safety and well-being of its citizens. The current limitations pose significant challenges to effective preparedness. By investing in research, enhancing collaboration, and prioritizing public education, BIH can mitigate the potential impact of earthquakes, protect lives, and build a resilient future for its communities. It is time to prioritize earthquake risk assessment and take proactive measures to safeguard the nation against this formidable threat. The current state of earthquake risk assessment in BIH reveals significant gaps and challenges that need to be addressed to ensure the safety and well-being of its population.

However, through a concerted effort and a commitment to the above-mentioned improvement, the country can enhance its earthquake risk assessment framework and build a more resilient future. The government needs to demonstrate leadership and allocate sufficient resources to earthquake risk assessment and management. Financial investment, policy integration, and long-term commitment are imperative for the sustainability and effectiveness of the framework.

By implementing these recommendations, BIH can significantly improve its earthquake risk assessment capabilities, minimize the potential impact of earthquakes, and protect lives, infrastructure, and the economy. As seismic events pose an unpredictable threat, it is crucial to act with urgency, resilience, and foresight to ensure the nation's ability to withstand and recover from future earthquakes. By embracing these measures, BIH can pave the way for a more resilient and secure future, where the risks of earthquakes are effectively managed and the well-being of its citizens is protected.

References

1. Ademović N, Kalman Šipoš T, Hadzima-Nyarko M (2020) Rapid assessment of earthquake risk for Bosnia and Herzegovina. Bull Earthq Eng 18:1835–1863. https://doi.org/10.1007/s10518-019-00775-1
2. Institute for Standardization of Bosnia and Herzegovina [BAS] (2018) National annex BAS EN 1998–1:2018—design of structures for earthquake resistance—Part 1: General rules. Seismic actions and rules for buildings
3. Bureau of Statistics [CBS] (2013) National population and housing census 2013 (national report). Agency for Statistics of Bosnia and Herzegovina
4. Ademović N, Hadzima-Nyarko M, Zagora N (2020) Seismic vulnerability assessment of masonry buildings in Banja Luka and Sarajevo (Bosnia and Herzegovina) using the macroseismic model. Bull Earthq Eng 18(8):3897–3933. https://doi.org/10.1007/s10518-020-00846-8
5. Ademovic N, Zagora N, Hadzima-Nyarko M (2021) Seismic vulnerability analysis in urban and rural regions of Visoko, BIH, 12th days of BHAAAS in Bosnia and Herzegovina advanced technologies, systems, and applications VI. In: Proceedings of the international symposium on innovative and interdisciplinary applications of advanced technologies. https://doi.org/10.1007/978-3-030-90055-7

6. Ademovic N, Hadzima-Nyarko M, Zagora N (2022) Influence of site effects on the seismic vulnerability of masonry and reinforced concrete buildings in Tuzla (Bosnia and Herzegovina). Bull Earthq Eng 20:2643–2681. https://doi.org/10.1007/s10518-022-01321-2

Enhancing Human Security Through Regional Cooperation in Disaster Management and Risk Reduction Policies in South East Europe

Arslan Umut Ergezer and Jovana Popova

Abstract South East Europe (SEE) is the most natural disaster-prone sub-region of Europe. The latest earthquakes in Türkiye reveal that major disasters cause multi-dimensional damage threatening the sustainability of life. Human security as a comprehensive concept that involves all the aspects of human life, which are of critical importance for its sustainability, needs to be introduced as a higher policy objective in all multisectoral and multiannual strategies. All other policies including disaster management and risk reduction might well be incorporated and coordinated with other policies affecting human security. The Official Development Assistance's responsiveness to human security aspects needs to be enhanced. Strategic interventions could be at the regional level in the SEE and South East Europe (SEE) 2030 Strategy is the best-placed framework to address those challenges.

Keywords Human security · South East Europe · Disaster management and risk reduction · Financing

1 Introduction

South East Europe (SEE) is Europe's most natural disaster-prone region. Droughts, floods, extreme temperatures, earthquakes, and landslides: the SEE region suffers from all. Moreover, the occurrence of natural disasters in the region is another incentive for taking urgent action. Only within a few years, several natural disasters of different strengths struck the region: from earthquakes in Türkiye, Albania and Croatia, recurrent seasonal floods in Bosnia and Herzegovina and Serbia, to wildfires in Greece, just to name a few. The latest catastrophic earthquakes in Türkiye

A. U. Ergezer
Regional Cooperation Council Secretariat, Sarajevo, Bosnia and Herzegovina
e-mail: umut.ergezer@rcc.int

J. Popova (✉)
Faculty of Political Sciences, University of Belgrade, Belgrade, Serbia
e-mail: jovana.popova@gmail.com

M. Arslanagić-Kalajdžić et al. (eds.), *Interdisciplinary Advances in Sustainable Development II*, Lecture Notes in Networks and Systems 804,
https://doi.org/10.1007/978-3-031-46692-2_3

in February 2023 were recorded by far as the biggest disaster to hit the SEE region. More than 48,000 people are reported to have lost their lives, 3.3 million people were displaced, and 2 million people had to be accommodated in temporary shelters as a result of the earthquakes. The damages and losses are estimated to be over 100 billion USD, or equivalent to 9% of Türkiye's forecast GDP for 2023 [2].

Multidimensional damage caused by the catastrophic earthquakes of Türkiye revealed once more the need to tackle disaster management and risk reduction-related challenges through comprehensive and holistic policy conceptualisation and instruments. In this regard, we are of the opinion that human security is an appropriate concept to cater sufficiently broad definition that may help cover all the critical aspects of human life. Incorporating disaster preparedness and risk reduction as an integral part of human security will offer a policy-level decision-making framework to manage disaster preparedness and risk reduction through holistic and well-designed multi-policy nexuses.

2 Background and Motivation

Human-centred security appears as a concept to safeguard the vital core of all human lives in a holistic manner. The objective of human security is "to safeguard the vital core of all human lives from critical pervasive threats, in a way that is consistent with long-term human fulfillment" [1]. With that definition which Alkire made, people-oriented security is all-inclusive in a way to accommodate all surrounding factors of human lives ranging from political, economic, cultural and social to health and environmental risks. The concept of human security[1] is mainly associated with Mahbub ul Haq's work [4], and UNDP's 1994 Human Development Report on Human Security [10]. People-centred security concept extends nation-state-based security understanding downwards from nation-states to individuals. According to Emma Rothschild, the extended security does not only go downwards from nations to individuals but also upwards from nation-states to the security of the international system [8]. Security is also extended horizontally as states, international system and individuals cannot ensure their security in the same way or with similar tools. Each level has its own security concerns and therefore security-ensuring instruments for each level should be different accordingly. In parallel to "extended security", the numbers of political responsible for ensuring security are proliferated. Thus, responsibility for ensuring security is dispersed from central governments to local governments, civil society, media, the private sector and nongovernmental actors [8]. Several other authors also tackle the concept of human security to include the factors of socioeconomic vulnerabilities and poverty. King and Murray explained human security from an individual's expectations point of view that each person hopes for

[1] The 1994 UNDP report defined human security as: (1) Safety from chronic threats such as hunger, disease and repression, (2) Protection from sudden and hurtful disruptions in the patterns of daily life—whether in jobs, in homes or in communities.

life without experiencing the state of generalised poverty. King and Murray accept security is based on the risk of severe deprivation. They took into account the interaction between security and poverty. However, they preferred to include a more general poverty definition, which reads as the deprivation of any basic capabilities. Thus, their generalised poverty includes the deprivation of different determinants of well-being with differencing thresholds corresponding to each sub-wellbeing determinant rather than a unidimensional factor of poverty, (i.e. income) [5].

The material and non-material dimensions of human security are the main focus of Caroline Thomas who puts democracy and human dignity at the qualitative core together with material sufficiency constituting the quantitatively measurable centre of human security [9]. The physical safety of the individual is the minimum level of ensuring human security but cannot be fully entailed unless the protection of basic liberties, economic needs and interests [3]. Thus, the concept of human security has come up as a notion that is comprehensive enough to involve all the socio-economic and political determinants of lives.

We believe it would be extremely valuable to analyse disaster preparedness and prevention, and risk reduction capacities of the countries from multi-policy perspectives. However, the current attention both academic and of policy makers mainly focuses on disaster response; referring to the activities undertaken after disasters strike. Mostly, security is considered as a concept of national or nation-state level security, and disaster management and risk reduction is considered as disaster response that needs to be tackled by civil protection agencies, while the quality of life determinants are dealt with by partial and sectoral policies. However, holistic strategies like UN Sustainable Development Goals and Agenda 2030, and the South East Europe (SEE) 2030 Strategy [7] require an explicit prioritisation of people-oriented security as a forefront concept.

3 Role of Official Development Assistance (ODA) and Level of Its Responsiveness to Human Security Challenges of SEE

The SEE region faces common challenges and types of human security-related challenges. The UNDP's Human Development Index (HDI) data between 1995 and 2019 disaggregated according to NUTS2 (Nomenclature of territorial units for statistics) reveals the divergence between SEE participants and their Western European neighbours.[2] According to the authors' computations based on the subnational data of UNDP's HDI, the gap between the SEE's regional average of human development index and the average of Austria, Germany, and Switzerland either remain the same in the period between 1995 and 2019 or widened in some years (Fig. 1).

[2] The Subnational Human Development Index: Moving beyond country-level averages | Human Development Reports (undp.org).

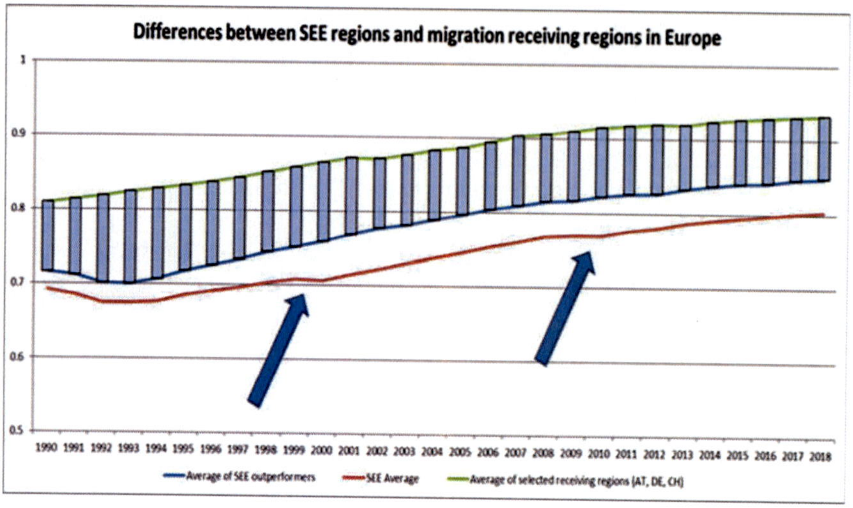

Fig. 1 Differences between SEE regions and migration receiving regions in Europe. *Source* the computations based on the UNDP's subnational HDI data by U. Ergezer in the presentation made in SEE2030 Monitoring Committee on 06 March 2023

In this regard, it would be of crucial importance to investigate the sources of financing human security. In generic terms, the financing of human security involves national budgets, private sector funding, and international donor financing (i.e. Official Development Assistance). For the current paper, the authors limit their intervention to focusing on official development assistance without undermining the importance of national budget responsiveness to human security aspects and private sector financing.

The latest Organisation for Economic Cooperation and Development (OECD) data reveal that democratic participation and civil society, human rights, and financial intermediaries are the first three sectors representing 18% of 3.7 billion USD disbursed in 2021 by the members of the Development Assistance Committee (DAC) of the OECD as ODA in the SEE recipient economies. Whereas, taking into account the total ODA disbursed in SEE since 2015, higher education has come up as the top sector with a share of 6% of the cumulated disbursed ODA of 26.4 billion USD.[3] The compartmentalized approach of donor financing in human security-related challenges in the SEE was observed. According to the OECD data, the ODA disbursed to the basic health infrastructure represent only 0.4% of all ODA disbursed to the SEE recipients since 2015. Despite the very high risk of natural disasters that the region has, the OECD DAC ODA donors allocated the exactly same share of their financial assistance to the disaster risk reduction sector in the SEE recipients. Thus, collectively both health infrastructure and disaster risk reduction related projects financed by OECD DAC donors could not even reach in the SEE recipients the amount that is allocated to the administrative costs of donors financed projects. Accordingly, the

[3] OECD DCD Statistics of 2021.

administrative costs represent 1.1% of the overall disbursed amount of ODA in SEE recipients since 2015 whereas the aggregated share of ODA disbursed to basic health infrastructure and disaster management reached 0.8%.

Since the Accra Agenda [6], aid effectiveness improved, but it still does not meet the set goals,[4] especially during the 2020s marked by severe crises: the COVID-19 pandemic and the war in Ukraine—the events that re-prioritised aid targets, changed its flows and occupied the attention of Official Development Assistance, international and private donors. The devastating earthquakes that hit Türkiye in February 2023 were a harsh reminder that disaster management and risk reduction is an area deserving of great aid attention, especially in the SEE region, which is highly vulnerable and prone to disasters, but consisting of (mostly) small economies. Moreover, it became ever clearer that disaster prevention, preparedness and resilience are the key to resisting natural disasters and to continue advancing towards sustainable and inclusive growth even if a disaster occurs.

4 Can a Regional Holistic Approach Be a Worthwhile Instrument to Respond to Natural Disasters as Part of Human Security?

Apart from national strategies for preparedness and prevention of natural disasters, the SEE region needs to reconsider enhancing and structuralising regional cooperation in this area. What is more, since natural disasters know no borders, joint actions towards disaster management and risk reduction are highly desirable, particularly in regard to disaster preparedness and prevention in order to tackle the challenges in the most efficient and effective manner.

This could be done in various ways. Firstly, human-oriented security needs to be introduced as a holistic concept that offers policy interconnections between all the relevant sub-sectoral policy instruments. In line with the people-first and leave no one behind principles of Agenda 2030, human security should be placed as the highest political objective that both national and regional strategies could be aligned with. Human security offers a hierarchy to all other policies including disaster preparedness and prevention-related policies and at which level they could be incorporated into the strategic action. In this regard, financing instruments will have to be adjusted to that higher-level policy objective, i.e. promotion of the human security situation in the SEE. In the context of the SEE a private sector pooling financing mechanism could be a valuable addition to the ODA and foreign direct investments (FDI) of recipients in the region.

For the SEE region, non-financial assistance instruments, called instruments of solidarity particularly among the SEE economies, may significantly enhance knowledge transfer, experience sharing, and information exchange.

---

[4] The majority of DAC donors did not reach the goal of 0,7% GNI in ODA.

In SEE, the SEECP and SEE2030 Strategy are the best-placed frameworks for further regional cooperation in the area of disaster management and risk reduction. The Strategy creates a SEE region-specific perspective towards implementing the SDGs and even has its own clear reference to this area through priorities and targets.

5 Conclusion/Recommendations

In conclusion, several recommendations could be put forward with the aim to help the SEE region in meeting SDGs, SEE2030 Strategy goals, and stability and resilience of the region in case of natural disasters. The primary approach of the region towards disaster management and risk reduction should be part of the collective efforts of the SEE economies to improve human security situation in the region.

The focus under the human security objective could be made on the following:

- Adopting a regional disaster prevention and preparedness strategy in line with the Sendai Framework;
- Prioritising preventive measures, resilience and disaster preparedness rather than post-disaster reaction and reconstruction;
- Attracting non-financial assistance for training, best practices exchange, and data collection coordination;
- Promoting regional risk sharing through promoting private sector financing mechanisms;
- Creating a regional early warning system;
- Better responsiveness of national budgets to natural disasters;
- Enhancing human security responsiveness of ODA, particularly increasing its shares dedicated to human security-related policies and instruments such as basic health, education infrastructures, and disaster management and risk reduction.

References

1. Alkire S (2003) A conceptual framework for human security
2. Strategy and Budget Office of Republic of Türkiye (2023) Earthquakes recovery and reconstruction assessment report
3. Hampson FO, Daudelin J, Hay JB, Martin T, Reid H (2002) Madness in the multitude: human security and world disorder. Oxford University Press, Ottawa
4. Haq M (1995) New imperatives of human security. World Affairs: The Journal of International Issues 4(1):68–73
5. King G, Murray CJ (2001) Rethinking human security. Political Science Quarterly 116(4):585–610
6. OECD (2008) Accra Agenda for Action. https://doi.org/10.1787/9789264098107-en
7. Regional Cooperation Council (2021) South East Europe 2030 strategy
8. Rothschild E (1995) What is security? Daedalus 124(3):53–98

9. Thomas C (2000) Global governance, development and human security: the challenge of poverty and inequality. Pluto Press, London and Sterling, VA
10. United Nations Development Programme (UNDP) (1995) UNDP human development report on human security. UNDP, New York

Energy Efficiency and Environment

How Accurately Energy Audits Estimate the Required Investment in Energy Efficiency Measures?

Jasmina Mangafić, Azra Smječanin, and Josip Nikolić

Abstract Energy efficient design is both a design philosophy and a practical technique that has been based on detail energy audit and relevant methodology. Based on the implemented subprojects in Bosnia and Herzegovina, over the last years, increasing effort has been put into the development of energy efficient audit and design. This paper aims to show how accurately have the energy audits predicted the actually realized investment in energy efficiency measures implemented in public buildings. The analysis is performed for different building size, type and location. The need for urgent revision of the current methodology of developing the energy audits has been identified in order to increase the quality of information disclosed.

Keywords Energy efficiency · Public buildings · Detailed energy audit · Design · Retrofitting · Financing

1 Introduction

The 21st Conference of Parties (COP 21) to the United Nations Framework Convention on Climate Change (UNFCCC), and the resulting Paris Agreement, was the global response to this problem, with a collective effort to limit the global temperature rise to well below 2 °C and pursue efforts to limit the further temperature increase to 1.5 °C [1]. The follow up efforts provided evidence for an even stronger consensus that only the collective global action can address this challenge and achieve the goals set out in the Paris. Energy efficiency is a highly effective way to reduce

J. Mangafić (✉)
School of Economics and Business, University of Sarajevo, Sarajevo, Bosnia and Herzegovina
e-mail: jasmina.mangafic@efsa.unsa.ba

A. Smječanin · J. Nikolić
Ministry of Spatial Planning of the Federation of BiH, Sarajevo, Bosnia and Herzegovina
e-mail: azra.smjecanin@beep.ba

J. Nikolić
e-mail: josip.nikolic@beep.ba

M. Arslanagić-Kalajdžić et al. (eds.), *Interdisciplinary Advances in Sustainable Development II*, Lecture Notes in Networks and Systems 804,
https://doi.org/10.1007/978-3-031-46692-2_4

GHG emissions, and it can make a significant contribution to combatting climate change [2]. According to the International Energy Agency (IEA), EE measures could result in 40% of the GHG emissions abatement required to achieve the goals of the Paris Agreement. EE measures also bring additional co-benefits, such as reduced air pollution, reduced spending on energy, enhanced energy security, and improved competitiveness [3].

Bosnia and Herzegovina is a country with very high energy inefficiency within the building, industry and service sectors. The average energy consumption of a public building in Bosnia and Herzegovina (BiH) is three times higher than the EU average. More than 70% of public buildings in BiH were built more than 30 years ago without much regard for their energy efficiency (EE). In addition, over 80% of public sector buildings are currently using fossil fuels for heating or are connected to district heating systems which are predominantly coal-based [4].

As one of the steps towards fulfilling the goals of the Nationally Determined Contribution (NDC) of Bosnia and Herzegovina [5] as the National action plan for energy efficiency [6], the Government of Bosnia and Herzegovina has started the implementation of multiple projects related to energy efficiency in public building sector. The main development objective is to demonstrate the benefits of improving energy efficiency in public sector facilities and to support the development of flexible financing models for energy efficiency.

Due to the complex system of government in Bosnia and Herzegovina defined by the Constitution, the field of EE in BiH is regulated at the entity level.[1] In the Federation of BiH, which is in the focus of this study, more than 300 buildings in total were refurbished between 2016 and 2023, financed from different sources as budget funds, grants and loans.

The implementation of energy efficiency measures requires a holistic approach that includes the procurement not only the construction works but also all other accompanying consulting services and goods. The first step is detailed energy audit that should provide a list of recommendations regarding energy efficiency improvements that should, inter alia, result in the decrease of utility bills. It also gives us an estimation how much do we need to pay to implement the suggested energy efficiency measures, together with the analysis of the cost efficiency for every individual measure as payback period. Based on the selected measures, the design has to be developed to be able to prepare the tender for construction works.

This paper studies the extent to which a Detail energy audit is a good indicator of the necessary energy efficiency investment. The goal of this paper is to compare the estimated investment value from Detail energy audit and Design with the Final Contract price. The contribution lies in assessing the factors that lead to differences in planned and realized energy efficiency investment.

[1] Federation of Bosnia and Herzegovina & Republic of Srpska.

2 Detail Energy Audit Versus Energy Efficiency Design Versus Contract for Construction Works

This section describes a link between detailed energy audit, energy efficiency design and final contract price.

The optimal scenario in the techno-economic analysis of the Detailed Energy Audit (DEA) provides the estimated value of the investment for individual facility. In accordance with Decree on energy audits and energy certificates [7] the DEA includes a detailed energy analysis of all construction and technical systems in the building. The guidelines [8] for conducting energy audits provide a methodology for conducting energy audits for new and existing buildings, residential and non-residential, with simple or complex technical systems (without industry and industrial processes). An energy audit consists of various works that can be carried out, which depends on the type of energy audit and the purpose of the energy audit facility. For existing buildings, depending on the purpose, in addition to the necessary calculation of the building's thermal needs, energy costs are analyzed for an optimal 36 months in order to model energy consumption and establish all energy needs in the building. The analysis can be completed with the necessary measurements of electricity consumption, heat losses, air permeability of the building, etc., which is important for determining energy losses in individual systems.

Proposed measures to increase energy efficiency are classified into categories according to energy, economic and ecological contribution, with the calculation of a simple return on investment period. After identifying potential measures to improve the building's energy properties, it is necessary to express energy savings for each individual measure, estimate investment costs and calculate a simple investment return period. An energy audit consists of various jobs that can be carried out, which depends on the type of energy audit and the purpose of the energy audit facility [9]. The main requirements from a detailed energy audit are to calculate current energy consumption in the organization, to conduct an interview with the relevant staff in the facility, to find out the most appropriate ways to improve energy efficiency measurements and consequently reduce the consumption and increase savings. Fleiter et al. [10] had coined a term "energy efficiency gap" which describes the difference between existing and potential energy use after implementation of energy conservation measures (ECM). Sorrell et al. [11] classified the barriers for implementing ECMs into the following categories: imperfect information, hidden costs, risk, access to capital, split incentives, and bounded rationality. Energy efficiency policies designed to overcome these barriers include energy management obligations, financial incentives, energy labeling schemes, minimum standards, audit subsidies and requirements, and energy disclosure laws [10, 12]. These energy disclosure laws can facilitate steps towards effective performance-based energy standards, as opposed to those based on generic ECM recommendations and guidelines [13].

While energy audits should be a valuable tool for adoption of ECMs, high ECM implementation costs and incomplete or incorrect energy efficiency reports can reduce final results defined by energy audits [14]. Fleiter et al. [10] determined that high initial investment costs influenced on adoption of cost-effective ECMs and results also showed that high-quality, more detailed energy audits, resulted in more frequent ECM implementation. Shapiro [14] conducted a detailed case study of 30 commercial and residential energy audits. The common problems that he found are as follows: inadequate review (3 or more mistakes in audit data), found in 30% of audits; overestimated savings derived from the retrofit, found in 53% of audits; inadequate billing analysis (monthly summaries, projected savings, benchmarking) found in 57% of audits; poor building description, found in 60% of audits; low and/or missing installed costs, found in 60% of audits; poor selection of improvement or payback, found in 63% of audits; no life-cycle costing (use of payback instead of life-cycle costing), found in 73% of audits; improvement life is too long or not provided, found in 73% of audits; weak improvement scope (missing 2 of 3 of the following conditions—location/quantity, energy rating, and testing requirements), found in 77% of audits; missed improvements, found in 80% of audits. In his paper Shapiro gave a set of suggestions in order to improve energy audits: create a standardized templates for auditors, ensure that all energy auditors are thoroughly trained, accredited and understand standards, increase the level of quality control, verify the amount of energy savings and implement an energy auditing software to increase efficiency and accuracy. Similar conclusions were given by Kobbina J.A. et al. [15]. Survey of 479 residential energy auditors in the United States conducted by Palmer et al. [16] showed that auditors rarely validated projected energy savings, customers were uninformed about energy audits and high initial investments and low energy costs had a significant influence on ECM implementation [13].

Studies like [17–19] have applied energy audit and benchmarking data to understand factors influencing energy use in buildings, ECM implementation decisions and ECM eligibility [13]. Results from an econometric analysis showed that energy savings, financial incentives, and implementation costs were important factors behind retrofit decisions [17]. The study [18] identified several limitations of current benchmarking methods and found building occupant rates and type to have important implications for energy efficiency. Results of the study [19] showed that there was significant variation in building energy usage intensity regardless of building properties, which implies that the energy consumption can largely depend on operational habits and that non cost intensive measures can be applied which will result in energy efficiency improvements—savings. It has to be stated that the energy audit has a "use by date" or limited useful life. Technologies, energy sources, energy tariffs, operational habits of a facility and other variables change over time, so calculations and recommendations in energy audits become out of date. Energy audit standards generally consider that energy audits should be repeated every 3 years [20].

The Design is created according to the adopted optimal scenario from Detailed Energy Audit. In addition to energy measures, the design includes all accompanying non-energy measures and an estimate of works based on detailed market research. The tender document for procuring the construction works is prepared based on the Design. Final Contract Price is mutually agreed upon total amount that an investor pays to a contractor on completion of the contract, in accordance with contract terms and conditions and their subsequent modifications as defined during the works commissioning.

3 Research Design and Methodology

For the analyses, we have selected public buildings that were retrofitted under Additional Financing for Bosnia and Herzegovina Energy Efficiency Project (AF BEEP) financed by the loan received from the World Bank. DEA's for these buildings were performed in different time periods. For some of the buildings Design and Implementation of proposed measures was carried out almost immediately after the DEA's were performed. Opposite to that, Design for some of the buildings was carried out on the basis of older DEA's. It is important to note that public institutions that want to take a loan for the implementation of energy efficiency measures do so on the basis of the investment defined by the DEA. This indicates that the investments proposed by the DEA should be as close as possible to the actual values of the works.

In the observed period from 2020 to 2023, in the Federation of BiH, 34 public facilities were renovated. The selected buildings included 21 educational facilities with a total heated area of 64,356 m^2, nine (9) were health facilities with a heated area of 29,375 m^2, and four (4) administrative facilities with a total heated area of 18,048 m^2.

The largest amount of total investment was allocated to the replacement of openings with new energy efficient ones (almost 30% of the total investment in energy efficiency measures), followed by thermal insulation of the facade (about 26%), installation of new heating systems (about 12%) and renovation/warming of roofs (about 9%). Less than 1% of all investments in measures to increase energy efficiency were realized in the lighting system, as well as for other measures, which are not financially intensive. An overview of investments, the total amount and the share in the total investment, by type of measure, is given in Fig. 1.

From the total number of retrofitted objects we have selected a representative sample of 19 public buildings renovated under the AF BEEP in the period 2021/2022 to assess the cost increase, to recalculate the energy savings and accordingly the payback period. The selected sample buildings are located in three cantons on the territory of the Federation of Bosnia and Herzegovina. Data were analyzed for buildings of different purposes, geographical distribution, number of employees and number of end-user. The selected representative sample of buildings includes dataset for (14) Education buildings, four (4) Health buildings and one (1) Public administrative building.

To perform the required calculations, we needed the detailed energy audit, the design and the final contract price for each selected object. The indicators from the detailed energy audit were recalculated in accordance to the energy efficiency measures actually implemented in the buildings. For that purpose, we have used the KI EXPERT BIH software developed by Knauf Insulation d.o.o. [21]. This software is intended for designers and energy certifiers and is fully compliant with the Technical

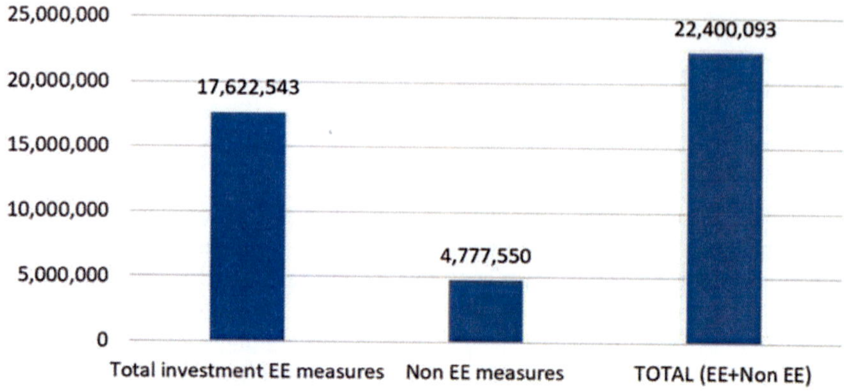

Fig. 1 The total amount invested per EE and Non-EE measures

Regulations on the Rational Use of Energy and Thermal Protection in Buildings, Algorithm for calculating the required energy for heating and cooling the building space and the Decree on energy audits and energy certificates, Rulebook on minimum energy performance of buildings and Guidelines for energy audits in new and existing buildings with simple and complex technical systems.

That enabled us to compare the planned and actually realized investment cost, savings of heating and electricity costs as the payback period in order to determine how accurately DEA's estimated required investments.

4 Results and Discussions

The first step was to provide an overview of the investment values for the population stemming from the different data sources, as presented in the table below.

Given the enormous differences between the estimated value of the techno-economic analysis in DEA and the estimated value of the Design, there was a reduction in the scope of the planned measures, and therefore it was impossible to achieve the planned energy savings, economic savings and reduction of greenhouse gas (CO_2) emissions. If the investment costs estimated by DEAs and actual final costs of retrofits are compared, it can be seen that there is an increase in actual capital expenditures (CapEx) of nearly 55% compared with estimated CapEx from DEAs. When expressed through unit costs, DEA analysis estimates unit investment costs to be around 144 BAM/ m^2, while the real unit costs for retrofits is around 205 BAM/ m^2. Detailed energy audits for retrofitted buildings were conducted in the period from the 2018 to 2019, while the design and civil works were carried out in the period from the year 2020 to 2023. If the EE investment is planned based on the DEA it is obviously that, the existing methodology proved to be inadequate and there is a need to change the guidelines for creating DEA.

Taking into account the difference between the planned and actual investment costs, we wanted to see how the difference has impacted the perceived savings of cost as the payback period. The analysis was done for 19 buildings, where all buildings are public. Table 1 shows all dataset including yearly data for energy savings, simple payback period (SPB) and index of savings, basically detailed energy efficiency measure data.

Using data collected for 19 facilities, equation was set to calculate the average indicator:

$$AverageSimplePaybackperiod = \frac{\sum InvestmentBAM}{\sum(energyandelectricitysavings)BAM}$$

$$AverageIndexofSavings = \frac{\sum(energyandelectricitysavings)kWh}{\sum InvestmentBAM}$$

It is evident from the collected data from the DEA that energy savings and savings in electricity for 19 facilities are 9,324 MWh in the amount of 1,428,651 BAM. The calculated average payback period is 6.4 years and the average Index of savings (kWh/BAM) is 1,02. The same method was used to analyze the data collected after the implementation of energy efficiency measures. Total energy savings and savings in electricity for 19 facilities are 9.101 MWh, in the amount of 1.603.757,00 BAM. The average payback period is 7,2 years and the Index of savings (kWh/BAM) is 0,78. Difference in energy savings in MWh is due to differences in slight changes in implemented measures and measures proposed by the DEA. Although savings in MWh based on implemented measures are smaller than the ones predicted by the DEA, there is an increase in financial savings. This is due the increase in energy prices, which is taken into account when savings based on implemented measures was conducted. Based on the data from these two tables, we can notice that the investment has increased significantly. There is also an increase in energy cost savings due to energy prices have also increased. The difference in the SPB is less than one year.

The next step was to elaborate in more deep the potential factors that lead to the increase of the estimated investment costs up to 55%. We have to start with the additional non-energy efficiency measures that were implemented but not predicted by the optimal scenarios in DEAs and as such caused an increase in total and average unit investments costs over 20% on certain facilities. For example, it is impossible to change the roof covering and install thermal insulation on the building without changing the vertical and horizontal gutters. Furthermore, as part of the aforementioned works on the roof, it is necessary to examine the existing lightning rod installation and replace it due to valid fire regulations and the safety of users in the building. Some energy efficiency measures were not proposed by the optimal scenario, but a field visit showed the need for additional measures in order to achieve energy performance certificate minimum B as required by the law.

This observed period of three years coincided with the period of significant changes in the entire world, including BiH. Special type of crises was caused by the Covid 19 epidemic. The market correction of 2020 is seen as one of the most

Table 1 Overview of the investment in EUR

No	Public facility	Total investments (EE + non EE) DEA	Total investments (EE + non EE) DESIGN	Total investments (EE + non EE) FINAL CONTRACT PRICE	Relative cost increase (Design to DEA) * (%)	Relative cost (Contract to Design) (%)	Relative cost increase (Contract to DEA) (%)
1	Physical *medicine and rehabilitation clinics mostar*	91,753	1,23,036	1,23,745	34	1	35
2	Directorate *of the clinical center mostar*	94,673	95,461	97,616	1	2	3
3	Clinic *for infectious diseases*	3,24,325	3,43,097	3,54,534	6	3	9
4	Clinic for *internal diseases* with *dialysis centre*	2,70,548	2,93,256	4,10,923	8	40	52
5	Police academy "Vraca ", Sarajevo	4,37,325	4,91,620	4,84,209	12	−2	11
6	Elementary school „Kulin Ban ", Visoko	2,77,007	2,41,965	3,28,630	−13	36	19
7	Elementary school „Safvet-beg Bašagić ", Visoko	2,10,747	2,15,201	2,40,711	2	12	14
8	Elementary school „Vareš ", Vareš	93,582	1,45,361	1,48,982	55	2	59
9	Elementary school „Safvet-beg Bašagić ", Breza	2,20,984	2,08,595	2,54,499	−6	22	15
10	Mixed secondary school ", Kakanj	2,48,531	2,54,304	2,32,517	2	−9	−6

(continued)

Table 1 (continued)

No	Public facility	Total investments (EE + non EE) DEA	Total investments (EE + non EE) DESIGN	Total investments (EE + non EE) FINAL CONTRACT PRICE	Relative cost increase (Design to DEA) * (%)	Relative cost (Contract to Design) (%)	Relative cost increase (Contract to DEA) (%)
11	"Mixed Secondary School ", Zenica	2,85,792	3,44,789	4,10,906	21	19	44
12	Elementary school „Rešad Kadić ", Brnjic-Kakanj	78,221	1,04,993	1,22,534	34	17	57
13	„*Medical School Zenica* ", Zenica	2,65,635	3,44,789	3,16,809	30	−8	19
14	Faculty *of Philosophy, Zenica*	2,29,471	2,93,752	3,31,841	28	13	45
15	„Faculty *of Mechanical Engineering ", Zenica*	4,50,557	5,00,506	5,55,260	11	11	23
16	„Faculty *of "Metalurgy and Technology ", Zenica*	4,30,664	5,69,076	6,29,988	33	11	46
17	Elementary *school „Mak Dizdar ", Zenica*	3,78,149	4,81,488	5,53,172	27	15	46
18	Elementary *School „Skender Kulenović ", Zenica*	1,94,498	1,85,453	2,37,355	−5	28	22
19	Elementary *School „Novi Šeher ", Novi šeher-Maglaj*	93,299	1,34,476	1,13,497	44	−16	22

Table 2 Table captions should be placed above the tables

No	Facility	Investment [BAM]	Energy savings for heating [kWh]	Saving energy costs for heating [BAM]	Saving electricity [kWh]	Saving electricity costs [BAM]	Simple payback period (year)	Index of savings (kWh/BAM)	Source
1	*Physical Medicine and Rehabilitation Clinics Mostar*	1,78,918.80	1,09,941.00	20,383.00	22,139.00	5,427.00	6.93	0.74	DEA
		2,41,303.06	1,09,941.00	25,478.75	1,200.00	294.16	9.36	0.55	FINAL CONTRACT PRICE
2	*Directorate of the Clinical Center Mostar*	1,84,611.59	91,802.00	17,020.00	20,382.00	4,997.00	8.38	0.61	DEA
		1,90,350.70	88,519.00	20,514.17	4,982.00	1,221.42	8.76	0.59	FINAL CONTRACT PRICE
3	*Clinic for Infectious Diseases*	6,32,434.00	5,07,914.00	1,40,208.00	63,550.00	15,887.50	4.05	0.9	DEA
		6,91,342.26	5,07,914.00	1,17,708.44	32,911.00	8,227.75	5.49	0.83	FINAL CONTRACT PRICE
4	Clinic for *Internal Diseases* with *Dialysis Centre*	5,27,569.00	5,11,473.00	83,882.00	32,316.00	8,102.00	5.74	1.03	DEA
		8,01,299.06	5,11,473.00	1,04,852.50	32,316.00	8,102.00	7.09	0.68	FINAL CONTRACT PRICE
5	Police Academy "Vraca", Sarajevo	8,52,784.00	18,94,141.00	1,93,769.00	26,249.00	4,919.00	4.29	2.25	DEA
		9,44,207.97	18,94,141.00	3,59,856.71	26,249.00	4,919.00	2.59	2.03	FINAL CONTRACT PRICE
6	Elementary school „Kulin Ban", Visoko	5,40,162.82	3,84,299.00	80,281.98	2,555.11	553.98	6.68	0.72	DEA

(continued)

Table 2 (continued)

No	Facility	Investment [BAM]	Energy savings for heating [kWh]	Saving energy costs for heating [BAM]	Saving electricity [kWh]	Saving electricity costs [BAM]	Simple payback period (year)	Index of savings (kWh/BAM)	Source
		6,40,828.53	3,84,299.00	1,00,352.48	2,555.11	553.98	6.35	0.6	FINAL CONTRACT PRICE
7	Elementary school „Safvet-beg Bašagić“, Visoko	4,10,957.50	3,03,006.00	1,23,169.85	4,037.00	928.65	3.31	0.75	DEA
		4,69,385.78	2,72,616.00	1,38,520.66	4,037.00	928.65	3.37	0.65	FINAL CONTRACT PRICE
8	Elementary school „Vareš“, Vareš	1,82,485.00	2,59,094.00	7,630.00	3,666.00	709	21.88	1.44	DEA
		2,90,515.60	2,59,094.00	9,763.55	3,666.00	709	27.74	0.9	FINAL CONTRACT PRICE
9	Elementary school „Safvet-beg Bašagić“, Breza	4,30,918.95	1,82,908.21	26,616.36	2,736.00	629.28	15.82	0.43	DEA
		4,96,273.14	1,73,952.00	25,313.07	1,272.00	292.56	19.38	0.37	FINAL CONTRACT PRICE
10	Mixed Secondary School“, Kakanj	4,84,636.12	7,67,497.00	57,186.66	6,415.00	2,303.05	8.15	1.6	DEA
		4,53,408.08	7,67,497.00	57,186.66	6,415.00	2,303.05	7.62	1.71	FINAL CONTRACT PRICE
11	“Mixed Secondary School“, Zenica	5,57,295.10	6,54,509.00	1,05,519.99	13,399.00	2,591.00	5.15	1.2	DEA

(continued)

Table 2 (continued)

No	Facility	Investment [BAM]	Energy savings for heating [kWh]	Saving energy costs for heating [BAM]	Saving electricity [kWh]	Saving electricity costs [BAM]	Simple payback period (year)	Index of savings (kWh/BAM)	Source
		8,01,265.92	6,42,902.00	1,03,648.71	8,600.00	1,663.00	7.61	0.83	FINAL CONTRACT PRICE
12	Elementary school „Rešad Kadić", Brnjic-Kakanj	1,52,530.00	2,49,271.00	8,883.00	2,284.00	457	16.33	1.65	DEA
		2,38,940.36	2,49,271.00	11,366.92	1,564.00	312.94	20.46	1.05	FINAL CONTRACT PRICE
13	„*Medical School Zenica*", Zenica	5,17,989.00	7,20,377.00	1,16,752.00	3,321.00	996	4.4	1.4	DEA
		6,17,777.88	6,54,623.00	1,06,095.20	3,321.00	996	5.77	1.17	FINAL CONTRACT PRICE
14	Faculty *of Philosophy, Zenica*	4,47,469.00	3,34,782.00	51,888.41	9,615.00	2,394.22	8.24	0.77	DEA
		6,47,090.39	3,20,790.00	49,719.77	9,615.00	2,394.22	12.42	0.53	FINAL CONTRACT PRICE
15	„Faculty *of Mechanical Engineering", Zenica*	8,78,587.00	5,10,261.00	79,090.00	15,116.00	3,820.00	10.6	0.6	DEA
		10,82,757.64	5,10,261.00	79,090.00	1,638.00	413.94	13.62	0.49	FINAL CONTRACT PRICE
16	„Faculty *of"Metalurgy and Technology", Zenica*	8,39,795.00	6,13,473.00	95,089.00	25,740.00	8,641.00	8.1	0.76	DEA

(continued)

Table 2 (continued)

No	Facility	Investment [BAM]	Energy savings for heating [kWh]	Saving energy costs for heating [BAM]	Saving electricity [kWh]	Saving electricity costs [BAM]	Simple payback period (year)	Index of savings (kWh/BAM)	Source
		12,28,476.44	6,13,473.00	95,089.00	25,740.00	8,641.00	11.84	0.52	FINAL CONTRACT PRICE
17	Elementary *school „Mak Dizdar", Zenica*	7,37,389.80	5,69,667.00	80,409.39	2,867.20	659.46	9.1	0.78	DEA
		10,78,686.02	5,69,667.00	80,409.39	2,560.00	588.8	13.32	0.53	FINAL CONTRACT PRICE
18	Elementary *School „Skender Kulenović", Zenica*	3,79,270.70	2,98,209.00	46,848.53	9,642.00	8,676.71	6.83	0.81	DEA
		4,62,843.05	2,98,209.00	46,848.53	9,642.00	2,892.60	9.31	0.67	FINAL CONTRACT PRICE
19	Elementary *School „Novi Šeher ", Novi šeher-Maglaj*	1,81,933.00	92,447.00	20,625.00	2,572.00	707	8.53	0.52	DEA
		2,21,318.79	92,447.00	25,781.25	2,572.00	707	8.36	0.43	FINAL CONTRACT PRICE

severe stock market crashes in the modern history [22]. The supply chain disruptions had severe operational and financial consequences on the construction sector leading to an increase in the prices of construction materials. Furthermore, in the period from the development of DEAs to the implementation of measures on the site, a certain period passed in which there were general changes in the labor market in BiH, with a very significant impact on the construction sector. In the last few years, the departure of trained labor force outside B&H has increased, thus reducing the number of experienced workers, which has created a problem for domestic construction companies. In order to mitigate this trend, there has been a significant increase in salaries in the construction sector in B&H. According to statistics from the state statistics agency, the average salary in the construction sector has increased from 525 BAM in 2015 to 624 BAM in 2019, which is an increase of 19%, but salaries are still generally very low, especially when compared to the market in the EU. With the increase in labor prices, the share of labor costs in the total price of construction services also increased.

All this has led to an increase in the prices of construction services in the market. It also has to be stated that the real estate market in the last few years is recording a significant increase in real estate prices with a causal relation to construction sector market.

5 Recommendations

The energy audit is essential for providing an investor an estimated value of required investment to reduce energy billing and to decrease energy wastages. Based on the information received form the starting document, the investor makes a decision whether or not to invest in energy efficiency project. From the presented analysis, it is evident that several factors affect the difference between the planned and realized investment. Some of them refer to existing legal solutions and some to extreme factors that are out of control, such as Covid 19, the rise in prices of construction materials that started during the pandemic and continued with the conflict in Ukraine. It also has to be stated that the real estate market in the last few years is recording a significant increase in real estate prices with a causal relation to construction sector market. All of the above mentioned can be seen as the main factors which lead to the increase of the estimated investment costs. Most of the identified determinants are caused by external factors on which investors have little or no influence.

Therefore, our recommendations go exclusively in the direction of updating the existing methodology on the development of the Detailed energy audit in order to increase its accuracy, complexity and functionality and ultimately the planned scope of measures and savings (financial and non-financial as GHG). Some of the recommendations that should give a positive impact on DEA accuracy are: external independent review of DEA's should be provided, accompanying non EE measures should be defined and valorized in the DEA, new EE measures besides the standard ones should be analyzed, DEA's should not be older than three years or they should be updated if any changes in building construction or system occurred or there were significant changes in energy and construction market. Government of the Federation of BiH must take some steps to make energy audits compulsory for different purposes and implementations. Alongside a given structure of such an audit, a few imperative undertakings are underlined to achieve a reliable energy audit. For that purpose the Government of the Federation of BiH needs to draft the amendments and additions to the Regulation on conducting energy audits and issuing energy certificates, including Annexes and to the Rulebook on minimum requirements for the energy characteristics of buildings, including Annexes [23]. In order to fulfill the obligations under the Agreement on the Energy Community sign by Bosnia and Herzegovina, the proposed changes have to including to include also the transposition of the relevant following Directives: (i) the Directive on the energy performance of buildings 2010/ 31/EU EPBD; (ii) the Directive on energy efficiency EED2012/ 27/EU; (iii) Directive (EU) 2018/844/EU on the amendment of Directive 2010/31/EU on the energy performance of buildings and (iv) Directive 2012/27/EU on energy efficiency. This refers primarily to modification of the Methodology for conducting energy audits [24], i.e. the algorithm and procedures. In the Rulebook on minimum requirements for the energy characteristics of buildings, it is required to revise and update the algorithm for calculating the energy characteristics of buildings, which refers to the definition of the reference value of primary and final energy consumption.

It is to be noted that the construction sector is one of the most important economic activities and yet there appears to be a dearth of purposeful effort in identifying and defining existing solutions applicable to various building typologies to serve as a reference for building designers and developers on the possibilities available [25]. In most of the retrofit projects, energy efficient retrofit strategies are not applied due to a lack of knowledge about the amount of investment required and the efficiency of the potential energy saving strategies [26]. For many, it is the complexity of retrofit and financing that present a barrier to intervention and uptake [27].

References

1. United Nations, Climate Change—United Nations Sustainable Development. https://www.un.org/sustainabledevelopment/climatechange-2, last accessed 21.01.2023
2. IEA, Energy Efficiency (2018) Analysis and outlooks to 2040. Mark Rep Ser 1–143. https://doi.org/10.1007/978-3-642-41126-7
3. Franfurt School, Certified Expert in Climate & Renewable Energy Finance. Unit 10: Financing Energy Efficiency Projects (Frankfurt am Main), p 4
4. Improving Energy Efficiency. https://fmpu.gov.ba/en/conference-improving-energy-efficiency/, last accessed 13.01.2023
5. National renewable energy action plan of Bosnia and Herzegovina, 23092021_NREAP_2016_BH_ENG.pdf (mvteo.gov.ba), last accessed 16.02.2023
6. PARIS AGREEMENT, United Nations 2015. https://unfccc.int/files/essential_background/convention/application/pdf/english_paris_agreement.pdf, last accessed 20.01.2023
7. Decree on energy audits and energy certificates, Official Gazette of the Federation BiH, no. 87/18). https://fmpu.gov.ba/en/legislation/decrees/, last accessed 16.02.2023

8. Guidelines for energy audits in new and existing buildings with simple and complex technical systems, (2009). https://fmpu.gov.ba/en/legislation/rulebooks/, last accessed 16.01.2023
9. Reddy SS, Pankay S, Seong-Cheol K (2021) Energy audit: types, scope, methodology and report structure. Indonesian J Elect Eng Comp Sci 22(1):45. https://doi.org/10.11591/ijeecs.v22.i1.pp45-52
10. Fleiter T, Schleich J, Ravivanpong P (2012) Adoption of energy-efficiency measures in SMEs-An empirical analysis based on energy audit data from Germany. Energy Policy 51:863–875. https://doi.org/10.1016/j.enpol.2012.09.041
11. Sorrell S, O'Malley E (2004) The economics of energy efficiency
12. Kontokosta CE (2013) Energy disclosure, market behavior, and the building data ecosystem. Ann NY Acad Sci 1295:34–43. https://doi.org/10.1111/nyas.12163
13. Marasco DE, Kontokosta CE (2016) Applications of machine learning methods to identifying and predicting building retrofit opportunities. Energy Build 128:431–441. https://doi.org/10.1016/j.enbuild.2016.06.092
14. Shapiro I (2011) 10 Common Problems in Energy Audits, ASHRAE J53
15. Kobbina JA, Frimpong EA, Okyere PY (2018) Energy auditing: an overview and key issues. In: Proceedings of the World Congress on Engineering and Computer Science, vol I, San Francisco, USA, ISSN: 2078–0958 (Print), ISSN: 2078–0966 (Online)
16. Palmer K, Walls M, Gordon H et al. (2013) Assessing the energy-efficiency information gap: results from a survey of home energy auditors. Energy Effic 6:271–292. https://doi.org/10.1007/s12053-012-9178-2
17. Gamtessa SF (2013) An explanation of residential energy-efficiency retrofit behavior in Canada. Energy Build 57:155–164. https://doi.org/10.1016/j.enbuild.2012.11.006
18. Kontokosta CE (2015) A market-specific methodology for a commercial building energy performance index. J Real Estate Finan Econ 51:288–316
19. Hsu D (2014) How much information disclosure of building energy performance is necessary? Energy Policy 64:263–272. https://doi.org/10.1016/j.enpol.2013.08.094
20. Rowse B (2023) Advantages and Disadvantages of an Energy Audit. https://8020green.com/advantages-disadvantages-energy-audit/, last accessed 20.02.2023
21. KI Expert BIH. https://www.knaufinsulation.hr/preuzmite-dokumente/ki-expert-bih, last accessed 27.04.2023
22. Reisner R (2021) 8 of the biggest stock market crashes in history—and how they changed our financial lives. In Galijasevic A, Tegbaru J (eds) Decision-making In Mutual Funds During the COVID-19 Pandemic , Master of Science Thesis, Industrial Engineering and Management, Stockholm
23. Rulebook on minimum requirements for the energy characteristics of buildings, including Annexes, Official Gazette of the Federation of BiH No. 81/19 & No. 85/19, Rulebooks | Federalno ministarstvo prostornog uređenja (fmpu.gov.ba), https://fmpu.gov.ba/en/legislation/decrees/, last accessed 20.02.2023.
24. Methodology for calculation and reporting of energy performance of buildings with the calculation algorithm. https://fmpu.gov.ba/en/legislation/decrees/, last accessed 18.02.2023
25. Tatang N, Ray T (2022) Design and planning of energy efficient building: a critical review. Int J Eng Res Appl 12(2):28–31. https://doi.org/10.9790/9622-1202012831
26. Basarir B, Diri BS, Diri C (2023) Energy efficient retrofit methods at the building envelopes of the school buildings. In: Proceedings Retrofit. http://www.academia.edu/Ellipsis/Energy_efficient_retrofit_methods_at_the_building_envelopes, last accessed 12.02.2023
27. Domestic UK Retrofit challenge: drivers, barriers and incentives leading into the green deal (2012). Energy Policy 50:294–305

Energy Efficiency in Public Buildings in BiH

Asmin Veladžić, Aida Veladžić, and Azra Kapetanović

Abstract With the potential to reduce the average energy use by about 60%, public buildings have been identified as the sector with the largest potential for cost-effective energy saving in BIH, by applying interventions on buildings to increase their energy efficiency and thus enhance energy performance. Considering the critical amount of power consumption in public buildings, increasing energy efficiency can present an opportunity to tackle the climate change and air pollution problems, while having a positive impact on the overall energy stability and sustainable growth of BIH. In addition to this, in order to achieve membership to the European Union, BIH must ensure complete compliance with EU standards and regulations and has to eventually achieve the EU energy and climate goals. This paper provides a brief overview of energy efficiency in BIH with an emphasis on EU targets and visions related to EE. Furthermore, it describes the related EE legal framework, the public buildings sector and major initiatives and EE projects in BIH. Finally, the paper shortly presents the technical implementation process for public buildings EE projects and provides an estimation of potential energy savings and benefits.

Keywords Energy efficiency · Public buildings · Bosnia and Herzegovina

A. Veladžić · A. Veladžić (✉) · A. Kapetanović
AVCON d.o.o. Sarajevo, Branislava Nušića bb, 71 000 Sarajevo, Bosnia and Herzegovina
e-mail: aida@avcon.ba

A. Veladžić
e-mail: asmin@avcon.ba

A. Kapetanović
e-mail: azra@avcon.ba

M. Arslanagić-Kalajdžić et al. (eds.), *Interdisciplinary Advances in Sustainable Development II*, Lecture Notes in Networks and Systems 804,
https://doi.org/10.1007/978-3-031-46692-2_5

1 Introduction

Energy Efficiency (EE) is a broad term that refers to topics related to energy savings, consumption, sufficiency and transition. However, simply put, EE can be defined as the use of less energy to perform the same task or produce the same result. Bearing in mind that less energy means lower energy costs and less pollution, EE is one of the easiest and most cost-effective ways to combat climate change, reduce energy costs for consumers, and is a vital component in achieving net-zero emissions of carbon dioxide.

The EE measures have several advantages: they can be implemented rather quickly, and when implemented correctly, they ensure long-term energy and financial savings, improve the building structure and design, and increase comfort for the occupants. In addition to this, EE measures can help facilitate the transition to a sustainable future by unlocking the potential for an all-renewable energy supply for everyone significantly faster than without EE measures.

Having in mind that improving of EE measures in BIH can have a significant impact on reducing environmental pollution and improving the overall energy stability and sustainable growth of Bosnia and Herzegovina, this paper aims to respond to the question: are the potentials of energy efficiency in public buildings in BIH properly used and which shortcomings should be addressed to reach the full effect of implementation of EE measures on public buildings in Bosnia and Herzegovina?

Considering the above-mentioned, EE is a cornerstone of the European energy policy and addressed in different strategies and policies developed by the EU Energy Community, as one of the necessary components of building decarbonization. The Energy Community Treaty was signed on 25 October 2005 in Athens between the European Community on the one hand and nine parties, countries of South East Europe, including Bosnia and Herzegovina (BiH), on the other. On 29 May 2006, the Energy Community Treaty was approved on behalf of the European Community with the Council's Decision 2006/500/EC. This Treaty, which came into effect on 1 July 2006, commits all parties to establish a common energy market that will function according to EU energy market standards and to transpose and implement the appropriate EU directives and regulations in the area of electricity, gas, energy infrastructure, oil, supply security, renewable energy sources, energy efficiency, environment, market competition, energy statistics and social issues [1].

The first document of this type was the Kyoto Protocol, which aimed to operationalize the United Nations Framework Convention on Climate Change by committing industrialized countries and economies in transition to limit and reduce greenhouse gas emissions in accordance with agreed individual targets. Following Kyoto Protocol, the Energy Community deepened this approach with the so-called 20–20–20% targets for 2020. The 20–20–20 Climate and Energy Package, adopted in 2008, was introduced as a binding strategy that sets three targets for the members of the Energy Community. The first set target requires EU Member States to cut greenhouse gas emissions by 20% relative to 1990 levels, for the year 2020. The second target sets an EU-wide share of 20% of gross final energy consumption from Renewable Energy Sources (RES), and the third target is to cut gross primary energy consumption by 20% by 2020 compared to projections published in 2007 and in 2009. In October 2014, the European Council agreed on a new target approach and framework for 2030: to reduce greenhouse gas emissions by 40%, produce 27% of energy from RES and cut energy consumption by 27% [1].

2 Energy Efficiency in Bosnia and Herzegovina

Bosnia and Herzegovina, as one of the Contracting Parties of the Energy Community Treaty, is legally committed to adopting the EU energy legislation (so-called "acquis communautaire") [2] and has the obligation to implement the energy acquis in force. Despite the long-term commitment, according to reports from the Energy Community Secretariat, Bosnia and Herzegovina has achieved only limited progress in most of the energy sectors during the last years [2–4]. The 2021 Annual Implementation Report [2] prepared by the Energy Community Secretariat states that, since BIH energy legal framework remained fragmented along entity lines, which blocks much-needed reforms, many commitments under the Treaty have jet not been completed. In addition to this, the gas sector legislation is outdated, and the environmental standards remained low with levels of sulfur dioxide, nitrogen oxides and dust from large combustion plants above the ceiling. However, EE targets and policy measures are met by up to 46%. The most significant effort was made in the buildings sector, where amendments to primary legislation and long-term building renovation strategies, to align with the energy efficiency provisions under the Clean Energy Package, have been drafted on state and entity levels. Both entities remain focused on working on updating existing by-laws, including improvements in energy audits and certification of building procedures. In addition to this, many initiatives toward improving EE in the building sector have been launched and are being implemented successfully [2].

3 Legal Framework Related to Energy Efficiency in BIH

Due to Bosnia Herzegovina's constitutional framework, the entity ministries are in charge of developing legislation related to energy, and therefore Energy Efficiency. The regulation of the legislative framework in the field of energy efficiency in the BIH is the result, among other things, of the obligations pursuant to international agreements which BIH signed as the Energy Community Treaty member. One of the main such agreements is the Treaty on the Establishment of the Energy Community from 2005, which directly prescribes the activities and legislation in this area and gives the obligation to transpose relevant EU directives into local legislation. The Directive 2010/31/EU on the energy performance of buildings (EPBD) is one of the most complex energy efficiency directives for implementation in the Energy

Community and requires cooperation between various stakeholders and various other activities, besides work on the development of legislation [5].

The basic document that established the legal framework in the Federation of Bosnia and Herzegovina (Federation of BIH) is the Law on Energy Efficiency in the Federation of BIH (Official Gazette of the FBiH No.22/17), which regulates the EPBD requirements. Further harmonization was achieved through ordinances based on the mentioned Law, as follows [6]:

- Rulebook on minimum requirements for energy performance of buildings;
- Rulebook on Energy Certification of Buildings in the Federation of BIH;
- Methodology for calculation and declaration of energy characteristics of residential and non-residential buildings;
- Rulebook on Technical Requirements for Heat Insulation of Buildings and Rational Use of Energy;
- Rulebook on regular energy audit of heating system and air conditioning system;
- Rulebook on the Energy Efficiency Information System of the Federation of BIH.

Starting from 2017, pursuant to Article 24(1) of Directive 2012/27/EU which commits each Contracting Party to set national energy efficiency targets for energy consumption reduction, BIH has a reporting obligation to the Energy Community Secretariat to gather and submit the data on energy efficiency targets, indicators, consumption trends and key Energy Efficiency Directive implementing measures. In addition to this, starting from 2019, BIH has a reporting obligation pursuant to Annex VIII.B of the LargeCombustion Plan Directive 2001/80/EC as amended by Decision 2013/05/MC-EnC [5].

Although BIH has a legal framework for EE, many of the laws and regulations have to be amended and updated in accordance with the latest guidelines and initiatives of the Energy Community Treaty. BIH adopted the compliant Energy Efficiency Law in Brčko District in July 2022, and drafted by-laws on the entity level. Both Federation of BIH and Republic of Srpska have established an Energy efficiency and environmental Fund [7]. Unfortunately, despite the fact that EE laws of both entities and Brčko District recognize ESCOs (Energy service companies) and energy performance contracting, the ESCO market is not functioning due to the big implementation gaps which exist in public procurement, multi-year budgeting and adoption of model ESCO contracts.

Generally, there are several legal and regulatory barriers that may pose a challenge to the implementation of EE projects, especially in the public sector such as complex government structure, limited implementation of the Laws on Energy Efficiency, lack of institutional focus on EE, restrictions on borrowing by public agencies, restrictive procedures that may prevent retention of energy cost savings, public procurement rules and procedures, inability to sign long-term contracts, no provisions for performance contracting and lack of comprehensive Public–Private Partnership law.

4 Public Buildings in Bosnia and Herzegovina

Despite the initiatives to decarbonize the building sector, buildings are still big energy consumers worldwide: they consume 40% of global energy, 60% of the world's electricity, 40% of waste, and 40% of material resource use. Furthermore, buildings are accountable for 25% of global water consumption and are responsible for up to 33% of the world's greenhouse gas emissions [8].

In Bosnia and Herzegovina, due to the construction characteristics and the lack of legal regulations related to energy efficiency, one of the biggest energy consumers are particularly public buildings built in the period from 1950 to 1980. For example, the average older building annually consumes 200–300 kWh/m^2 of energy for heating, but in cases of public buildings in BIH, due to dilapidated heating and cooling systems and lack of adequate thermal insulation, the energy consumption, primarily for heating and cooling, can be significantly higher [9]. In addition to this, many public buildings throughout Bosnia and Herzegovina were partially or completely destroyed during the war, after which they were inadequately renovated and maintained. In total, there are 92 buildings in Bosnia and Herzegovina owned and used by central state institutions at all four administrative levels, which makes a total of 396,226.39 m^2 of heated and/or cooled building area. The BIH state level occupies 43 buildings with an area of more than 250 m^2. These 43 buildings occupy an area of 244,993.86 m^2 of heated/cooled area. Central government institutions of Federation of BIH use 46 buildings, with the total area of these facilities being 71,593.07 m^2. Central government institutions in the Republic of Srpska (RS) are in 6 buildings with a total area of 60,000 m^2 and the central government institutions in the Brčko District occupy a total of 17 buildings of 19,631.46 m^2. The total area of non-residential buildings in Federation of BIH is 12,730,000 m^2, of which 7,568,713 m^2 are covered by commercial buildings and the area of public buildings is 5,161,287 m^2 which is nearly 57% of the overall heated surface of the public buildings in BIH. Buildings intended for education have the largest area of 1,726,842 m^2, while buildings intended for kindergartens have the smallest, i.e. 73,796 m^2. The total area of non-residential buildings in RS is 5,890,000 m^2, of which the area of public buildings is 3,614,839 m^2 and the area of commercial buildings is 2,275,161 m^2. Educational buildings have the largest area, 1,084,867 m^2, and buildings intended for kindergartens have the smallest, 46,602 m2. The largest number of public buildings in RS was built in the period from 1974 to 1987. In Brčko District, buildings intended for education have the largest area of 105,492 m^2, while buildings intended for kindergartens have the smallest, i.e. 2,311 m^2 [5].

As to detailed energy audits conducted in public facilities by UNDP [8], the average energy use in a building can be reduced cost-efficiently by about 60%, assuming a given comfort level in the building (e.g. 20 °C) before and after retrofitting. Thus, public buildings have been identified as the sector with the largest potential for cost-effective energy saving in BIH. As to International Energy Agency data, the energy intensity of BIH is 0.5 tons/1000$ of GDP. For comparison purposes, it is 4 times higher than the average in the European Union, which means that an average

European country earns 4 times more national income than BIH for the same amount of consumed energy [9]. By applying different interventions on buildings to enhance their energy performance, investments made to increase EE in public buildings have a large potential to reduce the enormous economic and environmental damage caused by high energy consumption, which might, in the long run, help BIH to achieve the targets set forth in the Nationally Determined Contribution (NDC) under the Paris Agreement.

Although it is a cost-effective solution, budgetary constraints of public buildings owners, who are often municipalities or other city bodies, lead to challenges in providing financing for these projects. Due to the fragmented and complex inter-authority jurisdictions, especially in the Federation of BIH, authorities and line ministries do not possess a clear overview of public buildings under their jurisdiction, or energy and water-related consumption and the costs that incur on a monthly basis. In addition to this, official data on the energy intensity of public building stock do not exist. Although draft plans for improved energy performance in buildings (Operational Energy Efficiency Action Plans of public sector buildings in several cantons in Federation of BIH and Energy Efficiency Action Plan of RS) are being laid down, a comprehensive policy implementation platform and monitoring framework for public buildings is still missing. Furthermore, public building occupants and authorities/entities in charge of public building management are often not aware of the importance of energy-saving mechanisms, nor properly trained, prepared or equipped to control or manage the systems for heating and cooling. There is a lack of essential capacities, human and technical resources to identify, prepare and implement low-carbon investment projects [8]. Furthermore, there are several other financial barriers that slow down investments in EE in public buildings. Firstly, the unwillingness of the lenders to provide debt financing to public agencies, the lenders' perception of high risk and the lack of customized financial products. Moreover, project financing often requires collateral, and many public agencies may find it difficult to provide it. Finally, EE projects are generally smaller in the terms of the amount of financing, which leads to high transaction costs.

5 EE Funding Programs in BIH

Despite the challenges which may arise prior and during the implementation of EE projects, the Federation of Bosnia and Herzegovina made a significant effort with the implementation of so-called direct EE measures by implementing several programs for increasing energy efficiency in final energy consumption. These EE measures are mainly related to the energy renovation of public buildings under the jurisdiction of the Federation of Bosnia and Herzegovina. According to the Action plan for energy efficiency (APEE) for public buildings for the period from 2018 to 2021 [7] sector of public buildings in the Federation of BIH is deemed to be the leading example of achieving the EE goals, set in the Renewable Energy Action Plan of Bosnia and Herzegovina (NREAP BiH), adopted in line with the Decision on the

Implementation of Directive 2009/28/EC on the promotion of the use of energy from renewable sources by the Ministerial Council in October 2012 [5].

Both Federation of BIH and RS have established an Energy efficiency and Environmental Fund, covering monitoring of implementation and reporting on achieved savings. The Revolving Fund provides sustainable financing for the EE sector in Bosnia and Herzegovina because the budgets do not have enough funds for EE projects. The users of the EE Revolving fund can be legal entities, municipalities, cities, bodies and organs of the entity and cantonal administrations and other public institutions, non-profit organizations and non-governmental bodies (individual craftsmen, companies, facilities) which want to implement EE measures through a revolving fund loan.

The Environmental Protection Fund of the Federation of Bosnia and Herzegovina cooperates with governmental and non-governmental entities to provide funding for projects related to environmental protection, in particular: protection, preservation and improvement of air, soil, water and sea quality, and climate change mitigation and protection of the ozone layer; remediation, encouragement of waste avoidance and reduction; exploitation of valuable properties and waste treatment; protection and preservation of biological and landscape diversity. Every year, this Fund announces a public call for the allocation of funds for the implementation of programs, projects and similar activities in the field of environmental protection. In 2021, 1.5 million BAM was intended for co-financing of projects to increase energy efficiency in residential and non-residential buildings, in public services, and in industrial processes, whilst in 2022, this budget increased to 2 million BAM. In addition to this, in the past few years, the Federation of BIH Environmental Protection Fund implemented several energy programs focused on energy efficiency measures and renewable energy sources. One of the most important EE projects related to public buildings which the Federation of BIH Environmental Protection Fund implements in cooperation with UNDP and Federal Ministry of Physical Planning is Scaling-up Investment in Low-Carbon Public Buildings in Bosnia and Herzegovina project. The project goal is to support the efforts of BIH to address climate change challenges, reduce CO_2 emissions and adapt to climate change in accordance with the recommendations of the Nationally Determined Contribution (NDC) and the Climate Change Adaptation Strategy of Bosnia and Herzegovina. The expected results of this project are as follows: low-carbon retrofits in 430 public buildings will result in zero carbon emissions; conditions will be improved for 150,000 users, more than will be 2,500 persons trained in sustainable energy management; direct reduction in greenhouse gas emissions will reach cca. 2.02 million tCO2e [10]. Many other multinational and bilateral financial institutions are active in BIH and have launched significant projects related to energy efficiency:

The World Bank conducted a comprehensive assessment of the options for scaling up EE projects in buildings in the Western Balkan countries in 2013/2014 and started implementing the Bosnian Energy Efficiency Project (BEEP). Since 2016, within BEEP, energy efficiency measures have been implemented in the buildings of public institutions in the Federation of BiH. To date, more than 25 million BAM has been invested in the reconstruction of buildings, which resulted in reducing CO_2 emissions by 8,354 tons per year. In addition to increasing comfort for more than 456,500 users, the Project also results in creating of more than 1,960 green jobs. Buildings of high social importance, such as educational and health institutions, are the focus of the Project, along with other public buildings. The average investment per building is 492,080 BAM with more than 39.366.610,00 BAM invested so far. Since the start of the project, a total of 80 facilities have been renovated, 58 of which are educational, 17 are health, and 5 are other public institutions. The extension of the BEEP project is secured through the preparation of the Additional Funding (AF BEEP) project in 2021. The Federation of BIH Ministry of Spatial Planning realized multiple measures under the AF BEEP, in order to conserve and reduce energy consumption, while simultaneously modernizing living conditions and improving economic performance. In total, 33 public buildings and 2 roof replacements in the Federation of BIH are confirmed to be retrofitted. The selected buildings include 22 cantonal and 11 federal-level public buildings, with an estimated cost of 8,4 million EUR [11].

Since 2013, United Nations Development Program (UNDP) is implementing the Green Economic Development Project (GED), financed by the Government of Sweden, the Environmental Protection Fund of Federation of BIH and the Environmental Protection and Energy Efficiency Fund of RS. The Project partners are the Ministry of Foreign Trade and Economic Relations of BIH, the entity's ministries for spatial planning, cantonal ministries and other institutions. So far, within this project, energy management institutionalization has been introduced in 8 cantons, the Decision on obligatory EMIS data input is adopted in 6 cantons, 2 Energy Efficiency Action Plans have been developed, over 4,900 users have been trained to use the EMIS system, over 270 detailed energy audits on public sector buildings were conducted to identify the most effective measures to improve energy efficiency, the Revolving fund within the Environmental Fund of Federation of BIH which enables investments into energy-efficiency projects on public, business and industry buildings was established, energy efficiency retrofits on 229 public buildings across the country are conducted, modernization of public lighting system is implemented in nine municipalities and over 67,700 citizens have participated in educational events and workshops. In the last four years, through various projects, UNDP has managed about 70 million USD of funds in the Federation of BIH of which about 20 million USD of combined funds have been directed to energy efficiency measures and renewable energy sources. Only in Sarajevo Canton, through the Green Economic Development (GED) project the measures to increase the energy efficiency of public buildings have been implemented in 70 public buildings. In addition to a significant improvement in the quality of living and working conditions for about 30 thousand users of these public institutions, it was estimated that energy consumption was reduced by approximately 59%, and that CO_2 emissions were reduced by about 4,100 tons per year, which directly contributes to the improvement of air quality [8, 12, 13].

European Bank for Reconstruction and Development (EBRD) established the Western Balkans Sustainable Energy Financing Facility (WeBSEFF), to provide debt financing for renewable energy and industrial energy efficiency projects to small and medium-sized enterprises (SMEs) in the Western Balkans. WeBSEFF is working closely with local partner banks which conduct a full credit assessment of each borrower and their project proposals. Partner banks under WeBSEFF in BIH are Raiffeisen Bank and UniCredit Bank. Overall, the Facility's investments in projects have contributed to saving more than 103 million kWh of energy and more than 32 million kg of CO_2 emissions per year. Furthermore, EBRD established a legal framework to support Energy service companies (ESCOs) that can successfully implement EE projects in the public sector. In November 2018, Sarajevo Canton joined EBRD Green Cities. This framework is designed to help cities to articulate a sustainable development vision and their strategic objectives as well as define actions and investments necessary to address priority environmental issues. EBRD and the European Union provided a financial package for improvement in the EE of public buildings in Canton Sarajevo under the Green City Action Plan. The investments cover up to 40 public buildings such as: schools and related facilities, kindergartens, student dormitories and health centers. The energy savings are estimated to be around 13.7 GWh and would result in 4,774 tons of CO_2 savings annually [14]. Energy efficiency measures on the public buildings within this project include the implementation of a series of interventions, such as the replacement of openings, thermal insulation of facades, installation of LED lighting, installation of thermostatic valves, etc. The project is being funded by long-term loan from EBRD in the amount of 8 million EUR, while the European Union, through the Regional Energy Efficiency Program, provided a grant of 2 million EUR. The EBRD has also mobilized additional grant funds for technical assistance from the Government of Austria in the amount of 492.240,00 EUR and from the EBRD Special Share Fund (Shareholder Special Fund) in the amount of 600.000,00 EUR. In 2022, the rehabilitation of the 13 public buildings in Canton Sarajevo has been finished. Currently, the works are being done on one elementary school and one health center. The implementation of the EE measures on the rest of the 25 public buildings is expected during 2023 [14].

U.S. Agency for International Development (USAID) launched the Energy Policy Activity to help BIH coordinate, manage and improve transparency in the gas and electricity sectors while providing targeted technical assistance to improve and simplify the energy policy environment and legal framework. Throughout this project, USAID develops and/or recommends legislative and other measures at all levels of government to ensure that the BIH energy sector is compliant with EU requirements. USAID also assists the BiH Ministry of Foreign Trade and Economic Relations and relevant entity ministries in improving energy strategies and action plans, including the development of the National Energy and Climate Plans. Additionally, USAID is helping BIH develop a cyber security roadmap for the energy sector [16]. The project also supports a strong public outreach and awareness program to promote a market-based energy sector and educate the general public about the benefits of the changes taking place in the energy sector.

6 Technical Implementation Process for Public Buildings Energy Efficiency Projects

Based on the legal and technical requirements, the EE project has to follow the following implementation phases:

6.1. Preparation of Detailed Energy Audit Reports;
6.2. Preparation of the Conceptual Design;
6.3. Preparation of the Detailed Design;
6.4. Procurement phase.
6.5. Contract execution.

These phases are explained in the next sections.

Preparation of Detailed Energy Audit Reports

The first step in the implementation process for EE public building projects is to perform detailed energy audits. The basis for the preparation of the detailed energy audits is the conduction of all on-site measurements of the selected facility(ies) to determine energy losses and the current energy efficiency of the heating and lighting system and building user pattern. **Detailed Energy Audit Report** includes, inter alia: *Site visit and data collection, Analysis of energy characteristics of the buildings and characteristics of energy consumption and cost management, Identification of the presence of potential asbestos-containing materials (ACMs), Analysis of energy costs and modeling of energy consumption, Analysis of possible measures for increasing energy efficiency of the buildings, Defining measures for increasing the energy efficiency of the buildings, Analysis of energy, economic and environmental-related effects of proposed measures, Comparative analysis of energy consumption indicators of the building in comparison with other buildings of similar purpose, Defining recommendations for energy management within the building, Measures for the safe management of ACMs and Optimal scenario for selected building.* Prepared and finalized detailed energy audit will be base for the **preparation of the project documentation—Conceptual and Detailed Design**.

Preparation of the Conceptual Design

For the purpose of preparation of the project documentation, detailed measurements of the selected public building will be taken (current state drawings that will cover the situation, ground and floor plans, sections, facades etc.). During the site visit, the engineering team will gather all the relevant information and documentation from the end users such as permits, building drawings, project or technical documentation, information regarding the history of the building, performed works during the time, etc. If the client has existing project documentation, urban permit and construction permit for the building, he shall present all documents and drawings to the engineering team. The engineers will determine if the documentation is useful, and in such case, the relevant documentation will be temporarily given to the Design Company t and it will serve as the basis for the design.

The team of engineers shall take the dimensions of the interior and exterior spaces with the characteristic elements. When measuring windows and doors, both their dimensions and their characteristics (frame material, number of windows, mode and opening scheme, door filler) shall be determined. The engineering team shall inspect the condition of the complete envelope/facade and its water tightness, the roof structure with the layout of the attic spaces and the condition of the support structure and its ability to carry loads. Moreover, the engineering team shall inspect heating, cooling and electrical installations in detail. If the building has centralized hot water preparation for individual local boilers, the appointed engineer shall inspect its condition, take all the necessary dimensions, measures, and capacities to prepare the analysis and make a proposal for improving or changing the hot water preparation system. The engineers who examine the position of the luminaires and the installations are obliged to inspect all security systems in order to identify any security problems. The electrical engineer shall propose the project documentation to reduce the peak load (reactive power) and thus the consumption of electricity. If the roof is changed or works on the installation of the thermal façade are performed, it is necessary to dismantle and reinstall the lightning protection installation on the part treated with energy measures in order for these to be operational. All this will be ground for the preparation of the Conceptual Design that will be used for obtaining of an urban planning permit and the main design that will be used for obtaining the construction permit. In practice, if the architectural design remains unchanged, the implementation of the EE measures is being done through the ongoing maintenance of the building so the permits for this scope of the works are not mandatory. **The Conceptual Design** contains a *technical description, floor plans, basic sections and facades of the existing state and the newly planned state.* It is especially important to highlight the intervention plan that will show the planned measures for improving EE. Depending on the complexity and technical structure of the building, it may contain other drawings and documents.

Preparation of the Detailed Design

Preparation of the Detailed Design for obtaining a construction permit and performing of works begins after obtaining the urban permit. **The Detailed Design** is a set of mutually harmonized phases of projects that provide a technical solution for the building, and prove the fulfillment of essential requirements for the building, other requirements stipulated by law, special regulations, norms and technical standards. The Detailed Design is made based on the conditions given in the location information, i.e. urban permit. The detailed design, depending on the technical structure of the building, can be made of one or more parts—phases and must include *the name and registration of the legal entity that performs the design, signatures of responsible designers for all parts of the project, terms of reference certified by the investor, general and special conditions, standards, norms and regulations for an object or operation, technical description for all works with a summary of the area of the premises, Bill of Quantities of works, materials and equipment, all the necessary phases of the project such as: the architectural project, construction project, hydro installation project, electrical installation project, the mechanical project, technological project, technical details, study of fire protection, the study of safety at work, and other contents in accordance with special regulations.* Special emphasis should be placed on the preparation of technical description, Bill of Quantities and drawings details. This will be the basis for the preparation of tender/bidding documentation for the execution of works. When all project phases are completed and the estimated cost is calculated, the design documentation is submitted to the client for comments and revision, after which, if needed, it should be submitted to the competent municipal authorities for **obtaining a construction permit** in accordance with the Law on Spatial Planning.

Procurement Phase

Depending on the needs of the project, the relevant legal framework, the market situation and the needs of the contracting authority, it is necessary to decide on the type of public procurement framework. Generally, four procurement options are possible: open tendering, restricted tendering, negotiation with or without prior publication of a contract notice and direct method. The choice of which procurement procedures will be applied depends on whether the procurement is being done in line with the local legislative framework or if the procurement process is conducted in line with the international financial institution's procedures. After choosing the procurement type, it is necessary to create **tender documentation** which contains the minimum relevant information in relation to the selected contract award procedure: *procurement notice, invitation to submit requests for participation/offer (initial and final), technical specifications, qualification criteria for selection of the most favorable bid, draft or basic elements of the contract and other relevant documents and explanations.*

The procedure for review and **evaluation of bids** is to be performed by the procurement commission with the support of external experts who are authorized by the contracting authority, on the basis of the conditions and requirements from the tender documentation. To ensure the integrity of the competitive process, the evaluation of proposals must be undertaken objectively, consistently & without bias toward particular suppliers and/or contractors. The bids submitted by qualified bidders should be evaluated by applying the award criteria set out in the tender documentation, choosing the most economically advantageous tender or the lowest price. An evaluation team will examine each tender received and make recommendations as to which bidder is the most qualified one. Once the final decision has been made on the **tender award** to the particular bidder, the tender administrator creates the tender results notification which is in letter form and then sent to all participants. The next step is the **preparation of the final agreed form of contract** and its schedules and appendices so that it can be signed and that the works can commence. The Contract forms are usually prepared in line with the local legislation. In cases where the procurement process is done in line with the international financial institution's procedures, the usual contract form is FIDIC Conditions of Contract for Construction.

Along with the procurement of a Contractor for the work execution, it is necessary to do the **procurement of the Supervision of works**. Supervising engineer is one of the professional participants in construction, whose duty is to control if the construction work is built in accordance with the design and regulations. The Supervision of works includes inter alia: *conducting regular site visits, coordinating and regular monitoring of the status of project activities through preparation and updating of implementation plans and schedule-liaising on a regular basis with the Contractor and the Client, check and determine if the quality of the works, installed products and equipment is in accordance with the project requirements, conditions prescribed by relevant construction law, special laws and regulations, valid technical regulations and standards, and that this quality is proven by prescribed tests and documents, confirm the orderliness of the works by signing the construction diary, enter changes made in relation to the project documentation, control the construction book and, if necessary, coordinate with the designer and the contractor for any changed circumstances in the technical documentation and on the construction site, preparation of monthly and periodic progress reports, review of design documents comprehensiveness, project closure and preparation of Final Project report.*

Contract Execution

Contract execution starts with the *Letter of acceptance and Contract signing*, after which its terms become binding for all the contracting parties.

7 Potential Energy Savings and Benefits of EE Projects

The potential of the energy savings from the typical EE measures is mainly achieved by [8, 15]:

- replacing the old joinery with the new more energy efficient. With this measure, the average amount of cca. 35% energy savings can be achieved in comparison to the baseline depending on the condition and surface of the building.
- thermal insulation of the façade walls. With this measure, it is possible to save the average amount of cca. 26% of energy.

- replacing old heat supply systems with modern boiler(s) for central heating systems is a measure that can save up to 20% of energy, and
- insulating of the ceiling/roof can save up to 21% of the energy.

Since the return period for the investment can depend on many parameters (heating fuel is one of the most important ones), the time required to recover the cost of the EE investments is relatively long: approx. 15 years for the replacement of windows and doors, 16 years for the building insulation, 28 years for the roof insulation and 4 years for the investments to replace old heat supply systems. The payback periods for EE investments can be long, but they provide several other benefits such as improved thermal comfort, extended lifetime of the building and overall visibility and image of the facility.

The main environmental benefit achieved with very high energy performance public buildings is the reduction of the primary energy demand, e.g. for space heating/ cooling, water heating, air conditioning as well as a reduction in the consumption of electricity. Using energy more efficiently means burning less fossil fuels, which reduces the emission of pollutants into the atmosphere, soil, and water while also reducing the need to procure additional fuels, allowing us to conserve the environment by leaving those resources in the ground. In addition to contributing to energy savings and global environmental benefits, EE projects improve the access of local communities, including vulnerable communities, to clean, safe and affordable energy.

8 Conclusion

The aim of this paper is to answer the question are the potentials of energy efficiency in public buildings in Bosnia and Herzegovina properly used and which shortcomings should be addressed to reach the full effect of the implementation of EE measures on public buildings in Bosnia and Herzegovina? As an answer to this question detailed research was carried out on the largest completed and currently implemented EE projects. Furthermore, the financing models and sources of financing were investigated, and a brief explanation of the technical and administrative implementation process of energy efficiency projects was given with an overview of the steps made toward the implementation of EE initiatives with an emphasis on EU targets and visions related to EE in BIH.

As part of the candidacy process, in order to achieve membership to the EU, BIH must continue harmonization of the energy legislation and policies and ensure that Bosnia and Herzegovina can participate fully in the EU's environmental policy. BIH has to achieve the EU energy and climate goals, which stipulate a reduction in CO_2 emissions of 40%, a share of renewable energy sources in energy consumption of at least 32% and an indicative energy efficiency target of at least 32% energy savings by 2030. Regarding particularly energy efficiency, the main obstacle for BIH is the above-mentioned lack of an energy strategy, as well as the appropriate institutional environment and mechanisms for the implementation of EE plans.

The European Union awarded Bosnia and Herzegovina with candidate status in 2022, more than six years after the country applied to join the EU. Although the candidate status will not lead to ensuring special or additional funds, there are numerous funds, for which BIH is already eligible. The European Union is the largest single donor in BIH, with more than 3.5 billion € invested in post-war reconstruction and various projects since 1996. In the period from 2014 to 2020, BIH allocated 552 million € from EU funds for different projects. However, to receive EU funding, BIH must assure legal compliance with the current EU standards and procedures and draft strategic documents on a state level, which is often challenging due to the country's constitutional framework. For example, after adopting the Transport System Development Strategy, in 2016 on a state level, BIH received almost 315 million € in the period from 2018 to 2020 for the construction of new and rehabilitation of existing (motor) roads and railways. In February 2023, the European Commission signed the Financing Agreement for the Energy Support Package worth EUR 70 million to be disbursed to BIH, which aims to provide assistance to the country's most vulnerable households in coping with rising prices and support energy transition. 50 million € has been earmarked to mitigate the impact of high prices, 10 million € will be disbursed to improve the energy efficiency of 4,000 residential buildings, and the remaining10 million € has been allocated for around 400 Micro, Small and Medium Enterprises that are facing challenges as a result of the energy crisis. However, the precondition to receiving these funds is a positive assessment from the European Commission on progress in implementing the Action Plan, approved by the BIH Council of Ministers in December 2022, which lists the measures that BIH authorities will undertake to use the funds from the Energy Support Package and enable the country's energy transition.

Significant steps such as the adoption of draft long-term building renovation strategies, secondary legislation on EE in buildings, the EE obligation scheme and energy performance contracting have to be made. In addition to this, Energy Efficiency criteria should be introduced in public procurement procedures. Some improvements regarding these challenges are already in process, with BIH being on track to adopt its first-ever countrywide environmental strategy and action plan. In 2022, Federation of BIH, the RS and District Brčko have adopted the Bosnia and Herzegovina Environmental Strategy (BiH ESAP), which will be operational for 10 years. This document will steer policy measures and domestic and international financing and will help to further align BIH's legal and institutional frameworks with the EU environmental laws and procedures. This step could be especially beneficial for the public building sector, which is identified as the sector with the largest potential for cost-effective energy saving in BiH, with an estimated rate of 20 to 60% potential savings, faces big obstacles due to fragmented and complex inter-authority jurisdictions and budgetary constraints of its owners.

Positive examples from developed countries have shown that the public sector is the best place to start implementing EE policies. Public institutions are known as large consumers of energy, and therefore the effects of EE measures are significant. Especially in BIH, where most of the public buildings were built before the demands for reduction of energy use and have been poorly maintained throughout the years, implementation of EE projects can lead to significant savings of energy and thus public money, reduction of negative effects on the environment, creation of new green jobs, economic growth and sustainable development. Bearing in mind that increasing energy efficiency has many benefits for the economy, and environment but also leads to the increased comfort of building occupants, it is necessary to have a systematic approach to implement the projects in the field of EE. The implementation process starts with the building analysis, and it aims to primarily reduce energy consumption. In addition to this, it aims to improve buildings' structural stability, integrity, and aesthetics, as well as to prolong buildings' life cycle, enhance indoor thermal and air comfort, and provide a healthier environment for the occupants. At the end, the following can be concluded:

- Bosnia and Herzegovina have one of the most energy and carbon-intensive economies in the region. Due to inefficient insulation, obsolete heating systems and inadequate lighting, buildings in BiH require more energy per square meter compared to countries in similar climatic zones. Some studies says that there are approximately 7,600 public buildings in BiH, requiring an estimated EUR 350–510 million in energy efficiency investments.
- During the past several years Bosnia and Herzegovina have recognized the importance of energy efficiency projects. Many projects have been or are being implemented and there is no question that energy efficiency is the single largest measure to avoid energy demand in the future and to achieve the goal of carbon neutrality. In the long term, the reduction trend will be shaped by the improved energy efficiency of new buildings and the renovation of existing buildings.
- In the process of joining the EU, Bosnia and Herzegovina must make efforts to ensure the overview of harmonization of the regulation and policies with the legal legacy and environmental, health & safety, energy and climate policies of the EU, such as the Energy Efficiency Directive.
- Furthermore, the Eurocodes must be used during the design of the new building. Eurocodes are a series of 10 European Technical Standards that provide a common approach to the structural design of buildings and other civil engineering works. The use of these standards increases safety in the construction.
- Bearing in mind that most projects aiming to improve the energy efficiency of public buildings are implemented in cooperation or supervision of international financial institutions (through credit or partly donor funds), in the forthcoming period it is necessary to strengthen existing institutional structures so that EE projects are regularly planned in local, entity and state budgets and implemented within the BIH institutions. In this regard, preparation of the integrated renovation plan on the local and state level will be considered as an asset.

Bibliography

1. Energy Community. About. https://www.energy-community.org/aboutus/whoweare.html. Accessed: February 2023

2. Energy Community Secretariat (2021) Bosnia and Herzegovina Annual Implementation Report. https://www.energy community.org/implementation/Bosnia_Herzegovina.html. Accessed: March 2023
3. Energy Community (2016) National renewable energy action plan of Bosnia and Herzegovina (NREAP BIH). https://www.energy-community.org/implementation/Bosnia_Herzegovina/reporting.html. Accessed: January 2023
4. Energy Community (2021) Bosnia and Herzegovina Fourth Progress Reports under Renewable Energy Directive 2009/28/EC as adapted by the Ministerial Council Decision 2012/04/MC-EnC. https://www.energy-community.org/implementation/Bosnia_Herzegovina/reporting.html. Accessed: January 2023
5. Energy Community (2020) Report on the implementation of the Energy Performance of Buildings Directive. https://www.energy-community.org/dam/jcr:4f4af944-0d87-4bdc-a525-9bf7337f8bb2/EPBD%2520implementation%2520Report_BiH.pdf&cd=1&hl=bs&ct=clnk&gl=ba. Accessed: December 2022
6. Official Gazette (Službene novine) Federacije BiH, br.145/2017 (2017) Pravilnik o dodjeli sredstava iz revolving fonda za projekte energijske efikasnosti
7. Energy Community (2017) Energy Efficiency Action Plan of Bosnia and Herzegovina for the period 2016—2018. https://www.energy-community.org/dam/jcr:d5da6e89-291c-4e97-b978-85804d98d040/BIH_NEEAP_2016_2018_042017.pdf Accessed: December 2022
8. Marko NišandžiUnited Nations Development Programme in BIH (2017) Studija energetske efikasnosti javnih objekata Kantona Sarajevo
9. Sustainable Development Goals Fund (2017) Case study: energy efficiency and renewable energy sources in Bosnia and Herzegovina. https://www.sdgfund.org/case-study/energy-efficiency-and-renewable-energy-sources-bosnia-and-herzegovina. Accessed: February 2023
10. Fond za zaštitu okoliša Federacije BiH (n.d.) About. https://fzofbih.org.ba/. Accessed: March 2023
11. Bosnian Energy Efficiency Project (BEEP) (n.d.) About. https://beep.ba/en/. Accessed: February 2023
12. United Nations Development Programme in BIH (n.d.) About. Scaling-Up Investment in Low-Carbon Public Buildings. https://www.undp.org/bosnia-herzegovina/projects/scaling-investment-low-carbon-public-buildings. Accessed: January 2023
13. UNDP in BiH: Projects—Green Economic Development (GED) (n.d.) About. https://www.undp.org/bosnia-herzegovina/projects/green-economic-development-ged. Accessed: January 2023
14. European Bank for Development: The EBRD in Bosnia and Herzegovina (n.d.). https://www.ebrd.com/ebrd-in-bosnia-and-herzegovina.html. Accessed: January 2023
15. Knežević A (2009) Poboljšanje energijske efikasnosti u stambenim objektima na području KS. REIC
16. USAID From the American People: Bosnia and Herzegovina (n.d.) Bosnia. https://www.usaid.gov/bosnia. Accessed: January 2023

St John's Wort as a Possible Tool for Remediation of the Soil Contaminated with Heavy Metals

Azra Suljić, Ervina Bečić, Maida Šljivić Husejnović, and Zahida Ademović

Abstract Heavy metal pollution is a very common problem in the modern society. When concentrated in the soil, heavy metals can easily intoxicate food and water. Therefore, there is a certain risk of a chronic heavy metal exposure, even in the areas with no industrial facilities or pollution. There are many soil remediation solutions in the scientific literature, including herb-mediated ones. This research is focused on the possibility of St John's wort to become a tool in the remediation of the soil from heavy metal pollution. It is a very widespread plant, which has no specific soil requirements and grows in various areas around the world. In this research, soil where the St John's wort grows has been intentionally polluted with arsenic (As) cadmium (Cd) and lead (Pb), in order to gain insight into the St John's wort's heavy metal absorption capacity. As, Cd and Pb concentration in soil and St John's wort samples were analyzed by ICP-OES. The final results showed that the St John's wort had the significant absorption level of Cd and Pb.

Keywords Arsenic · Cadmium · Lead · Soil remediation · St John's wort

A. Suljić (✉) · M. Š. Husejnović
Faculty of Pharmacy, University of Tuzla, Tuzla, Bosnia and Herzegovina
e-mail: sulji.azra@gmail.com

M. Š. Husejnović
e-mail: maida.sljivic-husejnovic@untz.ba

E. Bečić
Faculty of Pharmacy, University of Sarajevo, Sarajevo, Bosnia and Herzegovina
e-mail: ervina.becic@ffsa.unsa.ba

Z. Ademović
Faculty of Forestry, University of Sarajevo, Sarajevo, Bosnia and Herzegovina
e-mail: zahida.ademovic@untz.ba

M. Arslanagić-Kalajdžić et al. (eds.), *Interdisciplinary Advances in Sustainable Development II*, Lecture Notes in Networks and Systems 804,
https://doi.org/10.1007/978-3-031-46692-2_6

1 Introduction

Heavy metal soil contamination is an important problem regarding the environmental pollution [1]. This problem has been especially present in the last few decades, due to an increase in industrial production. Moreover, inappropriate waste disposal contributes a lot to heavy metal pollution. Although there are adopted regulations regarding waste disposal they are mostly not fully implemented in the developing countries [2]. Because of their well-known features, including high persistence and non-biodegradable nature, increased heavy metal concentrations in the air, soil, or water lead to adverse impacts on the environment. Additionally, due to the high degree of toxicity and capacity for bioaccumulation in various tissues, heavy metals represent a serious threat to human health [3]. Health risk of heavy metal intoxication may include: cardiovascular disorders, neuronal damage, immune dysfunctions, renal injuries, and risk of cancer and diabetes, and all of that mainly through the production of reactive oxygen species and oxidative stress [4, 5]. Due to their high degree of toxicity and bioaccumulation behavior, arsenic (As), cadmium (Cd), and lead (Pb) rank among the priority metals that are of environmental and public health concerns.

Heavy metal bioaccumulation in medicinal plants is becoming more important, because plants are often used as an alternative medicine, and therefore they present a potential cause of metal poisoning [6, 7].

St John's wort (botanically known as *Hypericum perforatum* L.) is a widely spread herb, which grows in open, disturbed areas throughout many of the world's temperate regions. It is a native European plant that has spread to Asia, Africa, Australia, and North and South America [8]. Studies on heavy metals in St John's wort have mostly focused on its concentration in naturally occurring herbs rather than cultured in well-defined conditions, with limited data on the absorption ability of *Hypericum perforatum* L. [1, 9].

Over the years, numerous methods for removing heavy metals from the environment are developed, both chemical and physical. Some of these, however, are quite complicated and expensive. Therefore, there is a real need for finding affordable and effective alternatives. It has been found that some plants can adsorb heavy metals such as *Pteris vittata, Agrostis tenuis, Brasica juncea, Eleusine indica,* banana, cactus, rice straw, etc. [10–12], although their widespread application is constrained by their natural occurrence. Therefore, this study offers a slightly different perspective—what if St John's wort (taken here as a representative herb) could be used as a tool for remediation of the soil polluted with heavy metals? St John's Wort is a very common plant in Bosnia and Herzegovina. It does not have great requirements in terms of soil, and grows on neglected terrains—mountains, meadows, pastures [13]. It is taken as a representative herbaceous plant in this study, considering its prevalence in the territory of Bosnia and Herzegovina, and the fact that it is often used in Bosnian traditional medicine. This study aims to determine St John's wort uptake of three highly toxic heavy metals in the environment (As, Cd, Pb).

2 Materials and Methods

Sample Collection

St John's wort was taken from a wooded area in the municipality of Čelić, Bosnia and Herzegovina (44.7196248° N, 18.8040667° E) in June 2022. The area from where it was taken was far away from the road and industrial facilities (the purpose was to take St John's wort which was the least exposed to the heavy metal sources). Seedlings of St John's wort were taken and replanted into the seven pots, together with the pieces of its original soil, and then transferred to the laboratory. Seedlings were taken with only a part of their original soil, which was not removed due to possibility of damaging the St John's Wort roots. The pots with the replanted seedlings were then filled with the commercial soil for plants, with pH value of 5.5–6.5. The temperature was maintained at 25 °C during the experiment. The herbs had the natural source of the light in the room, and were exposed directly to the sunlight (Fig. 1).

Fig. 1 Pots with the replanted St John's wort. Seedlings of the St John's wort right after the replanting from the original habitat—on the left. Full grown plant after the treatment with heavy metals and right before sample collection—on the right

Chemicals, Reagents and Glassware

The soil samples were digested using ultrapure grade hydrochloric acid and nitric acid obtained from Lachner (Neratovice Praha, Czech). All dilutions were made with double-distilled deionized water. Analytical grade standard solutions of As (1000 mg/L with 2% HCl), Cd (10,000 mg/L with 5% HNO_3), and Pb (1000 mg/L with 2% HNO_3) were used (Sigma-Aldrich, Massachusets, USA). Prior to usage, all glassware and plastics were thoroughly cleaned by soaking them in dilute HNO_3 (10% v/v) for 24 h.

Soil Sample Treatment

For evaluation of the heavy metal uptake by *Hypericum perforatum* L., soil samples in which the plant was replanted were intentionally polluted soil with As, Cd, and Pb. Maximum allowed concentrations (MACs) in soil for As, Cd and Pb were used for the concentration selection.

According to the guidelines from the World Health Organization, MACs for As, Cd, and Pb in soil are 20, 0.8, and 85 μg/g, respectively. So, the soil in the pots with St John's Wort was treated with the following:

a. MAC of the specified heavy metals in the soil (per gram of soil) (pots 1, 2): 3 ml of As, 0.12 ml of Cd, 12.75 ml of Pb.
b. 150% of MAC of the specified heavy metals in the soil (per gram of soil) (pots 3, 4): 4.5 ml of As, 0.18 ml of Cd, 12.75 ml of Pb
c. 200% of MAC of the specified heavy metals in the soil (per gram of soil) (pots 5, 6): 5 ml of As, 0.24 ml of Cd, 25.5 ml of Pb

The soil in the pots was treated twice, starting on June 17th, 2022. The first treatment consisted of adding a combination of As, Cd and Pb, in concentrations presented in the Table 1. After the first treatment the samples of the soil and St John's wort from the each pot were collected and analyzed using ICP-OES. Four days exactly passed from the heavy metal treatment to the sample collection and analysis, meaning that the St John's wort in each pot had 4 days to absorb the heavy metals. After 5 days (including one watering of the seedlings in the meantime), the pots were retreated only with Cd (second treatment), using the same concentrations of Cd as in the first treatment. Cadmium retreatment was performed in order to ensure that this metal is present in the soil samples in higher concentrations so the evaluation of possible uptake from the plant would be much easier. Moreover, Cd has the lowest MAC per gram of soil, but despite of that, it is regularly found in higher concentration in the soil, food, and water in different regions [14, 15]. Seventh pot was used as negative control. Both heavy metal treatments occurred during the St John's wort blooming, and timed to be finished at the end of the blooming.

The standards of heavy metals used for treatment of the soil in the pots were:

Table 1 The first and second heavy metal treatment of the soil used for St John's Wort cultivation

POT	1st Treatment	2nd Treatment	Heavy metals and their concentrations (As/Cd/Pb)	Heavy metals and their volumes (As/Cd/Pb)
1, 2	Low As, Cd, Pb	Low Cd	20 μg/g/0.8 μg/g/85 μg/g	3 ml/0.12 ml/12.75 ml
3, 4	Mediate As, Cd, Pb	Mediate Cd	30 μg/g/1.2 μg/g/127.5 μg/g	4.5 ml/0.18 ml/19.12 ml
5, 6	High As, Cd, Pb	High Cd	40 μg/g/1.6 μg/g/170 μg/g	6 ml/0.24 ml/25.5 ml

a. Solution of Arsenic (As) concentration 1000 mg/L (Matrix: 2% HCl)
b. Solution of Cadmium (Cd) concentration 10,000 mg/L (Matrix: 5% HNO_3)
c. Solution of Lead (Pb) concentration 1000 mg/L (Matrix: 2% HNO_3).

3 Results

The ground state of the soil from the pots was analyzed first. Every single sample of the soil contained Pb. The level of Pb in each sample did not exceed maximum allowed concentration of Pb per gram of soil (Table 2). Cd and As were not present in any sample of the soil from the pots (measured concentration was below the limit of detection).

Control St John's wort which (not treated with heavy metals, but taken from the same area) was analyzed too, in order to see its heavy metal intoxication prior to the heavy metal treatment in this study. The analysis showed that it only contained Pb at 197.93 μg/g (as expected, because the soil samples prior treatment only contained Pb, as shown in Table 2). Cd and As were below the limit of detection.

After the 1st treatment with the As, Cd and Pb concentrations from Table 1 (Sect. 2), only one soil sample contained As, while in the five remaining pots, the level of As was below the limit of detection. Both Cd and Pb were abundant in the soil samples. The Cd concentration in every soil sample was higher than the MAC. As for Pb, only pots 5 and 6 (treated with 200% MAC) contained a lead concentration higher than the MAC (Table 3).

As stated in the Sect. 2, the second treatment consisted of Cd addition only (Table 4). Although As and Pb were not used in the 2nd treatment, their concentrations in the soil were also analyzed again. Contrary to the first analysis, after this second treatment, all soil samples except pot 3 contained As (which was above the

Table 2 The ground state of soil from the pots

Pot number:	1	2	3	4	5	6
Pb (μg/g)	9.33	3.26	5.5	8.36	8.3	10.46

Table 3 Concentration of As, Cd, Pb in the soil after the first analysis, from the pots 1–6 which were treated with the combination of As, Cd and Pb

Pot number	Concentration (μg/g)		
	As	Cd	Pb
1	0.000	8.63	47.20
2	0.000	6.46	17.73
3	0.000	30.60	58.56
4	0.000	27.40	53.86
5	0.000	94.16	254.73
6	41.56	124.26	355.66

0.000 means below the limit of detection

Table 4 Concentration of As, Cd, Pb in the soil after the second analysis, from the pots 1–6 which were at the second time treated with Cd only

Pot number	Concentration (μg/g)		
	As	Cd	Pb
1	5.03	35.23	18.40
2	2.16	28.46	15.03
3	0.000	24.03	21.16
4	9.60	47.66	71.70
5	55.43	219.86	186.83
6	28.63	370.33	142.63

0.000 means below the limit of detection

MAC in pots 5 and 6 only). Cd was present, as expected, in higher concentration in all soil samples except pot 4, comparing to the first analysis. Interestingly, Pb showed lower concentration than in the first analysis (in all soil samples except the pot 4).

The St John's wort absorbed Cd and Pb, but did not absorb As (the As level was below the limit of detection in all samples). Cd was highly absorbed into the St John's wort, and its concentration in the plant increased as its concentration in the soil increased. Pb was also absorbed in all samples. Its concentration decreased in the St John's wort (in the most samples) after the second treatment, following the pattern from the soil absorption.

4 Discussion

Regarding the aim of the study, to evaluate uptake of three highly toxic heavy metals in the environment (As, Cd, Pb), this research found out that the St John's wort absorbs Cd and Pb from soil solution. Soil samples prior heavy metal treatment contained only Pb (Table 1) in the average value of 7.53 μg/g. Therefore, St John's wort from those control soil samples was collected as one plant sample, which showed Pb concentration of 197.93 μg/g.

Pb and Cd were detected after the first analysis of the soil samples, while As was detected in only one soil sample, which was treated with the 200% MAC. Although As and Pb were not added in the second treatment of the soil, there were some obvious changes in their concentration in the soil samples after this second treatment with Cd. Most samples showed higher concentration of As and Pb at the second analysis, meaning it probably took more time for those heavy metals to be absorbed into the soil on a greater scale.

Cd, As, and Pb may exist in soils as inorganic species, organic complexes, adsorbed on solid phases, or constituents of solid phases, with varying solubility levels, but plants absorb only the available fraction of those heavy metals [16, 17]. St John's wort was treated at the time of blooming, which means the level of its absorption from the soil should be the highest because the plant needs more different elements from the soil at that period. The results showed that As was not absorbed into the St John's Wort (it was below the limit of detection in every single St John's wort sample), although it was detected in the soil samples after the second treatment. Absorption of As, as well as other heavy metals, into the St John's wort, depends on the form of the heavy metal which is present in the soil. Arsenate is the main chemical species of As in aerobic soils. Phosphate and arsenate share the same root transporters, but those transporters show a higher affinity for phosphate than for arsenate [16]. This might be an explanation for absent As uptake by St John's wort in this study. Also, although arsenate is adsorbed more quickly, arsenite occurs primarily under reduced soil conditions and is more soluble, mobile, and toxic [17], which cannot be associated with the results from this study because St John's wort did not show As uptake. Another possible explanation is that the absorption of As into the St John's wort could take some more time than given here (8 days in total). This is also supported by the fact that As was detected in the majority of soil samples only after the second ICP-OES analysis, meaning that the absorption of As into the soil itself took more time.

Cd and Pb were absorbed by St John's Wort (Table 5). The results showed that the Cd and Pb uptake into the St John's Wort could be mutually dependent. The level of Cd could have some influence on the level of Pb in St John's wort and vice versa, because some samples where the Cd concentration was higher showed a significant decrease in the Pb concentration. This means that Cd and Pb might have some interactions during the St John's Wort uptake. Soil supplementation with cadmium and zinc affected manganese, lead and copper concentrations in *Melissa officinalis* L. [18]. Cd and Pb uptake was also mutually dependent in *Ficus parvifolia*, where it was proven that presence of Pb could reduce Cd phytotoxicity in plants by inhibiting Cd uptake and accumulation [19]. So there are some studies suggesting Cd and Pb interactions when it comes to the herb uptake from the soil, but there is not much evidence specific for St John's Wort. There are many studies where uptake of heavy metals and other elements in St John's Wort from different regions is analysed [1, 3], but so far, the mechanism of interaction of Cd and Pb in the St John's Wort uptake was not investigated in detail.

Table 5 Heavy metal concentration (μg/g) in the St John's wort samples

Pot number	A			B		
	Concentration (μg/g)			Concentration (μg/g)		
	As	Cd	Pb	As	Cd	Pb
1 + 2	0.00	4.56	70.4	0.00	33.23	8.00
3 + 4	0.00	4.28	25.88	0.00	83.30	8.39
5 + 6	0.00	6.22	10.92	0.00	13.61	20.38

St John's wort from the pots treated with the same heavy metal concentration was collected as one sample, i.e. St John's wort from pots 1 and 2 is one sample of the St John's wort (1 + 2). A—after first treatment. B—after retreatment with Cd

5 Conclusion

This study showed that St John's wort is certainly capable of absorbing Cd and Pb. This is very important because those two heavy metals are commonly found in the soil [5, 14]. Developing countries, like Bosnia and Herzegovina, are facing even greater problems with heavy metal soil contamination because not much is invested in environmental protection. Although there are regulations regarding waste disposal, they are mostly not fully implemented [2].

So, St John's wort has the ability to absorb those heavy metals and probably accumulate them. In that way, St John's wort could be contributing to the soil's remediation from heavy metals. But, this could be a "double-edged sword", because St John's Wort is used in a phytotherapy a lot, especially in traditional medicine. In that case, it would not be desirable for it to contain heavy metals. So, if the St John's Wort is going to be used as a tool for intoxicated soil remediation, it needs to be used in the controlled environment, so that any accidental possibility of it being used for therapeutical purposes, could be excluded.

Even if the soil is polluted with very low heavy metal concentrations, there is a significant risk of chronic intoxication, so it is of great value to have a plant that is capable of absorbing these heavy metals. Therefore, based on the results from this study, St John's Wort should be considered as a tool for remediation of a soil contaminated with Cd and Pb.

References

1. Fischer A, Brodziak-Dopierała B, Loska K, Stojko J (2017) The assessment of toxic metals in plants used in cosmetics and cosmetology. Int J Environ Res Public Health14(10):1280. https://doi.org/10.3390/ijerph14101280
2. Suljic A, Suljic N (2021) Degradation of the pharmaceutical and medical waste on landfills due to rainwater infiltration. Int J Res Appl Sci Biotechnol 8(Issue-1). ISSN: 2349-8889, IF: 3,715 for 2019, Vandana Publications India

3. Šljivić Husejnović M, Janković S, Nikolić D, Antonijević B (2021) Human health risk assessment of lead, cadmium, and mercury co-exposure from agricultural soils in the Tuzla Canton (Bosnia and Herzegovina). Arh Hig Rada Toksikol 72:268–279
4. Dautović E, Šljivić Husejnović M, Bergant M, Sabitović D, Srabović N, Smajlović A, Begić L, Softić A (2019) Lead and cadmium induced cytotoxic and genotoxic effects on HL-60 and Jurkat cell lines. Genet Appl 3(1):57–64
5. Glavač N, Djogo S, Ražić S, Kreft S, Veber M (2016) Accumulation of heavy metals from soil in medicinal plants. Arch Ind Hyg Toxicol 68(3):236–244. https://doi.org/10.1515/aiht-2017-68-2990
6. Mafakheri M, Kordrostami M (2021) Recent advances toward exploiting medicinal plants as phytoremediators. In: Hasanuzzaman M, Vara Prasad MN (eds) Handbook of bioremediation. Academic Press, Cambridge
7. Maharia RS, Dutta RK, Acharya R, Reddy AVR (2010) Heavy metal bioaccumulation in selected medicinal plants collected from Khetri copper mines and comparison with those collected from fertile soil in Haridwar, India. J Environ Sci Health Part B 45(2):174–181
8. Klemow KM, Bartlow A, Crawford J et al (2011) Medical attributes of St. John's Wort (Hypericum perforatum). In: Benzie IFF, Wachtel-Galor S (eds) Herbal medicine: biomolecular and clinical aspects, 2nd edn, chap 11. CRC Press/Taylor & Francis, Boca Raton (FL)
9. Bakirdere S, Yaman M (2008) Determination of lead, cadmium and copper in roadside soil and plants in Elazig, Turkey. Environ Monit Assess 136(1–3):401–410. https://doi.org/10.1007/s10661-007-9695-1
10. Lata S, Kaur HP, Mishra T (2019) Cadmium bioremediation: a review. Int J Pharm Sci Res 10(9):4120–4128
11. Chibuike GU, Obiora SC (2014) Heavy metal polluted soils: effect on plants and bioremediation methods. Appl Environ Soil Sci 752708
12. Al-Senani GM, Al-Fawzan FF (2018) Adsorption study of heavy metal ions from aqueous solution by nanoparticle of wild herbs. Egypt J Aquat Res 44(3):187–194
13. Barnes J, Anderson LA, Phillipson JD (2001) St John's wort (Hypericum perforatum L.): a review of its chemistry, pharmacology and clinical properties. J Pharm Pharmacol 53(5):583–600
14. Zhang X et al (2015) Assessment of cadmium (Cd) concentration in arable soil in China. Environ Sci Pollut Res Int 22(7):4932–4941. https://doi.org/10.1007/s11356-014-3892-6
15. Satarug S et al (2018) Cadmium, environmental exposure, and health outcomes. Environ Health Perspect 118(2):182–190. https://doi.org/10.1289/ehp.0901234
16. Zhao F-J et al (2022) Toxic metals and metalloids: uptake, transport, detoxification, phytoremediation, and crop improvement for safer food. Mol Plant 15(1):27–44. https://doi.org/10.1016/j.molp.2021.09.016
17. Asare MO et al (2023) The fate of secondary metabolites in plants growing on Cd-, As-, and Pb-contaminated soils—a comprehensive review. Environ Sci Pollut Res Int 30(5):11378–11398. https://doi.org/10.1007/s11356-022-24776-x
18. Adamczyk-Szabela D et al (2020) Combined cadmium-zinc interactions alter manganese, lead, copper uptake by Melissa officinalis. Sci Rep 10(1):1675 https://doi.org/10.1038/s41598-020-58491-9
19. Li Y, Cheng X, Feng C, Huang X (2023) Interaction of lead and cadmium reduced cadmium toxicity in Ficus parvifolia seedlings. Toxics 11(3):271. https://doi.org/10.3390/toxics11030271

Cities

Understanding the Linear City. (Mis)interpretation, Categorization, and Realization

Tijana Tufek-Memisevic

Abstract The Linear City topic is one that periodically appears in the world of architecture and urban planning through a variety of visionary plans, yet rarely comes to successful fruition in the built reality. Most recently the discourse on linear cities has been stirred up again relating to the controversial Neom project in Saudi Arabia. While abundant concepts and models of linear cities exist in literature, seemingly unrelated to one another, no real consensus has been established on a normative definition and the term is ambiguously used. Furthermore, some urban forms evolve naturally in a linear fashion without significant planning strategy raising the question of the concept's validity. With previous notions in mind the crucial enigma resolved with this paper is what defines the concept, design and success of a Linear City? And why do linear city ideas reappear if they rarely get realized? This research elucidates the reasons behind the concept's misinterpretation and provides a much-needed categorization of the general models, bringing to focus whether Neom is actually a liner city. Subsequently, it is explained how the planned concepts relate to organically evolved linear urban forms or current trends in sustainable urban development. The paper draws on desk-based research, and a comprehensive discourse analysis within the theoretical frameworks of urban morphology and critical theory. It builds upon the author's prior research groundwork, provides theoretical review of existing primary sources and available body of knowledge encompassing a wide range of linear city models, ideas, concepts, and existing linear urban environments. The paper contributes to theory building by providing reasonable inferences and clarification on a contested topic.

Keywords Linear city · Neom · Linear urban planning · Linear megastructures · Soria y Mata

T. Tufek-Memisevic (✉)
Candarc, LLC, Arlington Heights, USA
e-mail: tijana@candarc.com

M. Arslanagić-Kalajdžić et al. (eds.), *Interdisciplinary Advances in Sustainable Development II*, Lecture Notes in Networks and Systems 804,
https://doi.org/10.1007/978-3-031-46692-2_7

1 Introduction and Methodology

The vision of an ideal city expressed through linear form accompanying a transportation route has been a tempting one for many architects and urban planners throughout history. These visions, however, seldom translate into real developments in the built environment. Urban and architectural professional society is quick to dismiss the linear city concept pointing to historical instances as examples that never get realized [1, 2]. On the other hand, some urban forms evolve naturally in a linear fashion without significant planning strategy. "*Although the linear planning of cities has never won popular support among professional planners, it has, paradoxically enough, been the natural pattern of growth of our great urban regions.*" George R. Collins 1960.

The discourse on linear cities has been galvanized once again with the "The Line" project-announced by Mohammed bin Salman, the crown prince of Saudi Arabia and chair of the country's flagship company Neom in January 2021 (Fig. 1). The intriguing proposition of the 170 km long and only 200 m wide straight mega-city in the middle of the desert has been quickly propagated through social and conventional media channels. It is a promise of a "globally unmatched" smart city meant to house 9 million people, built upon five pillars—proximity, density, diversity, mobility, and sustainability [3]. The general sentiment shifted fast from intrigue to sharp criticism, deeming it a vanity project or "*the latest pie-in-the sky pet project*" of the crown prince [4]. Many planners and architects have quickly jumped on the critique train explaining in detail why The Line is destined to fail, inevitably referencing past examples of linear city proposals, in particular Soria's Linear City [5]. Abundant reasons made the The Line become a *piñata* for heavily swinging criticism sticks. **But the one question that is not posed is to what extent does Neom truly represent a linear city?** This paper aims to elucidate how the research problem of linear cities is much broader than the widely circulating mainstream notions propagate. Early academic research engaging in the topic recognizes that linear urban development concepts around the world appeared throughout different timespans, whether related to previous or not [6]. The variety of the design models and sizes over time, suggests a **commonly identified potential of transit-line based development in utilizing and prioritizing public transport vicinity** as well as tackling sustainability challenges such as urban sprawl, car dependency, pollution, social inequality, absence of urban health and disconnect from the natural environment. The fruitless attempts in execution and absence of a normative definition, however, points to a lack of deeper understanding of the processes and limitations of a linear city development.

This prompts several research questions:

1. **What exactly defines a linear city?**
2. **Why does the linear city concept periodically reappear if it rarely gets realized?**

Subsequently the following sub-questions are imposed:

- Should any future attempt in linear city planning be forgone?

Fig. 1 The 170 km-long Line city in Saudi Arabia (*Source* Dezeen)

- What role does design form play in the linear concept?
- How do liner city concepts relate to organically evolved linear developments?

To answer the posed research questions, this paper covers the topic of linear cities through a timeline of discourse analysis as an interdisciplinary approach that draws on various theoretical frameworks. The frameworks include urban morphology, critical theory, and sustainable urbanism. In the context of linear city, discourse analysis examines how terminology, language territory and communication shape the approach to the concept and its perception in professional and public opinion. The analysis helps uncover values and wrong assumptions that determine the linear city concepts' success in realization.

While many examples of linearity in urban planning exist in literature, there is no consistent understanding of relevant criteria that classify a model as linear, as opposed to a concentric city model with much clearer definition and zone distribution. This paper does not consider the role of institutions, governance structures, and power relations in shaping the development of linear cities. It does, however, provide a basis for further research on those subjects within critical urban theory.

In the structure of this research historical overview on the topic distinguishes contribution of philosophical discourse from contributions in design proposals. The research builds upon the prior groundwork established in the author's earlier publications and thesis on the topic. The paper draws on a desk-based theoretical study of the existing body of literature, primary sources, and a comparative analysis of linear concepts, models, projects, and organic developments. The study dives deep into the linear discourse over different time spans providing a distinction of several historical phases of the discourse within the professional community and public perception. The analysis encompasses journal articles, project drafts, models and press articles ranging from the start of nineteenth century to contemporary sources. Apart from underlining the reasons for the concepts (re)appearance despite continued failure of real-life execution, it elucidates the relevant attributes and elements of linear urban developments, provides a categorization based on defined criteria and underlines the impediments.

2 The Linear City Discourse

Literature review points to several significant phases in the discourse of Linear Cities over the last century. The first phase covers the genesis of the concept, its worldwide promotion at the end of the 19th and early twentieth century and perception at the time. The latter phase covers its mid-century height in popularity, with outspoken figures engaged in philosophical argumentation in addition to fruitful design outputs by prominent architects. The prevailing sentiment in the consequent period is one of apparent saturation with failed linear proposals that is expressed through harsh critique in the 1970s–80s and a general dismissal afterwards. In contemporary literature the topic of linear cities is mainly covered through historical overviews with the underlined notion of unsuccessfulness of the concept. This section will focus on the understanding of the motivation behind the concepts, hindering factors in the past, general understanding and discourse at the time and the issue of understanding scale level, design and planning processes among the proponents.

Genesis of the Concept and Global Dissemination (End of 19th–Early 20th Century)

By now, it is well established that it was Spanish developer, intellectual and entrepreneur Don Arturo Soria y Mata in 1882 who first introduced the Linear City concept in urban planning, with his plan "La Ciudad Lineal de Madrid", published in the periodical *El Progresso.* Arguably though, earlier instances of linear urban concepts exist in literature, although not necessarily articulated as such. An example would be Leonardo Da Vinci's reconstruction plan for Milan motivated by the black plague outbreak in the fifteenth century [7]. Soria was a progressive political figure, inventor, and owner of a streetcar company that established one of Madrid's earliest tramways. He believed that it was **more feasible to build a new city** than to reconstruct existing ones in the effort to tackle urban problems of rural to urban migration, congestion, and pollution of industrial cities at the time [8]. His intention was to facilitate a well-connected, quality urban living environment while addressing the strong class inequalities he observed in Madrid. La Ciudad Lineal de Madrid, which Soria sometimes referred to as a "vertebrate animal," was envisioned as a 55 km development oriented towards an arterial street provisioned by tramway line and utilities, running around a circumference of Madrid. The arterial street was 40 m wide and was considered **an axis that represents the backbone** of the city. All functions were placed along that axis, crossed at regular 300 m intervals by subsidiary perpendicular streets. Large blocks of residential buildings were surrounded by greenery, with commercial and public structures located at junctions [9–11]. Soria proclaimed: "*To every family a house; to every house a garden*" A. Soria y Mata 1883 [8, 12].

Ultimately, Soria was not able to secure large funds to finance his whole project as envisioned. With his company Compania Madrilena de Urbanization he managed to build a 5 km long section of the Ciudad Lineal in the vicinity of Madrid and housed over 6000 people by the time he died [11, 13]. "*Obstacles of inertia, skepticism, and active hostility had to be overcome.*" stated Woods [8], whose article in the Journal of the American Institute of Architects in 1921 provided the concept more gravity in North American literature at the time. G. R. Collins highlights the fact that Soria was a republican and antimonarchist as likely hindrance in securing large government funds. Despite efforts to make the Linear City concept a worldwide recognized and acknowledged movement at the time, in professional literature it frequently appeared in perfunctory, inexact, or dismissive context and recurrently it has been misunderstood as "ribbon development" [6]. At the beginning there has been much comparison between Soria's Ciudad and Lineal Ebenezer Howard's Garden City concept, introduced in 1898. Although the Linear City concept occurred first and anticipated several features of the Garden City, the latter one fared better in professional circles than its predecessor, especially in the Anglophone world. The two concepts appeared to become intertwined in literature to the point that both were deemed versions of garden cities, as French architect-planner Georges Benoit-Lévy suggested. His devotion to the concept resulted in the creation of the *Association Internationale des Cités Linéaires* in Paris in 1928, endorsed by CIAM [6, 14]. The Linear City's main criticism revolved around its radical decentralization, rigidness and dominant focus on efficiency and functionality. As a great proponent of Ciudad Lineal and Soria's supporter, Hilarión González del Castillo summarized the linear city concept in several principles in 1913 [6]. Among other points the principles underline motion as the driving issue in urbanization, the need for planning to precede construction, preference of regular geometric forms of streets and blocks and a connection to nature. It is suggested that linear cities should be designed either as connections between existing cities, to form suburbs around larger cities or to cover the countryside. This speaks to a lack of clarity in the scale level and type of development at the time.

Over the years, the constructed portion of the Ciudad Lineal by Soria's Compania Madrilena de Urbanization has been caught up by urban sprawl and today looks profoundly different. The conditions set in place by the CMU ceased to be valid during the 1960s, complying with a new Mortgage Law. Most of the lots initially dedicated to residential housing became mixed-used developments with multi-family housing and accompanying facilities. The modification also included the removal of the central tram line, that is replaced by fast lanes for cars [15]. Outside of Spain, a linear model that came close to implementation was a proposal drafted by Carlos Carvajal Miranda of Chile. In the 1910s, he received official permission to construct a ten-kilometer-long linear development near Santiago based on Soria's ideas, with whom he was in constant correspondence throughout most of his career. A series of coeval linear theories and different models of linear development emerged at the beginning of the twentieth century as a result of the promotion of The Linear City. In contrast, other projects appeared without ever referring to the concept of Soria y Mata, such as Edgar Chambless' Roadtown in 1910 [16]. Considering the issue of land value and its dependence on transportation proximity, Chambless, who was also a patent investigator, developed an innovative idea for urban living. His proposal introduced the concept of an infinitely populated linear structure connected to a transportation system spanning across the United States. The idea was so intriguing that it received support from two prominent innovators in the form of patent donation—the Boyes Monorail by William H. Boyes as main transportation mode, and Thomas Edison's single-pour concrete house as main method of construction. The concept Soria proposed reached Soviet urbanists most likely through French economist Charles Gide or the Industrial City plan of Tony Garnier [6]. This fusion of thought, which arguably influenced Le Corbusier, known for his role in distribution of the concept among modernist architects [14] subsequently determined a more form-related interpretation of Soria's initial idea in urban planning history. In 1930s a linear development plan was implemented for Magnitogorsk and Stalingrad grounds by Soviet urban planner Milyutin [17]. This may be the only linear development model of that period that has been implemented on a city-wide scale. The system was implemented in the cities of Stalingrad (Volgograd), Magnitogorsk and several more in the Siberian region. Stalingrad's evolution process throughout time left little trace of the initially planned pattern. In its contemporary urban layout and contrary to the envisioned strip layout, mixed-use and industrial edifices are arranged along the riverbank and urban sprawl of residential housing is evident. Notably, almost every instance in overall literature has Soria's concept presented next to Milyutin's, treating them as representatives of the same utopian approach (Fig. 2).

The analysis of the concept's interpretations in this period underlines a few important conclusions: the concept strongly resonated within professional circles worldwide, whether it was accepted or not. It gained more popularity in Spanish-speaking territories, where the emphasis was on the ideals behind the concept. It has received less recognition in other areas of the world, where the focus leaned more towards the design aspects, such as geometry and scale. Overall, there was a lack of clarity in the scale level when referencing the concept at the time.

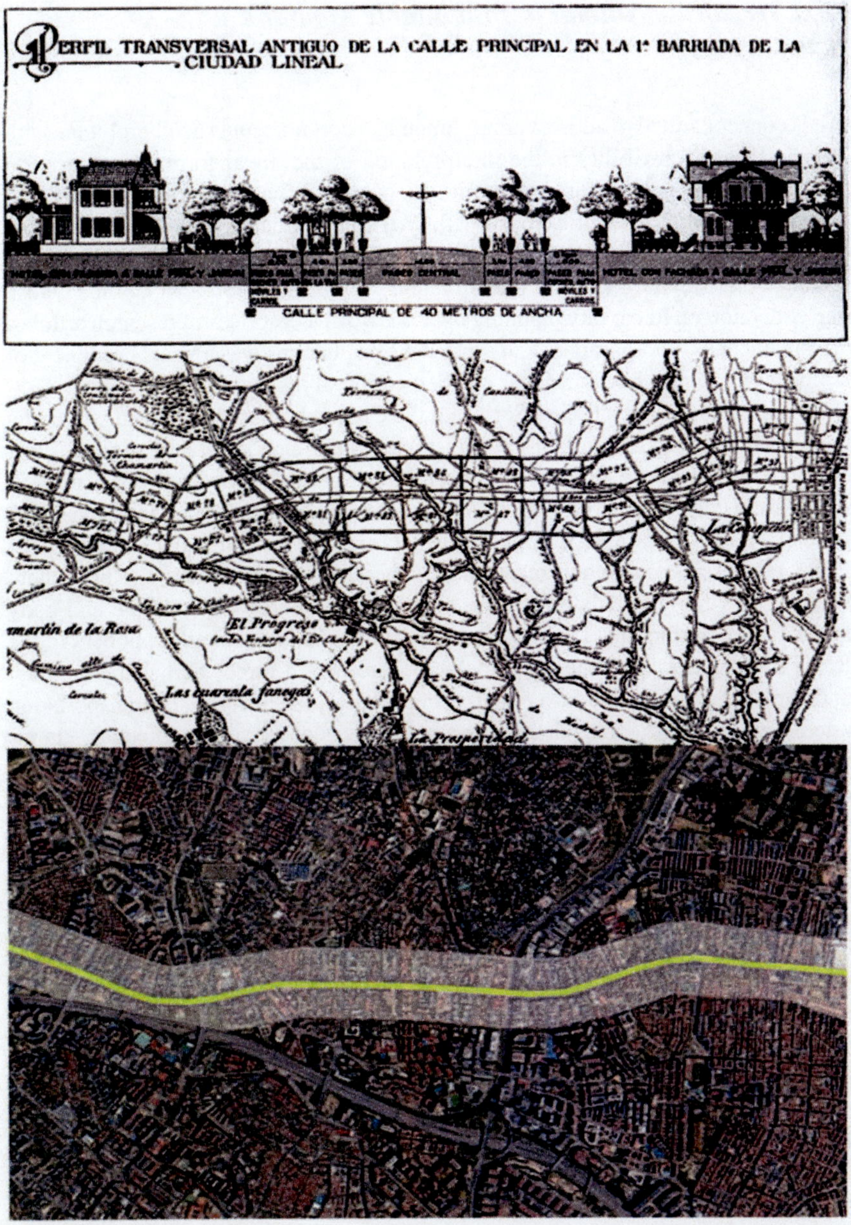

Fig. 2 Soria's drafts of Ciudad Lineal de Madrid in 1880s and the actual area of the plan's implementation today (*Source* Top and middle Soria y Mata, bottom author's annotation of the area over Google satellite image)

From Vertebrate Animal to Inhabitable Machine (1940s–1960s)

Soria's concept aimed to address environmental, economic and social problems with a certain level of flexibility in the interpretation of the linear form. He referred to it as a "vertebrate animal form" and with such natural characteristics it should also appear in the evolution of cities [6, 10, 18]. We gained valuable information about the concept's emergence, distribution of idea and development around the world by the middle of the twentieth century predominantly thanks to George R. Collins whose large collection on linear city planning over the world comprises of research articles, correspondences, photographic and printed works. Collins was a vocal proponent of the linear city whose level of detail and dedication in reporting on the topic is reinforced with deep appreciation of Soria and the Ciudad Lineal. Apart from criticizing the professional society on their skepticism and omittance in referencing the linear city concept in literature, his article "Linear Planning Throughout the World" in 1959 established a baseline historical overview encompassing ideas and plans from C. M. Carvajal, N. A. Milyutin, E. Chambless, M. Hastings to Le Corbusier, K. Tange, the MARS plans for London and many more. These are some of the most prominent examples inevitably circulating within contemporary linear city critique and body of literature. Constantinos Doxiadis, on the other hand, was a stark opponent to the idea of linear cities. He stated that there are three main forces that shape the city: concentric forces around areas of central importance, linear forces of transportation systems and aesthetic forces drawing people to areas of attraction. Doxiadis claims that Soria never designed a linear city and that La Ciudad Lineal is a corridor-like expansion connecting cities. He held a tenacious position on the matter, stating that the three-dimensional surface of the Earth simply does not allow a linear city development because any connection with the main axis would disrupt its linearity [1]. He then goes on to name two exceptions where we could have relatively linear cities—"*where the forces are not pointed in all directions, in a valley or on an island*" and in form of small, linear parts of cities between major axes. Collins raised concerns regarding the simplification of Soria y Mata's trademark notion of stretching the city "from Cadiz to St. Petersburg, from Beijing to Brussels." In his opinion, it diverted focus from the more significant features and values expressed in the concept. He states that "*it should be kept in mind that, regardless of the artists' renderings by which they may on occasion be presented, linear planning is primarily a schema, a process, a system, and not a physical or architectural actuality*" [19].

Collins thought of linear development as an ancient and naturally driven pattern of city growth. He suggests that turning this linear tendency of city deployment into a beneficial matter was essentially grounds for developing numerous linear planning theories. Adding to Soria's drive for efficiency, and the Soviet interpretation of radical decentralization, in 1930s the concept gained a new flavor of modernist values through the interest and promotion by Le Corbusier. His plan for the "Radial City" (Ville Radieuse) clearly abandoned the concentric city model in favor of axial development [20]. His ideas for Algiers and Rio de Janeiro, however, took it to another extent by expressing the form of a **linear megastructures.**[1] This chain of influence continued further to Bauhaus-linked German architect and urban planner Ludwig Hillberseimer, effectively prompting the MARS group's plan for London.

1960s geographer C. F. J. Whebell's contribution to the linear development concept reflects in his distinction of corridor agglomeration as large-scale linear system of urban places. He describes corridor development as an evolutionary process. As Collins did prior, Whebell also acknowledges the historical tendency of developments to form along corridors expressing a strong critique of the concentric city model presented by Christaller [13].

Failure of Linear Cities Becomes a Consensus Gentium *(1970s–)*

Perhaps the most abstract interpretation of linear cities came from two Florence-based architecture groups in the 1970s; Superstudio with their Continuos Monument and Archizoom with the Non-stop City. The groups produced dystopian renderings of monumental linear structures cutting through city and landscapes [21, 22]. These artistic experiments with their utopian overemphasis aimed to advance discussion on the role of design and form in shaping the future of cities.

Kevin Lynch in referencing linear city model examples ranging from Henard, Garnier, Milyutin, Leonidov, Le Corbuiser, to Paolo Soleri describes a futurist design primarily concerned with the physical environment rather than the social. Furthermore, he assigns the term "machine model" of cities, made up of undifferentiated parts linked together with clearly distinct functions and motions [2]. Particularly, to the works of Soviet architect Milyutin he compares the machine value of the scheme expressed as a production line. Notably, however, he singles out Soria's linear city as a "much more liberal and humane form" of machine model concerned about health, open spaces, accessibility, and affordable housing.

Contemporary literature on the topic of linear cities is less philosophical and predominantly shifted towards providing historical overviews of the various linear concepts that occurred after a couple of decades of intermission in significant discourse on linear cities. The author's contribution in the last few years was to shift focus on existing topographically conditioned linear urban developments [7, 18, 23] and distinction in categorizing planned city concepts from the radical form of linear megastructures [7]. The topic has been re-ignited with the proposal of Neom inevitably putting it into context of failed past examples. Although construction has already started, large skepticism about the project's success remains. Some of the main criticism of Neom range from questioning its promise of being 100% powered by renewable sources, while neglecting the embodied carbon during the

[1] Term coined by author in a previous publication, "*Linear Megastructures. An Eccentric Pursuit in Tackling Urban Sustainability Challenges*", referencing to a radical architectural interpretation of the liner city concept through a continuous built compact development.

construction process to the fact that the technology supporting this system is yet non-existent. Water supply issues, livability, concerns about social class inequality in vertical distribution, fragmentation of the natural landscape are just some of the main talking points. Additionally, there are human rights concerns about the relocation of local tribes from their historic homeland. In order to answer the question to what extent Neom is a linear city a clear categorization of linear urban developments based on their conceptualization and morphology is necessary.

3 Categorization of Linear Urban Developments

Based on the overview of historical discourse and the variety, abundance, difference in scale, design, and interpretation of linear urban developments within existing body of knowledge, breaking down the broad subject into understandable segments through categorization is necessary to streamline future research or implementation on the topic. The way this research approaches categorization of linear urban developments is within the framework of urban morphology and critical theory. This paper builds upon the author's prior categorization of linear urban development within theory of urban planning in accordance with the following recognized theoretical and morphological properties:

1. **Concept**—whether existing or only theoretical;
2. **Control of development**—naturally or conceptually imposed.
3. **Urban form inducing factors**—infrastructure, urban growth, topography, socio-economic drive.
4. **Setting**—a single compact urban setting or part of another (larger) urban fabric.
5. **Transportation**—ratio of transportation mode to size of development, single or multimodal transportation system.

Linear Villages Versus Ribbon Development

Linear villages are existing small-scale versions of linear urban developments, particularly common in central and eastern European countries. They represent the most basic elements and concepts behind the linear city idea. A linear village or *Strassendorf* as popular in Germany, is an independent settlement oriented along a single road with automobile and cycling transit. These villages are different from ribbon developments because they usually do not extend from cities as urban sprawl and they can be found as independent self-contained settlements. They did not emerge as a philosophical concept or strategic plan, but as an organically occurring sustainable urban growth. Linear villages retain a respectable size in relation to the transportation system of their main artery.

Ribbon developments, on the other hand, despite the linear appearance, typically originate as a form of uncontrolled city expansion along roadways. They first emerged

Fig. 3 Villages Valea Valor, Romania and Suloszowa, Poland, (up) and ribbon developments near Merthyr Tydfil, Wales and near Limburg, Belgium (down) (*Sources* Websites; valea-viilor.ro, World Urban Planning IG, ggat.org.uk, google satellite image)

in many industrialized countries worldwide during the 1920s and 1930s, attributable mostly to rising automobile ownership and a lack of planning control. Contrasting to other linear developments, ribbons are regarded as a form of urban sprawl. Although ribbon developments can increase accessibility to services and transportation, they essentially create negative impacts, like fragmentation of natural landscape, loss of agriculture land, increased traffic congestion, pollution and public transportation that is less efficient but more expensive. The adoption of the green belt policy is often implemented as mitigation strategy for such expansion (Fig. 3).

"In the USA, a pernicious type of uncontrolled "linear" or ribbon growth has occurred... Thus, linear growth receives acceptance as a fact, but not as a theory." G. R. Collins, 1960.

Topographically Conditioned Linear Cities

Apart from the road as a *force mobile* for the urban development in existing linear examples such as linear villages, the larger development determinant forcing linearity is topography. Even Doxiadis, who firmly opposed the idea of linear cities, does however acknowledge that there are exceptions in cases where the terrain provides a natural constraint on the urban form and allows for a linear settlement to form. He names two topographically constrained possibilities—an island, such as Miami Beach, where the ocean provides a natural boundary, and a valley. This is exactly the case of Sarajevo, Bosnia and Herzegovina [23].

Apart from being an exception, as per Doxiadis' notion, Sarajevo also demonstrates traces of the linear city concept in planning, although it was never officially acknowledged as a strategy. The dominant features recognized in the city's layout (not including informal hillside sprawl) are visible in the strong transit orientation towards the main growth axis with tramway transit, secondary perpendicular axes creating semi-modular sections, polycentricity and vicinity to the natural environment.

It would be interesting to consider whether Soria's concept would have been more successful if implemented in a setting deemed to have linear potential according to Doxiadis—namely, a coast, valley or island city. Would the Ciudad Lineal be more successful if implemented in Sarajevo?! The coastal region of Grand Tetouan (North Marocco) is another example of an area largely determined by a linear growth pattern, as also recognized by Soria's follower Hilarion Gonzales del Castillo, who proposed a "linear garden city" for the region [14]. This research emphasizes that topographically conditioned linear cities ultimately are conditioned to evaluate the principles and features of the linear city concept in pursuing and establishing a sustainable urban development.

Contemporary Development Trends Aligned with the Linear City Concept

A contemporary embodiment of the linear concept aligned with its essential premise of an arterial backbone as magnet force is found in a widely accepted trend of sustainable urban planning—the transit-oriented development (TOD). The concept is different in the sense that it is generally understood as an area incorporated into the existing urban fabric of a non-linear city. More specifically, it is an area of not more than ½ mile (approx. 800 m) radius from a transportation hub. Both the linear city concept and TOD prioritize the use of mass transit systems to reduce car dependency and increase the efficiency of public transportation systems.

Corridors, in general, are linear developments on a large scale and serve as a planned development connecting cities and regional centers. Whebell offers a tentative hierarchy of corridors distinguishing three levels; a continental or sub-continental corridor, a regional corridor, and a local commuter rail corridor. In terms of corridors and culture gradients he states: He proposes that the term corridor denotes a linear pattern of major towns joined by highly developed clusters of transport routes [13]. Contemporary use of the term corridors includes natural (waterways, greenways etc.) and man-made (highways, arterials etc.) developments that establish connections on a neighborhood, district as well as regional level.

The Planned Linear City

A somewhat linear approach in urban planning appeared in the US around 1919 in Milo Hastings' plan "Solutions to the Housing Problem." The dwellings were arranged in strips and oriented towards a U-shaped road loop. The Modern Architectural Research Group (MARS) presented a master plan for Greater London in the 1930s based on linear development concept. Along the Thames in the east–west direction a primary artery was to be laid out with railway, industrial, commercial and administrative functions. While the residential areas were to be placed across a series of transversal ribs intersecting the main artery. Each rib would be approx. 1.6 km wide with separating green belts double in width. Every important function would not be more than a 10 min walk away from any dwelling. The layout, however, is enclosed by an outer transportation line connecting to the main artery. The Finger Plan of Copenhagen is an internationally renowned regional plan developed in 1947. It was designed to meet the future challenges of the city by providing a development strategy that enhanced access between the city and green environment and a better connection with the Greater Copenhagen area [24]. The plan addressed the rising population growth, the centralized city form with gradually concentric, yet incoherent and less developed layers around it, which increased the distance from city to countryside and time-consuming daily commute. The plan represented the backbone structure of Copenhagen's development and it has been carried out by different bodies and different legal effects. Polish theorist, architect, and urban planner Oskar Hansen also expressed interest in the linear form, however his approach was based on an "Open Form" theory that amplified the need to engage the user of space as recipient of ideas and to abandon the concept of superior experts' decision-making that influences human living. Hansen introduced a linear urban plan called "Continuous Linear System" which was envisioned as elongated decentralized cities across Poland and Europe demonstrating the open form philosophy. The general idea behind the concept was to create a connection of places from one part of the country to another by establishing equal accessibility, as opposed to division of center-oriented cities and their surrounding environments. Together with his wife, Hansen was successful in realizing such a project on a smaller scale with the Slowacki Housing Estate in Lublin (Fig. 4).

Linear Megastructures

The radical and utopian interpretation of linear cities found its expression in a form called **linear megastructures**. They are socio-economically driven, compact yet large and continuous forms of linear urban development. Linear magestructures are designed as modular built structures, where zoning is usually vertically distributed below or above a transportation spine, different from other linear development concepts where form and structure are arranged in planar configuration [7].

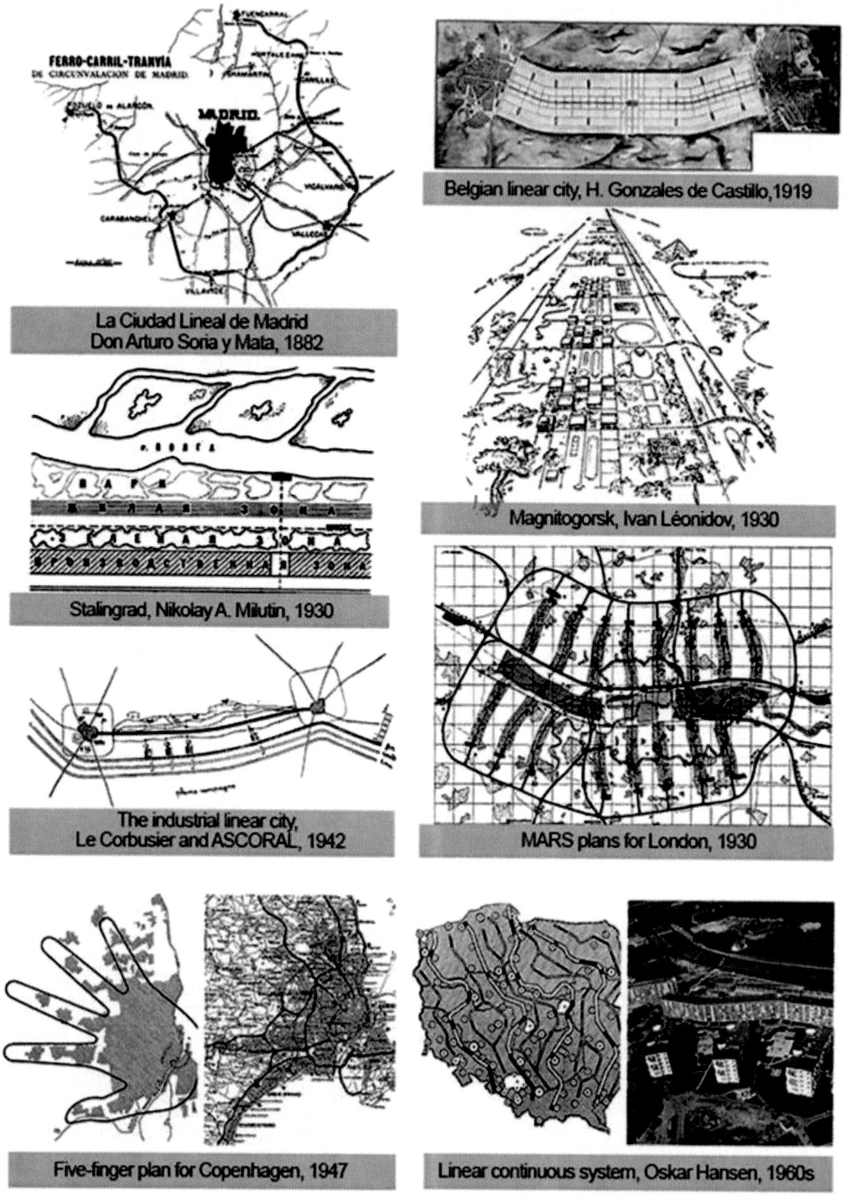

Fig. 4 Variety of linear city concepts in history of urban planning (*Source* Author's compilation based on various open sources)

"*…the idea occurred to me to lay the modern skyscraper on its side and run the elevators and the pipes and the wires horizontally instead of vertically…*" He then goes on to say: "*I would not cure the evils of congestion by perfecting congestion as is the case with the skyscraper—I would build my city out into the country.*" Chambless, E. 1910.

Although unique in certain design aspects, Neom's linear city project bares most significant similarity with other linear megastructure ideas. Hence, it's place in the large spectrum of linear urban developments belongs in this category. Starting with Chambless' Roadtown, notable examples include Le Corbusier's plan for Algiers, Graves and Eisenman's Linear city, P. Soleri's Lean Linear City and Gauthier's Linear city. Linear megastructures are driven by ambition, technological superiority and a sense of grandiosity. Their intrigue relies on technological and innovative promises, while their validity is promoted by the involvement of prominent "starchitects' and names in the industry. This is equally true for Roadtown as it is for Neom (Fig. 5).

4 Discussion

The Ideal City Aspect

Ideas of ideal cities in general emerge as strong **responses to critical urban crises with the intent to tackle all problems at once**. From the black plague outbreak that took tens of thousands of lives in the Middle Ages, to pollution, overcrowding, rural–urban divide and social inequalities of the early industrialized cities, towards analogous issues on a larger scale in contemporary cities with aspirations for a post-oil future and climate change mitigation. One common thread in these attempts is that all are based on the belief that it is more feasible to build new cities from scratch than to deal with the urban problems within existing cities. A noticeable design feature of most linear urban plans is **the repetition of modular segments**. This utilitarian design approach, sometime regarded as Fordist, originates from Soria's desire to create equitable housing opportunities. Later it was expressed and emphasized through Soviet ideology and values of the Modernist movement in architecture. Other elements emphasizing the modularity are expressed through secondary perpendicular axes (or vertical communication in linear megastructures). Such development pattern, however, requires significant planning constraints, not always suitable for real environmental and topographical settings. In linear "ideal cities" inevitably there is a great emphasis on the environmental aspect expressed through increased walkability and public transportation efficiency as opposed to car dependency. In linear megastructures, however, the plans are somewhat detached from the actual physical environment and surrounding land, while the boundaries are physically reinforced (Fig. 6).

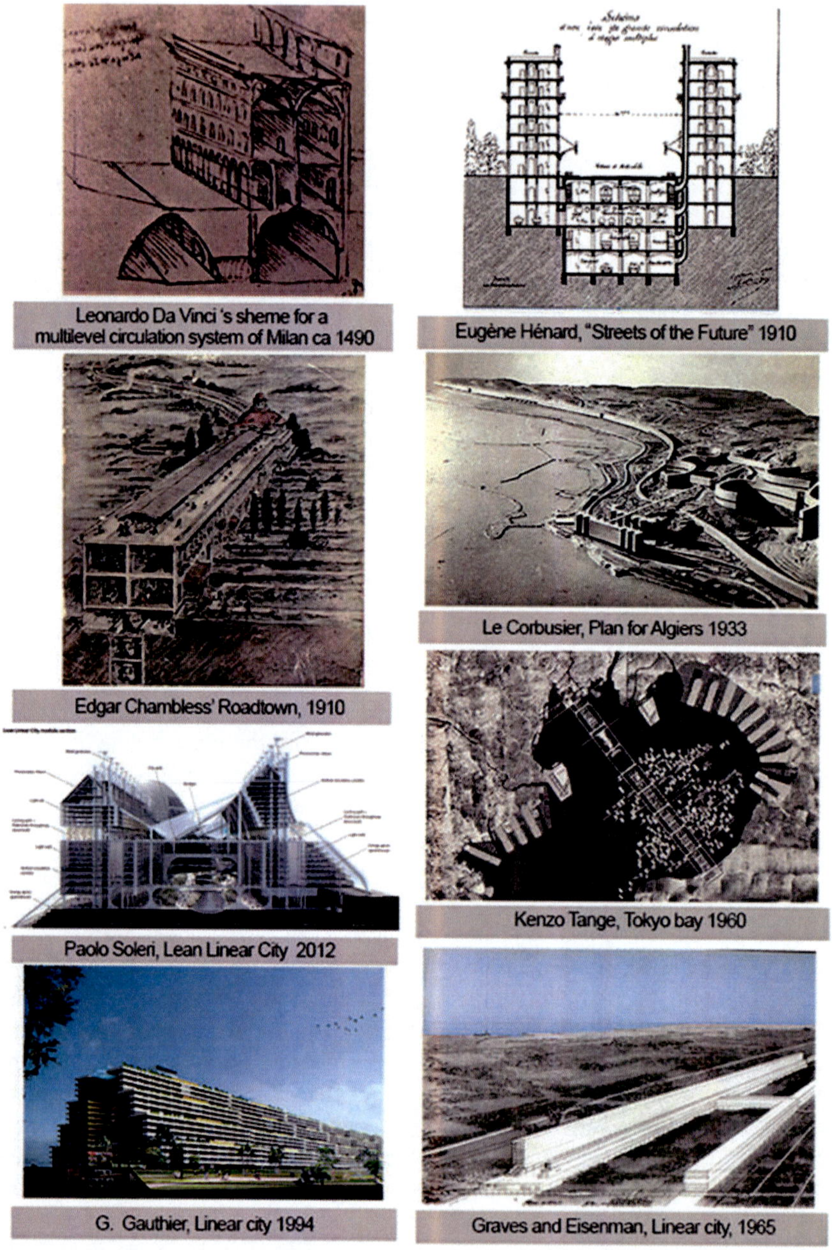

Fig. 5 Prominent linear megastructure concepts in history of urban planning (*Source* Author's compilation based on various open sources)

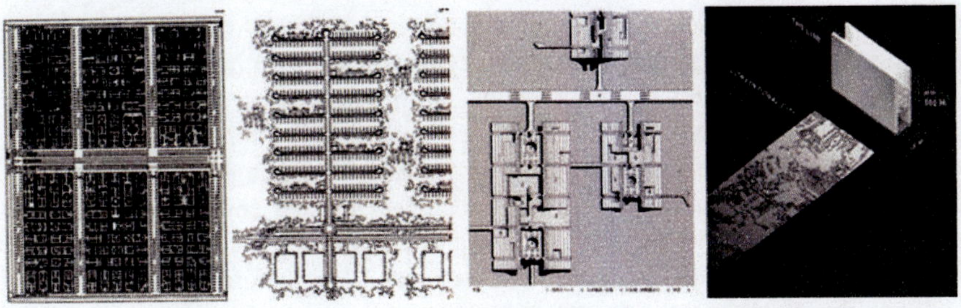

Fig. 6 Repetition of modular segments in different linear models—1. Soria y Mata, 2. Hilbersheimer, 3. Tange, 4. Neom. *Source* Author's compilation based on various open sources

Essentially, in the history of city planning, we can distinguish professionals, albeit a minority, with grand ambitions, who seek improvement of existing issues by significantly restructuring of the urban form and those who strive to optimize development under the existing rules [25]. When referring to Utopian planners, Wlliam H. Wilson underlines "*…a study of city planning is quite different from a study of planned cities*" [26].

Therein lays the key to why linear concepts are dismissed as unrealistic at the very start.

The Design Versus Planning Aspect—An Expression Through Form

The design versus planning aspect is perhaps most understandable when comparing Neom's to Soria's linear city concept. **The first one entirely revolves around a geometrically straight line; the latter provides a flexible linear layout following a certain route surrounding Madrid in a semi-circular plan** [11]. Moreover, the criticism around Neom's rigid form and strict geometrical definition of a line underlines the argument for its classification as a structure, rather than a realistic city plan. Soria's plan arose from the observed inequalities with a genuine intent to create a more livable, just, healthy living. To put things into the right perspective, the total human population on Earth in 1900 was estimated to be 1,6 billion people. The largest city at the time was London with an approx. of 6.5 million inhabitants. It is not unrealistic to conclude that Soria did envision a city in terms of what a city meant at that time. However, Soria's model was flawed in the sense that it did not consider the importance of boundary as crucial constraint factor, the natural landscape as magnet force and size to transportation mode ratio (a faster transit system would not work in the given layout as it would require less stops). Additionally, his concept was largely misinterpreted. The critiques range from concerns of size, to being deemed as futuristic, utopian and a machine model. To some extent through his own doing by self-sabotaging promotional exaggerations [10]. To other extents because the form was placed into central focus overshadowing the socio-economic and urban planning aspect of the proposal. The vaguely different design interpretations of the initial conceptual features is what fuels overall misconception about the notion of linear cities today. They largely depend on the creator's professional background, motivation, the period of architecture history, style and technological advancement of the time of their creation. The question of the Linear City is essentially also a question of difference in the fields of urban planning and architecture. Once architects started producing their interpretation of the concept, which by the nature of their profession relies greatly on the aspect of design, geometry and structural thinking—some of the concept's humanistic character has become ostensible. Planners on the other hand, focus on the processes and significance of locally specific development determinants. **Neom is therefore a contemporary linear city interpretation** driven by powerful ambition of creating "a revolution in civilization" through the utilization of latest technological advances and aspirations. In its promotional campaign it promises all the previously recognized social and human benefits of urban living, but the extreme approach to the design widens the disconnect from concept to reality.

Control of Development

Control of development plays a major part in the linear city and brings into focus the challenges of existing or imposed boundaries. **The issue of boundary in linear development over time has shown to be of utmost significance and one that cannot be resolved by administrative and legislative measures.** This is because it involves a mixture of social, economic, and political factors that are constantly changing and evolving. As such, it is difficult to draw definitive boundaries that will remain unchanged over time.

The width restriction is a common design feature and probably the most contingent issue in linear cities. While Soria relies strictly on planning policy to restrict width (unsuccessfully), Neom builds a structural entrapment of an unrealistic height and narrow width. Both underestimate the importance of locally specific conditions and the critical role of the natural environment. Even though The Line's is situated within a desert, the topographical conditions of the area vary from the vicinity of the sea on one side to elevations on another. The plans, however, show no architectural treatment, volumes or openings on the outer façades that would indicate an attempt of cohesion and connection with the natural surroundings. They strongly reinforce the issue of boundary by forced compactness. This neglect of the natural magnet forces driving economic and urban development cannot sustain and the differences in attraction between the coastal and inland areas would eventually influence changes in the development over time.

Planned Versus Organically Evolved

Terminology drives a lot of misconception in the discourse of linear cities which rarely includes existing linear development examples. In planning practice, we ought to evaluate place-specific conditions and processes in establishing a trajectory for sustainable urban development. Therefore, it is necessary to evaluate the possible benefits of implementing principles and features of the linear city concept onto existing organically evolved linear layouts. **The question is whether to adopt topographically conditioned linear developments within the umbrella term of "linear cities?"** Arguments speak in favor of such, considering that their main physical features align with the concept of linear cities to a larger degree than features of linear megastructures promise to do. Linear megastructures are widely promoted and accepted as linear cities in literature, despite their exaggerated interpretation of the concept and the fact that not even one has ever successfully been realized. On the other hand, existing linear cities demonstrate at the minimum the basic features behind the concept, a transit-oriented constrained development, a polycentric layout, often with strong walkability potential and the vicinity to natural environment.

Size to Transportation Mode Ratio

In 1967 Richard Llewelyn-Davis published an article primarily to apologize to C. Doxiadis for previously referring to his concept of Dynopolis as being identical to the linear city concept [27]. He further elaborates a clear but often neglected point—the relation between the transportation mode and the scale of the development. Essentially, what this means is that small towns need low-speed transit with frequent stops, while large cities should accommodate faster transit options and allow for larger distances between the transport hubs. Otherwise, the linear city concept is rendered obsolete. **As logical as this seems, it has not been an often-recognized principle in many existing linear city proposals.** In the case of Neom, the transportation system is supposed to stimulate 15-min walkability to main services and to comprise of flying taxis, self-driving cars, the modern metro and an underground ultra-high-speed transit that can transport passengers from one end of the 170 km line to the other in 20 min. With no clear indication about the amount of stops and their distribution nor accessibility from the enormous 500 heigh structures on both sides, it is questionable whether the size to transportation mode ratio is adequately proportional.

5 Research Outlines and Conclusion

After a thorough theoretical review, we have a better understanding of the motives that every so often spark another linear city idea and how (or if) they relate to other linear city concepts in history. **We understand that new linear concepts are predominantly triggered with the advancement of transportation technology and as a response to contemporary urban crises.** They are driven by the idea of tackling all urban issues at once in a completely new setting while utilizing transit as the most efficient development force. Additionally, we can recognize and determine that a categorization of linear urban developments can be made in relation to the **conceptual features, size of development and whether they are existing or envisioned linear phenomena.**

Within the framework of urban morphology, critical theory and sustainable urbanism, the analysis covers the factors that influence the emergence and development of linear cities within urban systems, such as the socio-economic drivers that lead to the creation, dissemination, and different interpretation of the linear city idea. Environmental and geographic constraints that affect their design and growth, point to a need for distinction between organically versus theoretically evolved linear cities, as well as linear planning tendencies. Both planned linear cities and organically or coincidentally evolved linear developments can be encompassed by a broader idea of linear urban development and this paper provides a categorization separating between concept, design and criteria for success. By providing a deeper insight into their emergence inducing factors and existence of variations, this research argues that certain features can be (and already are) extracted from the general conceptualization and applied in real-life settings and current planning trends.

The discourse analysis concludes that the climate of opinion within professional society, including most vocal proponents over the 150 years shows lack in clear understanding of the differences in scale level, existence of organic linear growth as part of the concept, the issue of control of the development, and the design versus planning aspect. The interchangeable use of different terminology speaks to the ambiguity of the concept. Misinterpretations of the linear city in existing literature range from urban sprawl (ribbon) to planned corridor developments. The most prevailing notion, however, is the interpretations of the linear city concept as a continuous large edifice, namely a linear megastructure. This paper argues that instead, existing elongated cities that demonstrate basic principles of the concept, should be the focus of attention and possible implementation of linear concept features.

Defining Features

Based on the discourse analysis, desk research and overview of historical models within existing body of knowledge, the paper concludes with providing a list

Table 1 Linear City common characteristics based on literature examples

Prevailing conceptual features	Prevailing design features	Success defining features
Tackling: • Urban congestion • Pollution and environmental issues • Social inequality **Promoting**: • Transit orientation • Public transportation • Utilization of transportation technology advancement • Walkability • Vicinity to natural environment or agricultural land • Urban health	• Main transportation axis • Secondary perpendicular axes/connections • Restricted width of development • Modular elements • Uniform design/repetition • Policentrality—Multiple centers/nodes • Starting and finishing points not strongly defined • Specific edge/boundary design	• *Force mobile*-transit-orientation • Type of development control (natural/structural/ administrative) • Size to transportation mode ratio • Planning process versus form design

of defining features of the Linear City, as a general concept. It differentiates between conceptual features, prevailing design features and success defining features (Table 1).

Categorization

Following is an overview of recognized types of linear developments specified by these characteristics. The author classifies linear urban developments as described in Table 2.

Main Misconceptions

The paper elucidates some of the most common misconceptions of the initial linear development concept. The following conclusion related to the original concept as envisioned by Soria, later subjected to various interpretations, but also including existing linear urban developments that demonstrate the concepts features. It is based on an objective observation of potential for future implementation of the recognized principles or features of the Linear City on existing urban layouts. It is not based on the promise of developing new cities from scratch.

Misconception 1: A linear city is a utopian ideal—Actually, linear megastructure interpretations are dominantly the utopian ideals. Other planned and organically evolved linear city concepts retain a level of realistic implementation. In

Table 2 Linear urban developments' conceptual categorization and characteristics

Category	Types	Features		
		Control/Barriers	Transportation	Scale level
Linear organic forms	Linear villages	Self-contained	Ground level, Single or multi-modal	Community level
	Ribbon developments	Unconstrained (if not mitigated through policy)	Ground level, Single or multi-modal	Segment of urban structure
	Naturally conditioned elongated cities	Topography, water body	Multi-modal	City level
Planned linear urban forms	Designed linear cities	Zoning, exclusionary zoning	Multi-modal	City level
	Linear megastructures	Structural	Highly advanced technology, primarily underground	Regional level
Linear development trends	Transit-oriented developments	Policy, based on 15-min radius	Multi-modal	Segment of urban structure
	Corridor developments	Infrastructural, natural or economic	Multi-modal, Rail dominated	Regional level

addition to organically evolved linear developments (not sprawls), there are transit-oriented development or corridor development trends in sustainable urban planning that demonstrate features of linear cities as successful real-life settings. These provide argument for the concepts validity against solely claiming it as utopian.

Misconception 2: It is a machine concept, deprived of actual human and social context—The genesis of the concept as demonstrated through Soria's vision was an attempt in creating socially just, humane, and healthy urban living. Latter interpretations of that concept demonstrate a detachment from these issues. Deeming it a machine concept is a superficial approach in understanding the idea behind linear cities and neglects the existence of organically evolved linear concepts.

Misconception 3: A linear city is deprived of centrality—Incorrect. It represents a polycentric development based on the efficient utilization of transit as dominant development driver.

Misconception 4: A linear city is a rigid line—Incorrect. It is a transit-oriented way of city planning that can be expressed on transit routes with curvatures.

Misconception 5: A linear city is a large inhabitable structure—Incorrect. Linear megastructures are a form-based architectural interpretation of a concept, not rooted in location-specific context nor significant understanding of urban conditions and long-term processes.

Realization

Some of the design features can truthfully represent the conceptual features (Table 1) if interpreted and implemented with respect to locally specific urban conditions.

The argument this study makes is that immediate dismissal and a lack of understanding of nuances within the linear city concepts eliminates discussion and the implementation of the common beneficial features on smaller and larger scales in reality.

The following features extracted from the conceptualization of linear cities can be applied on existing or potential linear urban settings:

- efficient use of transportation technology
- prioritization of mass transit systems
- walkability
- polycentricity
- sustainable linear compactness
- efficient intercity connectivity
- vicinity of natural environment/agricultural land.

Finally, with a projected urban population growth of almost 70% by 2050, the urban planning challenges recognized by linear planning concepts are ever more prevalent than before, on top of other interlinked problems, such as the effects of climate change, disasters, the aftermath of the pandemic, and ongoing armed conflicts. In contrast to the idea of creating new cities from scratch, contemporary understanding among planners and researcher is in favor of mitigating existing urban environment issues and developing global as well as locally specific toolkits and policies that improve the health, resilience and sustainability of existing cities. Despite the criticisms, the linear city concept continues to be influential in urban planning and design. Linear city ideas, whether largely understood or not, have already been adapted and modified to create other planning models which are aimed at enhancing the livability, accessibility, and sustainability of cities.

References

1. Doxiadis CA (1967) On linear cities. Town Plan Rev 38(1):35–42
2. Lynch K (1984) Good city form. MIT Press
3. Alzaibag F (2021) Saudi Arabia's mega projects. J Int Aff 74(1):365–372. https://www.jstor.org/stable/27169791
4. Cunningham E (2022) Saudi crown prince wants you talking about his 'city of the future'. The Washington Post, last update July 28th
5. Gordon A (2021) Saudi Arabia resurrects discredited "linear city" concept in bigger, dumber form. Vice Media Group. https://www.vice.com/en/article/m7a598/saudi-arabia-resurrects-discredited-linear-city-concept-in-bigger-dumber-form
6. Collins GR (1959) Linear planning throughout the world. J Soc Archit Hist 18(3):74–93. https://doi.org/10.2307/987932

7. Tufek-Memisevic T, Stachura E (2019) Linear megastructures. An eccentric pursuit in tackling urban sustainability challenges, Środowisko Mieszkaniowe = Housing Environment. Krakow Iss 29:54–59. https://www.ejournals.eu/housingenvironment/2019/29-2019/art/16810/
8. Wood EE (1921) The Spanish linear city. J Am Inst Arch
9. Revista La Ciudad Lineal: Legado De Arturo Soria. Legado De Arturo Soria, Página De La Asociación Cultural Legado De Arturo Soria. https://legadoarturosoria.es/revista-la-ciudad-lineal-1897-1997/. Last accessed 05 Feb 2023
10. Collins GR (1959) The Ciudad Lineal of Madrid. J Soc Archit Hist 18(2):38–53. https://doi.org/10.2307/987976
11. Ferrocarril-Tranvía de Circunvalación de Madrid Á Canillas: Hortaleza, Fuencarral, Vicálvaro, Vallecas, Villaverde, Carabanchel y Pozuelo. Datos y Noticias Referentes Á Su Construcción y Explotación, United States: Creative Media Partners, LLC (2018)
12. Boileau I (1959) La Ciudad Lineal: a critical study of the linear suburb of Madrid. Town Plan Rev 30(3):230–238
13. Whebell CF (1969) Corridors: a theory of urban systems. Ann Assoc Am Geogr 59(1):1–26
14. Brunfaut V Terlinden B (2020) The territory of the Grand Tetouan as linear city: between description and project. Urban Plan. https://www.cogitatiopress.com/urbanplanning/article/view/2863/2863
15. Brandis García D, Mas Hernández R (1981) La Ciudad Lineal y la práctica inmobiliaria de la Compañía Madrileña de Urbanización (1894–1931). Ciudad Y Territorio Estudios Territoriales (49):41–76. https://recyt.fecyt.es/index.php/CyTET/article/view/81588
16. Chambless E (2012) Roadtown, 1910. Nabu Press
17. Milyutin NA (1975) In: Collins GR, Alex W (eds) Sotsgorod: the problem of building socialist cities. MIT Press. https://doi.org/10.7551/mitpress/6353.001.0001
18. Tufek-Memišević T (2018) Contemporary linear city development: a case of spatial interventions in Sarajevo, Bosnia and Herzegovina. Doctoral dissertation (Unpublished), Cracow University of Technology
19. Collins GR (2005) Linear planning. Forum xx-5. March. pp 1–26 (1968) as quoted in Herwin Sap, Corridors and/or linear cities; a historic contribution to the contemporary discussion on corridor development. PhD thesis, Eindhoven University of Technology
20. Corbusier L (2019) The city of to-morrow and its planning (2013). Dover Publications Inc., New York
21. Superstudio, di Francia CT, Magris A, Magris R, Frassinelli P, Natalini A (1973) Superstudio on mindscapes. Des Q 89:17–31. https://doi.org/10.2307/4090788
22. Stauffer MT (2002) Utopian reflections, reflected Utopias: urban designs by archizoom and superstudio. AA Files 47:23–36. http://www.jstor.org/stable/29544275.
23. Tufek-Memisevic T, Stachura E (2015) A linear city development under contemporary determinants. Housing Environment by Cracow University of Technology. https://www.ejournals.eu/pliki/art/7426/pl
24. The Finger Plan, Ministry of the Environment, Denmark (2015)
25. Levy JM (2016) Contemporary urban planning, 11th edn. Routledge. https://doi.org/10.4324/9781315619408
26. William HW (2018) Moles and skylarks. In: Krueckeberg DA (ed) Introduction to planning history in the United States (1983). Taylor & Francis
27. Llewelyn-Davies R (1967) Some further thoughts on linear cities. Town Plan Rev 38(3)

Rethinking Urban Mobility Under COVID-19: Milan's *Strade Aperte*

Hugh Bartling

Abstract As the COVID-19 disease began to spread in cities throughout the world in 2020, local authorities were faced with a host of policy challenges to contain transmission of the little-understood SARS-CoV-2 virus. One important element of policy action in urban settings is the regulation of mobility. When the COVID-19 crisis emerged, cities had to re-think their mobility strategies and develop policies that accommodated movement of people while also limiting viral spread. This chapter looks at the experience of Milan and their effort to implement a slow streets scheme in the city by situating the analysis in a theoretical framework that integrates the built environment, the natural environment, and the social system.

Keywords Mobility · Milan · COVID-19

1 Introduction

As the COVID-19 disease began to spread in cities throughout the world, local authorities were faced with a host of policy challenges to contain transmission of the little-understood SARS-CoV2 virus. One important element of policy action in urban settings is the regulation of mobility. In recent years, many local authorities have been seeking to increase the viability of low-carbon modes of transport such as urban railways, buses, cycling and walking in the effort to reduce greenhouse gas emissions and improve the quality of urban life.

When the COVID-19 crisis emerged, cities had to re-think their low-carbon mobility strategies. With social distancing measures being applied to minimize viral spread, local authorities in numerous countries quickly closed or limited the operation of public transport facilities [1].

Although much economic activity slowed as a result of shelter-in-place and social distancing orders, urban mobility remained an important urban function as cities

H. Bartling (✉)
DePaul University, Chicago, Illinois, USA
e-mail: hugh.bartling@depaul.edu

M. Arslanagić-Kalajdžić et al. (eds.), *Interdisciplinary Advances in Sustainable Development II*, Lecture Notes in Networks and Systems 804,
https://doi.org/10.1007/978-3-031-46692-2_8

attempted to manage the social and economic impacts of SARS-CoV-2 spread. Essential workers still needed ways to reach health care facilities, grocery stores, and other vital centers of production and residents still needed to obtain essential items and engage in activities.

One particular response to address mobility during a pandemic has been to enhance street spaces to better accommodate cycling and walking. With the threat of COVID-19 disrupting public transport, cities such as Milan, London, New York, and Bogotá began closing streets to cars and opening them to bikes and pedestrians in April 2020. In some cases local authorities were looking at the crisis of mobility under COVID-19 as an opportunity to make biking and walking more viable in their cities for the long term [2].

This study is a case study analysis of a particular "slow street" initiative in the wake of SARS-CoV-2: Milan's Strade Aperte. This study focuses on the following research questions: How were slow street decisions made in Milan within the context of a set of uncertainties about SARS-CoV2 transmission and the efforts to control COVID-19 cases. What were the various constituencies and social conditions operating in a swiftly-moving and ambiguous decision making environment related to accommodating mobility while limiting viral spread? Looking to the work of Béland et al. this study starts from the assumption that policies relating to urban infrastructure transformation involve conflict between advocates, elected officials, and technical experts from a variety of fields [3]. Understanding how this conflict is navigated and resolved in a variety of different settings is one objective of the chapter.

This chapter situates the analysis within the framework developed by Campbell in her article, "A Comparative Framework for Analyzing Urban Environmental Policy" [4]. She discusses how natural, physical, and social systems can provide a heuristic structure for understanding the different ways cities respond to policy challenges. Given the nature of the SARS-CoV2 spread and the infrastructural changes that are part of the slow streets policies being adopted by cities throughout the world, Campbell's framework lends itself to analyzing this particular policy phenomenon.

The remainder of this chapter looks first to explore Campbell's understanding of urban policy that links natural and built infrastructural systems with governance processes and social systems. This framework, it is argued, is well-suited to understanding the challenges and possibilities of urban policy making within the context of an urban public health emergency. Furthermore, this particular approach will highlight urban infrastructure and its importance in terms of planning during public health emergencies.

Following the discussion of Campbell's framework the focus will shift to Milan's Strade Aperte scheme which will help establish the potential efficacy of an urban environmental policy framework for understanding urban health emergencies and the possibilities for effective urban policy responses.

2 The Urban Policy Environment and Urban Health Emergencies

A dominant strand in studies of environmental policy over the past twenty years has been to highlight the importance of subnational governments–and especially cities–in addressing some of the more challenging environmental problems facing communities across the world [5, 6]. For some issues such as water quality, air pollution, and waste management the role of more locally-focused governance arrangements is essential and logical as the impacts and sources of environmental despoliation are often regionally or locally situated and experienced. For environmental issues that have a more global and diffused set of sources and impacts, such as climate change, the motivations for urban policy responses differ but are often oriented around the necessity of a local response in the absence of efficacious policy making at the national or international levels [7].

While there is a long connection between environmental despoliation and the development and spread of pandemic events [8], the sustainability and climate action plans of cities often do not explicitly address pandemic management. Even the broader health impacts of climate change are often not emphasized in local climate planning. Mendez [9], in his study of local climate action plans in California, finds that co-benefits of improved air quality stemming from climate action are rarely discussed in climate plans.

Even though the centering of health concerns may not be evident in many local climate plans, public health professionals have emphasized the potential for negative health impacts stemming from climate change. In 2008 Frumkin et al. offered a public health approach to climate change by mapping climate-induced health impacts to the essential services identified by the American Public Health Association, arguing that key tasks of public professionals like monitoring community disease development and educating the public about emerging health risks needed to be informed by the impacts of climate change [10].

While it is reasonable for policy makers to frame and address the problem of COVID-19 spread as a public health problem and forefront health authorities to develop responses to the pandemic, expanding the understanding of pandemic response to a wider set of urban processes can be useful for both enhancing the effectiveness of managing viral spread as well as exploring the interactions between public health and the larger urban setting. To explore the potentialities for this wider, integrative analysis theoretical frameworks developed in the context of comparative urban policy can be useful. Specifically, it can be argued that the systems-based framework that Campbell developed for understanding urban environmental policy can be broadened to include urban public health concerns.

Key to the Campbell framework is the assertion that urban policy making should address the complex interactions between natural systems, social systems, and the built system. While it is important to recognize the interrelated nature of these systems, developing analytically distinct categories allows for the clearer discernment of how the policy process may operate in particular urban contexts.

For Campbell, the "natural" system refers to the biogeophysical setting of the urban space. Elements such as physical geography, flora and fauna, hydrologic cycle, atmospheric sinks, and climate would be part of this category. For the purposes of this study it is argued that viruses should be added to this category. Although viruses do not have cells like plants and animals, they are infective agents that bond to living cells and can thereby replicate and impact populations of flora, fauna, bacteria, and other microbes. Viruses are evident in the natural system and both impact and respond to environmental conditions. As the climate changes, pathogens such as viruses are expected to migrate and cause new waves of infection [11]. While viruses and microorganisms such as bacteria are not commonly discussed in urban environmental planning and policies, the SARS-CoV2 spread suggests that they should be recognized as important elements in natural systems.

The "social" system, in Campbell's model, refers to the economic relations, social institutions, and formal and informal instances of human-to-human interaction. As such, this system is primarily concerned with the networks of social interactions as well as the norms, laws, and habits that mediate human interactions. The social system can include traditional institutions such as corporations, governments, and non-governmental organizations and the patterns that emerge as humans interact with one another in these spaces and networks.

Campbell refers to the built system as the "concretion" layer in her model. This is taken to refer to the infrastructure and built environment: roads, utility networks, sewage systems, buildings, mass transit systems, and other human-induced impacts on natural spaces. Campbell prefers the term "concretion" over "built environment" to emphasize the layered nature of the constructed city. Construction and the physical disturbance of the urban setting occurs in a complex manner over time and the layered nature of the concretized environment impacts the natural systems and the social systems upon which it is reliant.

The systemic approach, in fact, emphasizes interactions within the three layers (social, natural, built) and between the layers. Within an urban natural system, flora, fauna–as well as viruses–migrate, thrive, or suffer based on climatic changes and the qualities of other natural elements such as air and water. Symbiotic relationships exist between flora and fauna with animals and plants providing food and habitat. For zoonotic viruses like SARS-CoV2, as well, adhering to animal vectors allows for viruses to spread to other hosts. In the case of SARS-CoV2, because of its genomic structure it likely migrated from bats to other mammals which were kept by humans in the wet market in Wuhan that is suspected to be the original site of viral outbreak [12, 13]. Natural ecologies in urban settings are complex and their dynamism impacts social relations and the ways in which humans can intervene to concretize environments.

The idea of a social system that embodies the relations between people ensconced in economies, cultures, and interacting under systems and processes of laws, customs, and norms is not unfamiliar to social science and policy theorizing. The importance of its presence in Campbell's model is to offer an expression of the complexity of the social environment and to highlight the myriad ways that the entities within society interact with one another as workers, citizens, residents and other relevant identities.

The built environment can also be thought of as a complex amalgam of human interventions that have accumulated over time that serves as a literal support structure for human activity. The development of sophisticated urban stormwater systems in the late nineteenth and twentieth century industrial cities, for example, has been essential in contributing to the reduction in particular water-borne illnesses. Decisions made on how to change and mold the built environment influence subsequent decisions. Mass transit infrastructure, for example, can increase land value and demand near stations which, in turn, can result in higher density building. More density creates further demand for certain types of utilities and wastewater infrastructure. The cycle of concretization generates demand and scarcity and influences the ways in which the built environment continues to be forged.

As alluded above when describing the three key layers of Campbell's model the social, concretized, and natural "layers" of the city are consistently in a state of mutual interaction. The built environment responds to and impacts ecosystemic processes as buildings are constructed to provide shelter, drainage systems are constructed to manage stormwater, and utility systems are established and erected to distribute energy to urban inhabitants. The establishment of the built environment, in turn, impacts wildlife habitats, hydrologic cycles, urban microclimates, and air quality. The dynamics of urban social interaction, of course, rely both on a concretized built environment and functioning natural system. Clean air, potable water, and biodiversity are biophysical requirements for social interaction to function. The built environment similarly ensures there are places for interaction, those places are accessible, and they are supplied with essential utilities like water, sewage, and electricity.

How can this model be useful for understanding and addressing particular urban policy problems in the midst of urban health emergency, such as COVID-19? As mentioned above, viruses are a phenomena of the "natural" system. They are pathogens that search for hospitable cellular environments for reproduction and, as such, are evident in animals and plants. As with other natural elements, viruses interact within ecological networks shaping and being shaped by these interactions. When a virus like SARS-CoV2 that is significantly infectious and potentially deadly to people emerges in an urban environment, its management is going to be occurring within the intersection of natural, social, and concretized systems.

From the social standpoint, viral management must contend with the mechanisms that underlie social interactions and contribute to their sustainability. These might include the social processes that allow economic and cultural activities to function. Local urban economies are predicated on workers and consumers congregating in spaces, moving around to domicile, shop, work, and recreate. Although in the early months of the COVID-19 epidemic, the novelty of the virus made it difficult for epidemiologists to be authoritative in their understandings of the SARS-CoV2 transmission and infection pathways, it was suspected that airborne aerosols and droplets and surface contamination were the main mechanisms for viral spread.

Thus, managing viral spread created a degree of incompatibility with the ways in which existing social systems were operating. The sites of work, commerce, leisure, and dwelling that fueled consumption, production, and leisure became potential zones of pandemic spread. For urban policymakers, balancing a threat from the "natural" world (SARS-CoV2) with keeping social life "normal" became an urgent public problem in the first months of 2020.

As implied above, the social system functionality threatened by the urban pandemic emergency has a specific built environment–or concretized–element of concern. Viral spread is less a product of the actual social, cultural, and economic transactions that make up urban market-based economies of the twenty-first century, and more about the spaces where those transactions occur. And, more specifically, the concern is with constructed spaces such as buildings, public transportation facilities, and transportation corridors. If viral spread takes place in enclosed spaces with poor ventilation, then what are the implications for moving people around the city via public transport? Could streets and the public right-of-way accommodate activities that normally occur in buildings? Can the concretized and social systems be realigned given new pressure from the natural system? How can governance processes within the social system react to the urban emergency presented by a pandemic within the constraints of a concretized built environment and a viral behavior that is not well-understood?

In the remainder of this chapter, these questions will be addressed with a particular focus on the re-imagining of the urban infrastructure of street networks in Milan. The city was one of the first in Europe to see SARS-CoV2 and was extremely hard-hit by the virus. It has also been undertaking a multi-year process to rethink its urban infrastructure in terms of various initiatives to reconfigure public space. By exploring a specific case study this article will assess the viability of Campbell's conceptual framework and highlight how it could be used to address future urban emergencies that emerge in the confluence of the built environment, the social system, and the natural world.

The next section first situates the case of Milan within the context of larger debates about streets as public spaces with a particular focus on the efforts that have been ongoing in major cities to shift street spaces from being the near-exclusive domain of automobiles to being able to accommodate multi-modalities and uses. The presence of these discourses and ongoing policy changes with regard to street space provides important context for understanding the political efficacy of street reconfiguration and highlights the ways in which the social element that is presented in the Campbell model can also be thought of in a layered fashion. In the same way that she points out that it is important to recognize the evolutionary development of the built environment (hence, the use of the term "concretized"), the social environment is similarly layered. Ongoing policy discussions within the social environment factor in to the ways in which urban pandemic emergencies are addressed through governance processes. The existence of a robust set of ideas around managing the multiple demands for street space as well as examples of possible enhancements shapes the contours of responses to the SARS-CoV2 outbreak in the case discussed in this article.

3 Slow Streets

In the North American context, the concept of "slow streets" has evolved as a response to over forty years of urban planning efforts that have resulted in the dominance of automobiles in urban environments. The slow streets movement has antecedents in planning movements from the late twentieth century that sought to "tame the automobile" and reorient cities to be more human-focused. Historians of twentieth century urban planning discuss the era as being one which solidified a professionalization of the planning enterprise that looked at the project of coordinating the social, natural and built environment as one built on a particular mechanistic and positivist rationality. In the United States, the establishment of zoning to regulate land use was one of the steps in this direction. The nineteenth century city that developed along with the industrializing economy created a cacophony of diverse land uses that created conflict among property owners [14]. For some early professional planners, the municipal powers to determine the spatial distribution of broad categories of land use could help militate against such conflict and result in a more harmonious environment. As zoning segregated functionality, the emergence of the automobile as a flexible tool of mobility was embraced as the way to integrate the various land use districts in a metropolitan area [15].

Segregated land use and mobility via the single-passenger vehicle became the defining factors that shaped the built environment. In the new suburbs on the metropolitan fringe, low density development proceeded under the assumption that automobiles would be essentially the only viable mobility option. Zoning codes evolved to reflect this as well with parking requirements being prominent elements in land use regulations. A built environment replete with streets constructed primarily to accommodate cars and spaces on urban parcels for their storage became normalized both in newer suburbs, but also in the older industrial metropolis where streets and buildings were gradually reconfigured to be oriented around automobiles. For observers of the social and cultural geography of cities, the ubiquity of the automobile in cities by the end of the twentieth century had resulted in the automobile dominating much of the urban experience [16].

While the automobile theoretically offers a great deal of flexibility in terms of convenience and the ability to access a wide variety of places, there are shortcomings as it dominates options for urban mobility. Most concerning is how the concomitant land use regime that emerges with automobility works to make alternate mobility technologies unfeasible. Low density land use patterns eschew the concentration of diverse destinations in favor of the dispersion of zones based on single land uses. Thus, spaces of consumption, employment, and residence are segregated and engaging in the activities associated with daily life for residents often requires traversing great distances which make mobility by walking, bicycling, or public transportation inefficient or dangerous [17].

With the built environment offering little incentive for people to travel using modes other than the automobile, alternative modes were ignored in planning as resources would be concentrated on managing traffic flow and dedicating spaces for car parking. Access to a car would, in turn, become nearly a prerequisite for urban life as public transport would be inconvenient and less intrusive modes such as cycling and walking become infeasible. Because of the automobile's size and the problems with congestion and road capacity, the limitations of the technology's promise as a flexible, efficient mode of mobility become evident and urban policymakers invest in road widening and lane addition. This creates what Todd Litman describes as a cycle of automobility whereby the transportation planning decisions made to accommodate the automobile set forth a pattern of behaviors and responses that reinforce the idea of the automobile as the only way to move around a metropolitan area [18].

To use the language of Campbell, this results in a particular automobile-oriented concretized space that is supported by social pressures, the assumptions of urban bureaucracies, and the laws and regulations that reinforce social norms. Given the dynamic nature of these social processes, it is important to acknowledge that there have always been efforts to resist or reform dominant practices and assumptions.

In the case of the automobile's dominant influence over urban infrastructure and public space, the rethinking of the automobile's hegemony can be traced in recent years to the emergence of the New Urbanism planning movement in the 1990s. Of particular interest to New Urbanism designers, architects, and planners, was the desire to reorient professional thinking in urban disciplines from an overwhelming focus on mechanistic understandings of the city towards ones that were more humanistic. Central to this goal has been to encourage municipalities to think more deeply about the quality of street space and its supportive infrastructure to enhance people's lives and the urban experience.

In their early iterations, the successful experiments that were constructed under the influence of New Urbanism precepts were largely relegated to construction projects that converted greenspace on the urban fringe into new towns such as Seaside, Florida or Poundbury, Dorset. However, the central argument of New Urbanism–that cities and public space need to be planned for a human scale–was quickly embraced by planners and municipal policymakers in a variety of contexts and countries [19]. The connection between rethinking the city at a human scale and environmental sustainability was an important product of New Urbanism's evolution as prominent urban practitioners such as Peter Calthorpe integrated justifications from environmental thinking into their argumentation for new urban forms [20]. A city that allows for people to move around without requiring the use of a private automobile is both more humane and environmentally benign.

In urban carbon accounting exercises, the transport sector has been identified as an essential sector for decarbonization. As cities have developed climate action plans over the past two decades, transportation and mobility have emerged as important sites for greenhouse gas mitigation policies [21]. The promotion of cycling, in particular, has been positioned as a mechanism for achieving emissions reductions from transport. In terms of policy initiatives to realize modal shifts to low-carbon technologies like cycling, there have been numerous examples that involve changes to the urban infrastructure.

The establishment of the Paris Velib bicycle share scheme in 2007 and parallel efforts in China influenced the adoption of similar schemes throughout Europe, North

America, South America, and Asia. Bike sharing allows cities to deploy at a relatively low cost a system of easily accessible bikes that can enhance mobility and improve connections with fixed route public transit. In many cities the implementation and operation of such schemes has been conducted in partnership with private corporations which keeps public costs minimal and offloads the responsibility for operations from municipal authorities [22].

While bike sharing has been a relatively quick way to increase urban bike usage, perhaps more challenging has been creating a sufficient cycling infrastructure that insures safe mobility for urban residents. Numerous studies in urban contexts have found positive correlations between improvements in the form of bicycle infrastructure that protects cyclists from automobiles and increasing numbers of urban cyclists. Constructing a network of safe cycling options can help achieve persisting modal shifts that contribute to a city's sustainability and emissions reduction goals.

Thus, as local authorities elevate carbon reduction in transport on the urban policy agenda, moving from public policy problem identification to policy action has emerged as a policy and governance challenge. The next section explores the intersection of the emerging policy goal of low-carbon urban mobility transition and the urban emergency stemming from the COVID-19 pandemic in Milan. In terms of situating this in Campbell's model, the concretized elements of Milan have been a focus of efforts to reconfigure in a fashion to encourage more forms of low carbon mobility. With the emergence of the SARS-CoV2 virus in the community, these efforts gained renewed urgency and rethinking in the context of a new threat from the natural world.

4 Methodology and Data

This research takes a qualitative and interpretive approach towards understanding the policy process of mobility transformation during the public health emergency of COVID-19. As implied by Campbell's theory, the integrative nature of the interactions between social, natural, and built systems are inherently complex and contextual. As such, qualitative analysis can help to discern and emphasize the important elements in the specific case that will be discussed. The interpretation of the various data is essential as it highlights key elements of meaning and the importance of context in the policy process [23]. Additionally, the qualitative approach taken here is one that that includes interviews with key stakeholders. Hearing first-hand, well-developed recollections of policy formulation and implementation can help understand motivation and preferences, which is useful for understanding policy outcomes [24].

This study combines analysis of planning documents, public statements from policy actors involved in infrastructure planning and advocacy, and interviews. Interviews were conducted with 15 individuals who occupied various positions in the *Strade Aperte* planning process: elected officials, appointed officials, civil servants, and public advocates. Interviews were conducted remotely using teleconferencing software. Interview transcriptions were analyzed using the QualCoder qualitative data analysis software which allowed for aggregating interview responses thematically.

5 Slow Streets, Mobility, Social Distancing and COVID-19 in Milan

Italy was one of the first European countries to be affected by COVID-19. The SARS-CoV2 virus likely arrived in late January as two Chinese tourists traveled from Wuhan to Milan on 23 January 2020 and then were admitted to hospital in Rome with COVID-19 symptoms on 30 January [25]. Being aware of the COVID-19 outbreak in Asia and cognizant that it was likely to spread to Italy, on 31 January 2020, the national government declared a state of emergency for the country which released funds to be used for emergency preparations [26]. The first reported case in the country was detected in the Lombardy region on 20 February 2020 and within 24 h 36 additional cases–none of whom had contact with the initial patient–were diagnosed [27]. It was clear that community spread of the disease had commenced.

The first Italian with no travel history to known COVID-19 outbreak areas was diagnosed on 21 February 2020 in Codogno, a town of 15,000 inhabitants located about 60 km south of Milan [28]. The number of infections grew exponentially and by 11 March 2020 the country reported 12,462 confirmed cases and 827 deaths. The growing rate of infection prompted the national government to implement travel restrictions in Milan's region of Lombardy on 8 March 2020 after more localized lockdowns proved ineffective at minimizing spread [29].

Ren notes that the governmental response in Italy to the pandemic was piecemeal, evolutionary, and inconsistent [30]. In Italy, the national government was reluctant to engage in ambitious measures until the situation had deteriorated significantly. Although the state of emergency was called quickly and it was concomitant with restricting direct flights between Italy and China, Ren notes that travelers could arrive into Italy via third countries without restriction. Furthermore, the multilevel governance system in Italy resulted in a patchwork of compliance activities and degrees of enforcement. According to multiple interview subjects both within government and outside, political rivalries between the northern regions and the national government also resulted in efforts to politicize COVID-19 responses in ways that were not necessarily conducive to viral containment.

Comfort et al. [31] contend that these political rivalries in Italy inhibited cooperation and, perhaps most crucially, detracted from the presentation of a unified set of communications and messages about behavior that are helpful in convincing populations to exhibit safe behavior and practices [32]. Contradictory communications from public officials occurred alongside a 'catch-up' public health response and health care system that had not been prepared for an onslaught of cases of a very virulent and contagious disease. Even as the virus spread throughout the country in February and March, the policies and procedures to care for patients, insure safety for health care workers, and establish testing eligibility protocols differed throughout the country [33]. By 23 March the national government initiated a complete lockdown that only excluded workers and activities deemed essential [34].

Slow Streets as COVID-19 Response: Milan

To discuss the Milanese response to COVID-19 pandemic, key events and the actions taken will be explored within the context of Campbell's tripartite classification. As mentioned above, the social system includes governmental institutions, non-governmental entities, and processes by which individuals in society interact and cooperate. From the standpoint of how the social system interacts with urban infrastructure policy, it is important to first contextualize the policy environment in which sustainable mobility efforts were being developed at the time of the pandemic. In this fashion, the processes of policy making, the peculiarities of the concretized environment, and the incursion of a natural menacing virus will be addressed simultaneously. This section draws on an amalgam of themes mentioned by various interview subjects that contribute to a composite view of the municipal response to the emerging pandemic.

Municipal authorities had been striving to shift modal share in Milan for a number of years preceding the pandemic. Milan has historically had a relatively high usage of private automobiles for transport when compared to other European cities with 61 percent of daily trips being attributed to cars and motorcycles [35]. Beginning in 2008, the city initiated a congestion charge scheme called Ecopass that encompassed an 8 km^2 area in the central city that charges vehicles a sliding fee based on the vehicle's emissions profile. An early analysis by Rotaris, et al. shows that Ecopass was successful in reducing emissions, congestion, and encouraging modal shift to public transportation [36]. The pricing for the scheme was simplified and subsequently re-branded as Area C with the goal to not only reduce automobile emissions, but also to explicitly further reduce congestion [37].

In 2007, the city joined ICLEI–an international network of cities–in their "World Mayors and Local Governments Climate Protection Agreement," which was a campaign to publicly assert cities' commitment to combating climate change. This was followed the next year with the city endorsing the "European Covenant of Mayors" climate initiative that reflected the city's commitment to European-wide goals of a 20% cut of greenhouse gases, 20% sourcing of energy from renewable sources, and a 20% improvement in energy efficiency by 2020. As part of these efforts, Milan developed a climate action plan that was published in December 2009 [38]. The goal of the plan was to reduce Milan's overall CO_2 emissions by 20% from 2005 levels by 2020. In the carbon accounting that was done with the 2009 plan, direct emissions from transport was second only to those from buildings as the leading sources of greenhouse gas emissions. Thus, the transport sector was highlighted in the plan and enhancing bicycle mobility was one of the key specific goals of the plan. Among the implementation options prioritized were changing the street infrastructure to improve existing bicycle path networks as well as expanding safe bikeways.

Planning efforts to forge low carbon transition in Milan continued through the 2010s, most notably for this discussion was the production of the *Milano 2030* plan. *Milano 2030* is an urban planning document that the city developed in 2019 to be in compliance with the Lombardy region's growth management legislation. It provides a vision for how the city should grow and develop over a ten year term. The planning process was a multi-year effort that was explicitly inclusive and brought in technical experts from the community as well as engaging the larger populace through surveys and public comments. One key element that emerged from resident preferences was a desire to see more robust public spaces built at a human scale [39].

The Milanesi had ready examples of how subtle shifts to street configurations could enhance public space. As part of the effort to accommodate the large numbers of visitors to the city during the 2015 world exhibition, the municipality experimented with *tactical urbanism* techniques which, in this case, involved temporarily transforming the historic Piazza Castello from a congested street to a pedestrianized space using removable barricades and planters. Surveys done of residents following the pilot period revealed that the intervention was viewed with approval and the transformation became permanent [40].

The use of tactical urbanism had been deployed extensively in New York City during the Michael Bloomberg administration of the first decade of the twenty-first century, with the most publicized project being the pedestrianization of Broadway at Times Square in 2009 [41]. The New York experience had influenced planners in Milan and during the time that the *Milano 2030* plan was being finalized in 2018 and 2019, the city partnered with Michael Bloomberg's philanthropic consultancy, Bloomberg Associates, to develop a network of *Piazze Aperte*, or Open Squares, which took the concept of transforming street space to particular intersections in the city. Paint and other low cost material were used to demarcate space for casual recreation and congregation on the street. The new space freed from automobiles could also be used for pedestrians and cyclists to move about the neighbourhood. These new *Piazze Aperte* spaces had previously been dedicated to car parks or congested intersections.

The locations of the four initial *Piazze Aperte* were determined by the municipality, but in 2019 the municipality opened up the program for more community input. Under the scheme the municipality invited community groups throughout the city to suggest sites, develop programming and amenities for the sites, and apply for support from the city for implementation.

As COVID-19 arrived in Milan a well-developed pattern of rethinking public space and infrastructure was present. Crucially, cooperative and collaborative mechanisms of interaction between municipal authorities, organized community groups, and the public at-large had been operational. In Campbell's terms, a social conglomeration of formal and informal actors, public workers in the municipal bureaucracy, and elected officials had emerged around an architecture of improving street space. Although the *Piazze Aperte* project is not quite an example of the concept of *fluid governance* that Certomà et al. describe in their study of urban gardening in Rome's *Parco delle Energie* whereby squatters and community members were the initiators of public space transformation, in the Milan case significant discretion and autonomy had been granted to non-municipal actors to determine the character of the space [42].

In early March, as the city was considering how to manage urban infrastructure during the COVID-19 pandemic, the experiences of *Piazze Aperte* influenced the municipal response, according to interview subjects working in planning positions with the municipality. The idea of extending the open plaza concept to the street network was a logical extension. As mentioned above, the city has historically had a high modal share of private automobiles and policies to manage congestion were given high priority by policymakers over the previous decade. In addition, Milan has an extensive public transport system that typically accounts for over half of the daily trips within the city [43].

Thus, in March 2020 municipal officials were facing a public health emergency stemming from a highly contagious virus that was thought to be spread through the air in compact, poorly-ventilated spaces. With lockdown in place the normal mobility patterns the city experienced were fundamentally different. Both public transport ridership and traffic were reduced due to decreased demand. There was concern, however, that once the pandemic subsided and the lockdown could be lifted that there would be community hesitation about using public transport given the issues with viral transmission and that a modal shift to private automobile could occur.

In the planning document describing the *Strade Aperte* plan, the municipality argues that these shifting mobility factors align with a set of circumstances unique to the pandemic. Dedicating more urban infrastructure to open space is necessary to better accommodate social distancing but also there is a need to "identify light, cost-effective, fast and reversible alternatives" which will be easier to accomplish because the health emergency insures the municipality would "encounter less resistance to change" [44].

Some of the rhetoric in the *Strade Aperte* planning document suggests the municipality is capitalizing on the pandemic to bring to scale urban infrastructure changes that would normally be implemented on a longer timeline. Explaining the justification for the project, the planning document reads: "the crisis is therefore an opportunity to make a decisive leap towards the widespread use of traditional pedal and powered bikes, scooters and all other forms of micro mobility, which effectively provide the necessary distance to prevent new infections" [44]. While opponents of *Strade Aperte* projects could argue that city planners are using the pretext of COVID-19 to push through changes without the normal public consultation process, the municipality consistently refers to them as "experiments", which is partially meant to assuage concerns that infrastructure changes cannot be reversed. Rather, the contingent nature of the open street experiment builds trust in the planners' ideas, knowing that the infrastructure is malleable. The fact that *Strade Aperte* was being implemented in the months following the deliberative, community-oriented *Piazze Aperte* also contributes to public acceptance.

The *Strade Aperte* has several elements. First, it increases space not only for cyclists and pedestrians, but also for other recreational and even commercial uses. Second, it creates an extensive 35 km network of bikeways. Third, it establishes a network of slow streets that limit automobile speed to 30 km per hour. These elements are interrelated as the conversion of street space from automobiles creates more space for safe walking and cycling. In addition, many of the bike lanes are demarcated by colored paint which can slow down automobiles [45]. With more bike infrastructure there is evidence from other cities to suggest that safety perceptions are enhanced and the infrastructure compels more people to choose cycling as a preferred mode [46, 47]. Thus, the tactical urbanism interventions stemming from an emergency response to the COVID-19 epidemic may accelerate Milan's goal to decrease congestion and reduce carbon emissions from transport.

6 Conclusion

While Milan's *Strade Aperte* scheme is still in a state of evolution, the rapidity with which it was implemented, its acceptance by residents, and the extent to which it fits within a well-developed history of policies seeking to increase the use of low carbon mobility technologies suggests that it has been a successful program.

To contextualize the discussion of Milan within Campbell's framework, the existence of a social system that had been seriously engaged with improving public space and encouraging modal shift away from automobiles was important for understanding how the city reacted to the new challenges faced from the natural system in the form of SARS-CoV2. After the virus arrived in the city, its rapid spread required quick reflexes on the part of municipal policy makers. The virus disrupted the viability of the existing mobility infrastructure as public transport was perceived to be a space for easy transmission and having large numbers of people shift to automobiles would be infeasible from an air quality and congestion standpoint.

The need to move quickly to develop appropriate and safe socially distanced forms of mobility was compatible with the tactical urbanism approach the city had been using for months with the *Plazze Aperte*. Unlike a typical urban infrastructure project that is capital and labour intensive and which requires months of engineering and construction, to implement the *Strade Aperte* all that was needed was design creativity, temporary barriers, and paint. The implementation could be done quickly and cheaply in a time of urgency and economic collapse.

Finally, the urban infrastructure transformation to allow for socially distant mobility could theoretically have a dramatic impact on the virus itself, showing ways the concretized and social environments can impact the natural environment. In the absence of a vaccine, one purpose of social distancing is to drive down virus transmission rates until it is not widespread in the community.

It is indeterminate whether the *Strade Aperte* will actually be successful in managing to eliminate significant viral spread of SARS-CoV2 in Milan. However, the case offers insight into how a focus on the interrelationship between social, natural, and built systems can help policy makers understand the complex interplay of these elements in the event of other urban health or environmental emergencies.

References

1. Tian H, Liu Y, Li Y, Wu C-H, Chen B, Kraemer MUG et al (2020) An investigation of transmission control measures during the first 50 days of the COVID-19 epidemic in China. Science 368(6491):638–642
2. Laker L (2020) Milan announces ambitious scheme to reduce car use after lockdown. The Guardian
3. Béland D, Howlett M, Mukherjee I (2018) Instrument constituencies and public policy-making: an introduction. Policy Soc 37(1):1–13
4. Campbell HE (2010) A comparative framework for analyzing urban environmental policy. J Comp Policy Anal 12(4):373–394
5. Angelo H, Wachsmuth D (2020) Why does everyone think cities can save the planet? Urban Stud 57(11):2201–2221
6. Bulkeley H (2011) Cities and subnational governments. Oxford University Press, New York, NY
7. St-Louis E, Millard-Ball A (2016) Cap-and-trade, crowding out, and the implications for municipal climate policy motivations. Eviron Plann C Gov Policy 34(8):1693–1715
8. Ashton J (2020) The pandemic of coronavirus: tackling the latest plague. J R Soc Med 113(3):123–124
9. Mendez MA (2015) Assessing local climate action plans for public health co-benefits in environmental justice communities. Local Environ 20(6):637–663
10. Frumkin H, Hess J, Luber G, Malilay J, McGeehin M (2008) Climate change: the public health response. Am J Public Health 98(3):435–445
11. El-Sayed A, Kamel M (2020) Climatic changes and their role in emergence and re-emergence of diseases. Environ Sci Pollut Res 27:22336–22352
12. Wu Y-C, Chen C-S, Chan Y-J (2020) The outbreak of COVID-19: an overview. J Chin Med Assoc 83(3):217–220
13. Zhu N, Zhang D, Wang W, Li X, Yang B, Song J, et al (2020) A novel coronavirus from patients with pneumonia in China, 2019. N Engl J Med
14. Schwieterman JP, Caspall DM (2006) The politics of place: a history of zoning in Chicago. Lake Claremont Press
15. Hirt SA (2015) Zoned in the USA: the origins and implications of American land-use regulation. Cornell University Press
16. Taylor N (2003) The aesthetic experience of traffic in the modern city. Urban Stud 40(8):1609–1625
17. Newman P, Kenworthy J (1999) Sustainability and cities: overcoming automobile dependence. Island Press
18. Litman T (2018) Parking management best practices. Routledge
19. Grant J (2006) Planning the good community: new urbanism in theory and practice. Taylor & Francis
20. Calthorpe P (2010) Urbanism in the age of climate change. Island Press
21. Creutzig F, Agoston P, Minx JC, Canadell JG, Andrew RM, Le Qur C et al (2016) Urban infrastructure choices structure climate solutions. Nat Clim Chang 6(12):1054–1056
22. de Chardon CM (2019) The contradictions of bike-share benefits, purposes and outcomes. Transp Res Part A: Policy Pract 121:401–419
23. Ritchie J, Spencer L (1994) The qualitative researcher's companion. In: Bryman A, Burgess RG (eds). Routledge, pp 305–329

24. Rathbun BC (2008) Interviewing and qualitative field methods: pragmatism and practicalities. In: The Oxford handbook of political methodology
25. Giovanetti M, Benvenuto D, Angeletti S, Ciccozzi M (2020) The first two cases of 2019-nCoV in Italy: where they come from? J Med Virol 92(5):518–521
26. del Ministri C (2020) Dichiarazione dello stato di emergenza in conseguenza del rischio sanitario connesso all'insorgenza di patologie derivanti da agenti virali trasmissibili. Gazzetta Ufficiale della Repubblica Italiana 161(26):7–8
27. Livingston E, Bucher K (2020) Coronavirus disease 2019 (COVID-19) in Italy. JAMA 323(14):1335
28. Spina S, Marrazzo F, Migliari M, Stucchi R, Sforza A, Fumagalli R (2020) The response of Milan's emergency medical system to the COVID-19 outbreak in Italy. Lancet 395(10227):e49–e50
29. Remuzzi A, Remuzzi G (2020) COVID-19 and Italy: what next? Lancet
30. Ren X (2020) Pandemic and lockdown: a territorial approach to COVID-19 in China, Italy and the United States. Eurasian Geogr Econ 1–12
31. Comfort LK, Kapucu N, Ko K, Menoni S, Siciliano M (2020) Crisis decision making on a global scale: transition from cognition to collective action under threat of COVID-19. Public Adm Rev 80(4):616–622
32. Van Bavel JJ, Baicker K, Boggio PS, Capraro V, Cichocka A, Cikara M, et al (2020) Using social and behavioural science to support COVID-19 pandemic response. Nat Hum Behav 1–12
33. Romagnani P, Gnone G, Guzzi F, Negrini S, Guastalla A, Annunziato F, et al (2020) The COVID-19 infection: lessons from the Italian experience. J Public Health Policy 1–7 (2020).
34. Consiglio del Ministri (2020) Ulteriori disposizioni attuative del decreto-legge 23 febbraio 2020, n. 6, recante misure urgenti in mate-ria di contenimento e gestione dell'emergenza epidemiologica da COVID-19, applicabili sull'intero territorio nazionale. Gazzetta Ufficiale della Repubblica Italiana 161(76):1–4
35. Croci E, Melandri S, Molteni T (2011) Comparing mitigation policies in five large cities, London, New York City, Milan, Mexico City, and Bangkok. In: Hoornweg D, Freire M, Lee MJ, Bhada-Tata P, Yuen B (eds) Cities and climate change: responding to an urgent agenda. The World Bank, pp 55–85
36. Rotaris L, Danielis R, Marcucci E, Massiani J (2010) The urban road pricing scheme to curb pollution in Milan, Italy: description, impacts and preliminary cost–benefit analysis assessment. Transp Res Part A: Policy Pract 44(5):359–375
37. Trivellato B, Mariani L, Martini M, Cavenago D (2019) Leading knowledge mobilization for public value: the case of the congestion charge zone (Area C) in Milan. Public Adm 97(2):311–324
38. Milan, Mo.: Sustainable Energy and Climate Action Plan: Agenzia Mobilità Ambiente e Territorio (2009).
39. Comune di Milano (2019) Milano 2030: Visione, costruzione, strategie. Comune di Milano, Spazi
40. Arup (2016) Cities alive: towards a walking world. Arup
41. Sadik-Khan J, Solomonow S (2017) Streetfight: handbook for an urban revolution. Penguin
42. Certomà C, Chelleri L, Notteboom B (2020) The 'fluid governance' of urban public spaces. Insights from informal planning practices in Rome. Urban Stud 57(5):976–995
43. Woods R, Masthoff J (2017) A comparison of car driving, public transport and cycling experiences in three European cities. Transp Res Part A: Policy Pract 103:211–222
44. Consiglio del Ministri (2020) Milan 2020 adaptation strategy: Agenzia Mobilità Ambiente e Territorio
45. LaMondia J, McGhee J, Fisher M, Cordero F (2019) Evaluating the safety and behavioral impacts of green bike lanes in suburban communities. Transp Res Rec 2673(11):671–679
46. Aziz HMA, Nagle NN, Morton AM, Hilliard MR, White DA, Stewart RN (2018) Exploring the impact of walk–bike infrastructure, safety perception, and built-environment on active transportation mode choice: a random parameter model using New York City commuter data. Transportation 45(5):1207–1229

47. Goetzke F, Rave T (2011) Bicycle use in Germany: explaining differences between municipalities with social network effects. Urban Stud 48(2):427–437

Cultural Landscapes of Post-industrial Heritage as a Drive for Placemaking: Eleusina Case Study

Elena Douvlou and Ewa Stachura

Abstract Rapid urban growth leads to a shortage of vacant land for development and results in its fragmentation into plots of commercial value (Gzell, *Urbanistyka XXI wieku.* PWN, Warszawa, 2020 [1]). Consequently, areas for creating and enlarging public spaces in cities shrink. Such conditions do not favour the challenges and aims of a resilient city/neighbourhood, such as social pluralism and universal participation in culture. One way to solve this problem is a comprehensive reorganization of inefficient industrial zones. Involving local communities in shaping and improving the quality of a living space can be gratifying for the community, and takes an intersectional approach which values the lived experience and a diversity of perspectives. Culture is increasingly recognized as a driving force for urban development. Many former industrial buildings are considered as significant historical monuments that impose certain restrictions for private developers and public authorities (Bandarin and van Oers, The historic Urban landscape. Managing heritage in an Urban century. Wiley Blackwell, Chichester, 2012 [2]). However, preservation of heritage values produces various environmental and socio-economic benefits as strength of place identity, boosting tourism and attracting the creative sector (Graham et al., A geography of heritage: power, culture. Routledge, 2016 [3]). This paper aims to explore the role of industrial heritage in urban regeneration and placemaking processes in the complex and challenging environment of a city with a rich ancient and post-industrial cultural landscape, Eleusina [4], Greece.

Keywords Placemaking · Post-industrial heritage · Cultural landscapes

E. Douvlou (✉)
School of Architecture, Metropolitan College, Sorou 74, 15125 Marousi, Greece
e-mail: edouvlou@mitropolitiko.edu.gr

E. Stachura
Cracow University of Technology, Warszawska 24, 31-155 Cracow, Poland
e-mail: ewa.stachura@pk.edu.pl

M. Arslanagić-Kalajdžić et al. (eds.), *Interdisciplinary Advances in Sustainable Development II*, Lecture Notes in Networks and Systems 804,
https://doi.org/10.1007/978-3-031-46692-2_9

1 Introduction

Cultural heritage and contemporary development may bring synergies and conflicts. Areas of rich cultural heritage challenge architects, urban planners and city authorities to find optimal scenarios of sustainable revitalization and spatial change responding to the current combination of factors influencing sites, areas and districts. The in-depth study and diagnosis of phenomena and processes occurring in such areas should be provided and confronted with the needs of residents and other users of space. Tracking the processes mentioned above may bring space-user relation models useful in formulating variant scenarios of spatial interventions.

A city/area's identity and image are essential drivers in shaping strategies for urban interventions. Placemaking may enrich and strengthen the revitalization process and make a revitalized unit more resilient.

The following research methods will be used in the paper: literature research, urban analysis, expert research, and a questionnaire analysis as a basis for a student work in order to yield a set of strategies vision for the future regeneration of a city. The research results will contribute to the broadening and deepening knowledge about the functioning of a contemporary city of high heritage value. Such knowledge can provide valuable support for decisions regarding urban interventions within multi-contextual locations. It may also enrich the catalogue of sustainable design guidelines in heritage areas facing social, economic and technological challenges.

2 Placemaking—A Theoretical Approach

The various concepts and definitions reflect how many individual strategies and approaches have been implemented over the past years [5]. The definition of placemaking as "making better places" is very general and does not give us greater clarity on placemaking as a concept or process. Placemaking in urban design, planning and regeneration require a qualitative, contextual approach providing an understanding of how placemaking works as a process [6]. Understanding placemaking contextually explains how the strategies used in placemaking work and what outcomes improve the quality of community spaces [7]. The literature review presented by the Authors indicates three themes that underlay placemaking as a process:

- Placemaking is a process of reshaping space to make it more appealing and useable and to generate a sense of place, enabling an attachment or connection between the community member and the place where they perform life activities [8].
- "Sense of place" for an individual can be either positive or negative. Placemaking needs to strengthen a positive sense and overcome existing negative or placeless perceptions of the place.
- Placemaking starts as a process leading from environmental change imposed upon an individual to environmental change created by the individual. Understanding the process of placemaking as a continuum that implements overarching actions,

involves various groups of actors, and discusses outcomes helps to formulate effective strategies, scenarios and measures for the short- and long-term impact on community development.

Place Identity and the Meaning of Place

It is worth stating that urbanity has no clear character in the era of mobile technologies, individualism and the crisis of community values. The term "urban" refers to both well-embedded local communities and smaller/larger groups of city users or temporary communities where people gather to solve a problem, participate in an event, and disperse after. However, due to human nature, the need to interact with other people and create social bonds on other than ad hoc basis will not disappear [9]. Providing scenes for those as mentioned above, rapidly transforming interactions, and meeting individual, personalized aspirations and expectations of groups/communities are one of the most critical challenges in managing cities today [10]. These scenes are "places", and a helpful tool to create and transform them is placemaking. Places, buildings or localities of particular significance are elements of space. They play a special role in social consciousness and have a unique identity and image.

Place identity is a combination of memories, opinions, interpretations, ideas, and feelings that relate to areas of various scales called: "places". Place identity consists of memories, values, thoughts, ideas and settings and links to other places: like home, neighborhood, work, etc., and may be considered a substructure of social identity (like gender and social class) [11]. Based on tangible and intangible components and interactions with various groups/individuals, a place may have an unstable identity and image that evolves [12]. Creating a place or changing its multidimensional characteristics to improve quality means intervening both at spatial and social levels providing opportunities to meet the users' needs.

3 Cultural Landscapes in Towns Today—The Case Study of Eleusina in Greece

Eleusina, situated 21 km west of Athens, has a privileged geographic location, opposite the mythical Salamina and at a crossroads that connects the north Greece and the Peloponnese with Attica. It has been an important node from ancient times, it is the birthplace of Aeschylus, but for some 12 centuries, the historic city was home to one of the most important religious centers of both ancient Greece and Rome. Until the fourth century AD hosted the Eleusinian Mysteries, secret religious rites for the cult of Demeter, the goddess of harvest, agriculture and fertility, and her daughter, Persephone, wife of Hades and queen of the underworld. Among those initiated were the Roman emperors Marcus Aurelius and his son Commodus. In a large hall that could seat up to 3000 people, participants at nocturnal torchlit ceremonies were told of the endless cycle of life and death, and attended a re-enactment of the myth of Persephone. Being one of the five most important sacred cities of antiquity, Eleusina has a rare archaeological wealth, and until the first half of the twentieth century was a seaside resort with remarkable natural scenery.

Because of its natural harbor and strategic location, at the northernmost tip of the Saronic Gulf, it was transformed from the nineteenth century onwards into a "productive machine" of Greece as one of its largest industrial centers [13] (Figs. 1 and 2).

Fig. 1 Bay of Eleusina (*Source* Authors' own)

Fig. 2 View of a post-industrial site in Eleusina neighboring the archeological space (*Source* Authors' own)

The Past and Present of the City

Eleusina was the third and newest industrial center of Attica, with those of Lavrio and Piraeus leading the way. The industrial activity of the city dates to the end of the nineteenth century and by 1930, 10 industries had already been established [14]. The first units were relatively small processing agricultural products. After World War II, workers from all over the country arrived in the city to work in the industries of the region. The pressing need for work and the lax political policy resulted in an anarchy of industrial development at the expense of the cultural history of the city as well as the environment. The rapid pace of industrial development has made Eleusina one of the largest industrial centers in the country resulting in extreme environmental pollution and damage to its ecosystems.

Today, the wider area of Eleusis, the Thriasian Plain, comprises a total of 552 industries, while 50 of them are in the city of Eleusina. The industrial center of the city includes bauxite and manganese mines, steel mills, oil refinery, metal processing industries, cement industry, food processing factories, agricultural products, etc. The signs of industrialization dominate the body of the city, overlapping it and often creating the perception of a place, which keeps its cultural wealth well-guarded, as a timeless mystery.

The post-industrial now meets the archaeological in an offending case of de-territorializing thus of spatial loss of a symbolically, visually, communally and archaeologically significant field. This situation had not left indifferent thinkers and artists of that time, like the poet Andreas Embiricos, in 1955, immortalizes through his photographic lens the "mutilated" statues in the courtyard the archaeological museum of Eleusis against the backdrop of thriving industries (Figs. 3 and 4). The poet makes a comment about the fact that these ancient artworks are looking to find their place within a context too different from what had been created a few thousand years ago, although that spatially have not moved more than a few meters [15]. Celebrating this year Europe's Capital of Culture for 2023 (the designation was postponed due to the 2020 coronavirus pandemic), it is today a major, yet de-industrializing center, visible from, and noxious to, the reverent ancient site (Fig. 5).

The Industrial Heritage of Eleusina: Anchors of Collective Memory

Eleusina is a typical example of a port city, where the development plans for the coastal front have been abandoned, despite the strong dynamics and the many strong cultural advantages that a city has. The former industrial areas of Kronos and Iris and the complex of Old Elaiourgio offer an opportunity to reactivate vast abandoned spaces, allowing recreational uses. Axes of upgrading Eleusis triptych antiquities-waterfront- modern monuments of industrial architecture, can be the poles of social and economic development of Eleusis.

Fig. 3 A photograph inspired by Embiricos' (by Douvlou E., May 2019)

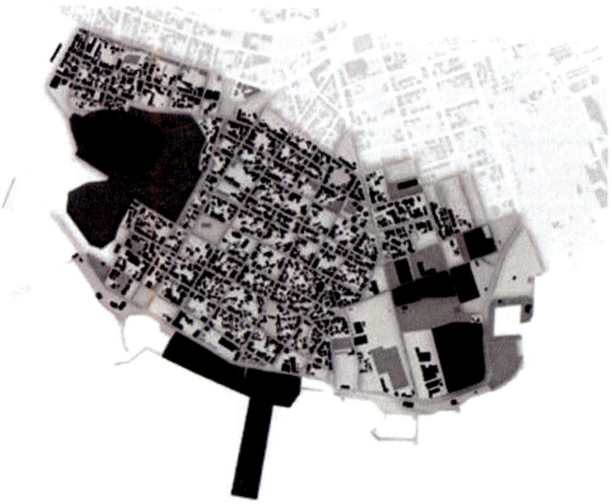

Fig. 4 A figure ground of Eleusina center demonstrating the large areas occupied by industrial activities compared to the residential areas (MArch_ AUD, Metropolitan College)

The abandoned Kronos industry (right)is located on the front of the Municipality of Eleusis along with other equally dilapidated ones such as Votrys, Old Elaiourgio, IRIS and together with TITAN, the Shipyards, the Xalyvourgiki Steel industry etc. occupy the most vital space of the city—the sea front! The study of industrial buildings is a matter of concern to the research world, as well as their restoration and maintenance, the protection of industrial heritage (Fig. 6).

The architectural quality and aesthetic value of the inactive industrial spaces as well as their value and contribution to the collective memory for the local communities undeniably dictate the protection and conservation of the technological and architectural heritage; reusing and highlighting the industrial landscape, the old factories and the former production buildings provides an opportunity for the utilization of the remnants of industrial activity [16]. The reuse of inactive industries for cultural and educational purposes offers a unique opportunity as these buildings are in themselves

Fig. 5 A map of Eleusina with indication of industrial activity, mainly located in the waterfront area (MArch_AUD, Metropolitan College)

living schools and museums. The proximity of some of them to the hill of antiquities indicates the necessity of preserving the industrial heritage while preserving the archaeological and historical cultural heritage.

Cradles for Placemaking: "Eleusinians Narrate"

Pendlebury & Porfyriou [17] mention heritage as a factor contributing to the quality of social space and placemaking. Indeed, heritage as a social asset can boost bottom-up initiatives and civic engagement in the creation of public places [18]. Placemaking (or place-making) is an overarching approach to planning, design and management of public spaces [19]. Placemaking capitalizes on a local community's assets, inspiration, and potential, intending to create quality public places that contribute to people's health, happiness, and well-being.

As noted at the artistic vision and strategic objectives of 2023 ELEUSIS [20], is to present the city as a raw museum: the substance of the city is composed of exceptional traces *of antiquity, modern stories of internal migration, industrial development, labor movements, but also de-industrialization and unemployment.* At the same time other equally important issues are intertwined such as the strong impact on the

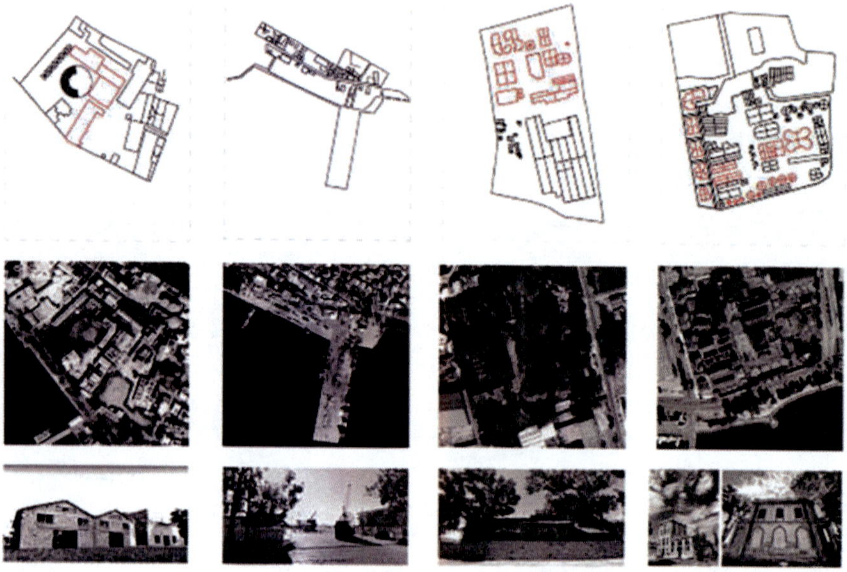

Fig. 6 (Left to right) Old Elaiourgio, the port of Eleusina, Iris and Kronos industrial complexes (MArch_AUD, Metropolitan College). One of the vital issues afflicting Eleusis is the fact that it has lost its status as a coastal city for years. From 13 km. of the coastline, the municipality does not manage a single measure, it has been entirely occupied by the industries and these are located in the commercial port, while the locals currently account for only 30 m of beach. New uses are spaced, completely excluding access to the beach and turning Eleusis into a seaside city without a thrift

environment, while the ambiguous relationship of the city towards the sea deserves attention, exploration, and exploitation.

The regeneration of Eleusina has been the case study for a studio workshop at the MArch Architecture & Urban Design, School of Architecture at Metropolitan College in Greece. The case study is based on the following sources:

- Documentation from the archives, local media analysis
- Guided field visits at the town center and the selected sites
- Interviews with activists and the office of the European Capital of Culture in Eleusina
- Questionnaires addressed to residents

In any act of placemaking, active citizens' participation as the key stakeholders, is an integral part of the process centered on people and their needs, aspirations, desires, and visions. Therefore, before embarking on strategic action proposals, a questionnaire was addressed to a group of citizens of Eleusina in two different periods of time, October 2021 and January 2023. A total of 30 questionnaires were collected and analyzed. The relatively small number of questionnaires does not allow for a meaningful statistical representation, but rather for a more qualitative analysis. It was identified that the most important issues needed to be addressed are shown in Table 1.

Most respondents commented that while Eleusina has an extended waterfront, the city is introvert. The main urban strategies developed, based on the above are shown in Fig. 7 and consequently summarized in Table 2.

The study also included the redefinition of the boundaries of the historical center of Eleusis, which includes the archaeological site, part of the Sacred Road (Iera Odos) up to Hadrian's bridge, part of the port and part of the modern city, for the expansion of the network of pedestrian and bicycle paths and for the establishment of conditions

Table 1 Issues to be addressed according to the residents of Eleusina

a. Reconnect the city with the sea-front	1. Transfer of the commercial port
b. Provide a thread of connection of Eleusina's landmarks-especially the archeological site	1. Designated route to landmarks
	2. Design a network of pedestrian friendly connections
c. Opportunities for regenerating the city's industrial heritage	1. Identify possible uses for the complexes of Kronos, Iris and Elaiourgiki complex
	2. Reinforce the Aischylia Cultural Festival to increase national and intenational appeal
d. Address sustainability issues (environmental, social, economical)	

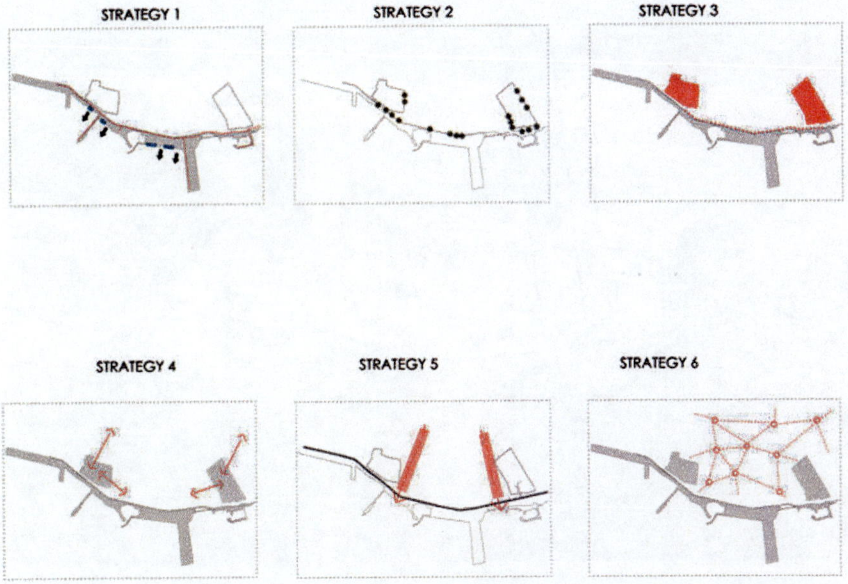

Fig. 7 Strategies as part of the development of a holistic intervention plan for the social and environmental regeneration of the urban area of Eleusina (MArch_AUD, Metropolitan College)

Table 2 The strategies developed that correspond to the diagrams in Fig. 7

1	Regenerate waterfront-new activities
2	Public spaces and increase of green areas
3	Connect fragments of landmarks- especially the archeological and industrial heritage
4	Reinforce social connectivity
5	Connect seafront area with the center of Eleusina
6	Pedestrian network to connect different neighborhoods of Eleusina

for the protection and promotion of the urban and architectural characteristics of the modern city. The proposed masterplan is seen in Fig. 8. The Cultural Capital of 2023 is a unique opportunity for residents, visitors, artists, scientists, and institutions to live unique experiential experiences through visits, outdoor activities, and discussions in the city's landscapes. In the current global context of the pandemic, ecological disasters and climate change, the landscapes of Eleusis are a unique world heritage site that unites natural, archaeological, and industrial heritage.

Figure 9 is a portfolio of proposed visions for the re-activation of the post-industrial complexes of Kronos, Iris and Old Elaiourgio as well as the waterfront. These include community centers, cultural centers for the celebration of the annual Aeschylean Festival mediateque/library, a museum for the battle of Salamina, artist's residents and studios, green parks, skate parks and pavilions alongside the waterfront complimented with shops for food and beverages.

Fig. 8 Proposed masterplan for the regeneration of the urban area of Eleusina (MArch_AUD, Metropolitan College)

Fig. 9 Proposed portfolio of activities for the post-industrial sites of Eleusina and the sea-front (MArch Architecture & Urban Design studio, Metropolitan College, 2019/20 and 2021/22 cohorts)

A network of pedestrian from the city center to the waterfront will provide new anchors of visual and physical connection with the waterfront in reference to the archeological hill. The waterfront is a united continuous public space of various qualities and scales, providing views to the bay of Eleusina, the island of Salamina and the urban landscape of Eleusina.

4 Discussion

Industrial heritage spaces could be key assets for constant growing cities that lack public spaces and communities that need venues for their gatherings, cultural events, educational, sports or care activities. In the past decades, many post-industrial heritage spaces became some of the most successful examples of adaptive reuse as a result of community-led development processes [21]. Learning from citizen-led, locally anchored heritage regeneration projects that have succeeded in securing spaces and services for communities, Eleusina's post-industrial sites have the potential for comprehensive regeneration and placemaking strategies to create new cultural landmarks and increase living standards.

Placemaking pays particular attention to the physical, cultural, and social identities that define a place and support its ongoing evolution, creating a sense of place and place attachment [19]. Therefore, the placemaking approach contributes to sustainable heritage management and social empowerment. At the same time, heritage sites and community activists and artists increase the quality of the environment, which in turn increases the value of the site and surrounding area [21]. Temporary use of abandoned space can offer freedom, unmodified social structures and social experimentation that benefits a broader community, potentially including non-local communities.

The interviews with the representatives of local activists' groups and the office of the European Capital of Culture in Eleusina reveal that there are opportunities to capitalize on the available land of the post-industrial sites to facilitate uses that serve the local community. Through a systematic process of evaluation of the existing buildings and open spaces, the city could benefit from a model of a contemporary sustainable economy, focusing on the pioneering linking of art with fields such as industry, science, technology, and social innovation.

Questionnaires addressed to residents also emphasized the need of the local people to reconnect with the seafront, re-establish their relationship with the fragments of the ancient and more recent past of the city that unfolds into contemporary life, through the regeneration of its post-industrial landscapes.

All of the above became tools of inspiration for the MArch_AUD students at Metropolitan College School of Architecture, that produced a portfolio of ideas, through which, one can visualize the potential of creating places for a thriving local community, anchors of collective memory, in a city that has reclaimed its identity, demonstrating that transition is a continual, a never-ending process. These are described in the following diagram in Fig. 10.

Fig. 10 Diagram summarizing the design driving tools for placemaking as a result of the multi-layered research carried out in Eleusina (by Elena Douvlou)

5 Conclusions

Only a holistic intervention plan for the social and environmental regeneration of the urban structure may bring quality change to the city. Such a plan must be based on a profound understanding of all the conditions and challenges of today and the past. Through research of archive material, interviews with the city of Eleusina stakeholders such as activists, officers of the Cultural Capital of Europe [4] organization and the residents, the current city identity and image must be re-established.

The meaning of "identity" content is very complex: it determines image; it covers "general issues", referring to a city as a whole and partial element of quality, prestige, landscape beauty etc. A study results on these values must be treated as a guidepost rather than a strict driver: this paper proposes such an approach.

Placemaking is a tool that connects all the goals for urban regeneration. Each of the strategies pointed out and graphically presented in Fig. 7 will need to be supported with placemaking actions, which proves a community's essential role in shaping the city's quality and maintaining the continuity of its lasting.

The paper also revealed that local communities are in favor of re-use of the large post-industrial complexes that can be used not only as tools to launch the redevelopment process but as the primary purpose of improving the quality of the urban environment.

References

1. Gzell S (2020) Urbanistyka XXI wieku. PWN, Warszawa
2. Bandarin F, van Oers R (2012) The historic urban landscape. Managing heritage in an Urban century. Wiley Blackwell, Chichester
3. Graham B, Ashworth G, Tunbridge J (2016) A geography of heritage: power, culture and economy. Routledge
4. Cultural Capital of Europe (2023) European capital of culture. https://2023eleusis.eu/en/. Last Accessed 11 Oct 2023
5. Ellery PJ, Ellery J, Borkowsky M (2021) Toward a theoretical understanding of placemaking. Int J Community Well-Being 4:55–76. Springer. https://doi.org/10.1007/s42413-020-00078-3
6. Hardy J (2022) Creative placemaking: recommendations from and impact of six advisory services panels. Urban Land Institute, Washington, DC. https://knowledge.uli.org/-/media/files/research-reports/2022/uli-aspr_creative-placema
7. Bohl CC (2002) Place making: developing town centers, main streets, and Urban villages. Urban Land Institute, Washington, DC, January 1, 2002
8. Edensor T, Kalandides A, Kothari U (2020) The Routledge handbook of place. Routledge Handbooks Online. Accessed 12 Feb 2023
9. White Paper on European Governance (2000) Enhancing democracy in the European Union. Work Programme. SEC. 1547/7 final. https://www.cpmr.org/pub/docs/57_contib_crpm_lb_gouv_gb.pdf. Accessed 11 Oct 2000
10. Schultz A-C, Lanzl C). Placemaking/making place. s.l.: https://www.urbancultureinstitute.org/uploads/1/1/4/6/11465358/placemaking_making_place__wentwortharchreview_v6_sm.pdf
11. Qazimi S (2014) Sense of place and place identity. Eur J Soc Sci Educ Res, Strony 1(1):306–310
12. Badach J, Dymnicka M (2017) City and urbanity in the social discourse. IOP Conf Ser: Mater Sci Eng 245:042058
13. Panagiotidis L (2019) The development of Eleusina in the 20th century—a process of industrialization and de-industrialization of an urban landscape. University of Patras
14. Papaggeli K, Chlepa E (2011) The transformations of the Eleusnian topography: antiquities and the contemporary city. Cultural Centre of the Municipality of Eleusina
15. Kassianou N (2008) Through the photographic lense of Empeirikos, 24 November 2008
16. Chrysa M (2019) The past and future of Elefsina's industrial buildings [os. udziel. wyw.] Douvlou Elena. May 2019
17. Pendlebury J, Porfyriou H (2017) Heritage, urban regeneration and place-making. J Urban Desi, Strony 22(4):429–432
18. Janssen J, Luiten E, Renes H, Stegmeijer E (2017) Heritage as sector, factor and vector. Eur Plan Stud, Strony 25(9):1654–1672
19. Project for Public Spaces (PPS) (2007) What is placemaking? https://www.pps.org/article/what-is-placemaking
20. Artistic Vision and Strategic Objectives of 2023 ELEUSIS, December 2022. https://2023eleusis.eu/wpcontent/uploads/2021/09/Artistic-Vision-and-Strategic-Objectives-of2023ELEUSIS_28092021.pdf
21. Patti D, Polyak L (2018) Funding the cooperative city from knowledge network to local. Tracce Urbane. Rivista Italiana Transdisciplinare di Studi Urbani, str. 2(3)

Understanding the Principles of the 15-Minute City Concept and the Repercussions on the City's Health: The Curious Case of the City of Sarajevo

Nasiha Pozder, Senaida Halilović-Terzić, and Andrea Pavlović

Abstract Climate change, pollution, profit driven construction, weakened quality of life and poor health of citizens, are cause-and-effect related problems that one can find all over the world. Developed, socially conscious societies entered the twenty-first century determined to dedicate themselves to the path of healing of their cities, so they put the fight against calamities of the modern times. Introduction of green jobs, clean energy, less waste and "novities", such is the 15-minute city, are often narrative of choice for different and resilient approach. In parallel, quality of life in transitional societies and underdeveloped countries, is severely under pressure, since the profit-driven paradigm is a priority. This is visible in the lower socio-economic status countries of Europe too, such is Bosnia and Herzegovina. Sarajevo entered the twenty-first century as an impoverished city destroyed by war. The necessary renovation, once based on the principles of equality and proximity of urban content, had been replaced by accelerated construction and homogenic zoning, with little or no attention towards healthy and functional city. Through the examination the possibilities and limitations of the principles of 15-minute and healthy city in specific circumstances and their comparison with today's approach, this paper aims for understanding of the specific needs of the planning approach in Sarajevo case, as in times of prosper, but also in times of crises. Methodological approach comprises out of several compatible methods, starting with historical overview and descriptive analysis, followed by comparative method of the findings, all combined in a case study for the city.

Keywords 15-Minute city · Healthy city · Resilience · Adoptability

N. Pozder · S. Halilović-Terzić (✉) · A. Pavlović
Faculty of Architecture, University of Sarajevo, Patriotske Lige 30, 71000 Sarajevo, Bosnia and Herzegovina
e-mail: senaida.halilovic@af.unsa.ba

N. Pozder
e-mail: nasiha.pozder@af.unsa.ba

A. Pavlović
e-mail: andrea.pavlovic@af.unsa.ba

M. Arslanagić-Kalajdžić et al. (eds.), *Interdisciplinary Advances in Sustainable Development II*, Lecture Notes in Networks and Systems 804,
https://doi.org/10.1007/978-3-031-46692-2_10

1 Introduction

Despite its recent topicality and relevance in the contemporary city planning, 15-minute city is not radically new nor innovative idea. In fact, the model which is very present in professionals' and general public narrative, and is often mentioned as a possible solution in terms of improving the conditions and health of today's burdened cities, relies on long-established principles of urban planning, the beginnings of which can be found as far back as the beginning of the twentieth century.

Clarence Perry [1], as well as Jane Jacobs [2], in their plans, works and reflections on the society of that time, already seriously considered and referred to the importance of organizing the neighborhood in a human-acceptable and, one might say, standardized way, in which preference to the resident over cars would have been prioritized. In addition, great attention was paid to the living and working environment, in the context of the availability of all the contents to be used on a daily basis.

Urban planning of the former Yugoslavia, which appeared as a formalized and formed discipline a decade after the devastation of World War II, often rely on the described ideas of Perry and Jacobs. Planning decisions and the postulates advocated by Yugoslav urban practice, introduced the organization of society and space in such manner it could be linked to the overseas movements of that time.

Acknowledging that these are completely different, even opposing political and social ideologies, and despite that one specific idea has equal, even the same significance, then the relevance and interest becomes even greater to observe.

According to Clarence Perry, an American urbanist, sociologist and writer, 15-minute city implies certain design principles and an urban standard for a functional and well-structured neighborhood organization. In order for the neighborhood to be successfully organized, one has to be equipped with the necessary elements of the core of urban contents, such as (schools, daily supplies, services), with the accompanying traffic network, as well as system of green areas and public spaces. All of the above must be incorporated into residential zones and all easily accessible within 15 min walking or cycling.

More precisely, in the essence of his ideological determination and methodological principle of reformation of existing urban units, as well as those that have yet to be created, Perry proposed a series of quantitative measures, which will emerge as the basic norm of urban space planning.

He introduced the interdependence of the number of inhabitants and the size of the area, standardized the density of housing in relation to the facilities that serve a certain area, and established an accessibility radius of ¼ mile, in which the spatial, functional and content needs of the residents' overlap. In summary, he brought the quantity and quality of life of people within one unit of space into an obvious and indisputable connection.

In our regional planning practice, Perry's principles had a huge impact and the urban planning started to be based on the unit of matrix of urban area, radius of 500 m (or ¼ mile).

In our regional planning practice, Perry's principles had a great resonance. Urban planning began to be based on the smallest unit/matrix of the urban area, that is, a radius of 500 m (equivalent to ¼ mile), in which Yugoslav urban planners planned and determined all necessary facilities for an average neighborhood.

Centralized political and social system of communism and socialism, made it easier to implement ideas and principles, which are related to the 15-minute city of today, than it was in its western counterpart societies.

The urban plans of Sarajevo in the past were based on the principles of this concept. In 1965, after a series of problems and delays, General Urban Plan of Sarajevo City [3] was finally adopted, followed already in the early 1970s by its first changes.

They were conducted through the document named Proposed Program of Construction and Spatial Development of the City for the Period 1971–1985, which reflected clear echoes of the principle of plurality of content, centers and roles that each part of the city and/or neighborhood must have in the overall picture of longitudinal urban morphology.

The aforementioned document pointed out the fact that previous longitudinal development of the city had brought a number of problems to the residents of the newly created settlements and the old part of the city as well, because both halves of the city remained deficient in terms of supporting spaces and functions.

Residents had to undertake long journeys every day in order to satisfy their needs for centrality facilities, which put an additional burden on the traffic network.

Therefore, systems and norms for equipping settlements, urbo-morphological entities and centers with the necessary superstructure, as well as infrastructure, were established. This concept created a clear network of social, commercial and service use in radii of 500–1000 m or 15 min of user movement.

The accepted concept of the formation and equipping of the settlements, proved to be exceptionally good in periods of crisis.

From the distance of nearly 30 years, it can be deducted that such a tight urban scheme contributed significantly to organization of life, or at least the appearance of a functional city, even in the period of unprecedented crisis brought by the war and the siege of 1990s.

The equipping of individual settlements and urban-morphological units with the necessary elements of social infrastructure, which was sufficiently adaptable and modular to adapt to the newly created situation, enabled reach within a radius of 500 m or 15′, which literally saved the lives of citizens. At the same time, and in such difficult circumstances, it made a city like Sarajevo miraculously functional and healthy.

Nevertheless, and one can say paradoxically, the idea of a 15-minute city in the period after the aggression against Bosnia and Herzegovina and the siege of Sarajevo, during and after the war reconstruction and construction, was completely abandoned and forgotten.

It gave way to the hyperproduction of capitalist products, which nor their purpose, nor for their physical structure, cannot replace the previous concept of forming functional city parts. Previous proved to be true the hard way—during the last global crisis.

The latest crisis caused by the COVID-19 pandemic has shown us how crises can be actuators of opportunities and that they are often a stimulus for the emergence of new utopias and reflections, in urbanism as well.

Previous experiences of organizing city functions in extraordinary circumstances, unfortunately, could not be applied.

Is the cause of this the loss of a clear network of social, commercial and service infrastructure elements within 500 m? Or did the change in living habits and the expansion of capitalist values make the inhabitants forget to use and appreciate the elements that make cities healthy?

It was the latest crisis that spurred changes on the global scene.

Contemporary crises, are putting the concept of the 15-minute city back into the focus of planning and political-management deliberations. The need to organize the city in the way that Perry advocated at the end of 1920, or as Yugoslav urban planners did half a century ago, arose this time under circumstances of necessity, at the moment when accessibility and movement were very limited, and the post-pandemic economy was looking for new incentives and solutions.

The return to previously recognized opportunities in terms of the organization of living, city space, occurs, *inter alia*, thanks to Professor Carlos Moreno who says:

"Were it not for COVID-19, I think that the conditions for deploying the 15-minute city concept would have been very hard to instigate, but the catastrophe of the pandemic has seen us drastically change how we live—it has forced us to reassess the nature and quality of our urban lifestyles" [4].

Prof. Moreno, as one of the most ardent proponents of the concept today, became the driving force behind the "*ville du quart d'heure*" [5] project, on which the mayor of Paris, Anne Hidalgo, based her re-election campaign in 2020.

It is evident from the above mentioned that the attractiveness of the model in wider aspects, and not exclusively within the closed professional and scientific community, as well as its application at the present moment, is what makes the 15-minute concept so desirable.

Discussing the new model of Paris, Hidalgo says: "My project is about proximity, participation, collaboration and ecology. In Paris we all feel we have no time, we are always rushing to one place or another, always trying to gain time. That is why I am convinced we need to transform the city so Parisians can learn, do sports, have healthcare, shop, within 15 min of their home. This will take Paris into the future" [6].

Although it sounds almost prophetic to announce "the entry of Paris into the future" (and let's remember Perry in 1920), everything that Hidalgo talks about a hundred years later, actually looks like a return to the past. Of course, in a good way.

Provoked by and unreconciled with non-functioning of the defragmented city, alienated people who waste long hours in everyday commuting, detached from their community, places of residence and work, which is undoubtedly affecting oil and fuel consumption and quality of environment and life in general, urban planners and today's politicians are returning to the postulates of the 15-minute city.

Do not forget Jane Jacobs, who in the 1970s warned about the conditions of the burdened cities and offered a solution, by saying: "The more successfully a city mingles everyday diversity of uses and users in its everyday streets, the more successfully, casually (and economically) its people thereby enliven and support well-located parks that can thus give back grace and delight to their neighborhoods instead of vacuity" [2].

But what is actually being said about this concept today and what are its features in a world that has drastically changed compared to the one in which Perry imagined the 15-minute city?

According to the United Nations Framework Convention on Climate Change (UNFCCC), the 15-minute city is accepted as an idea that will contribute to achieving the goals of the Paris Agreement and reducing global temperature, as one of the most significant indicators of changes in the way we use resources on Earth.

The same organization recognizes the advantages of the concept as propagated today by Professor Carlos Moreno and others, and evaluates the idea as simple, applicable, nature-based solutions oriented. It is also praised for its green and blue infrastructure sensitivity and for being aligned with reducing heat in cities, as well as reducing the effects of droughts and floods initiatives. Also, it is supported for its benefits for the residents of the cities, in terms of psychological and mental health, and not just benefits which would be obvious for the planet Earth.

Practically, the idea is implemented through the above described equipping of smaller spatial units with the necessary facilities, ensuring their availability within 15 min of pedestrian/bicycle movement, and reducing or completely banning car use.

The latter is also the reason why the concept of the 15-minute city has encountered its opponents, perhaps even where urban planners and decisionmakers least expected.

Namely, in certain circles of society, mostly small right-oriented groups, the 15-minute city model is perceived as a challenge, even an open attack on their guaranteed freedoms and democratic rights. And this is specifically aimed towards using a private car, undertaking long-term daily migrations between work and place of residence, and meeting the needs of a commercial and social nature beyond the nearby community resources. It even goes so far, that the model is viewed critically as a means of control and imposition of a new world order, in which draconian laws of movement and confinement of residents in their homes will be imposed [7].

Despite occasional controversies, which are not related to the quality of the model, but to the state of society today, the 15-minute city continues to develop and is accepted in an increasing number of world metropolises and cities. In addition to the aforementioned Paris, Milan also announced its commitment to changes in the direction of the 15-minute city, and at the C40 Cities Climate Leadership Group summit, a large number of mayors from different parts of the world promoted the concept.

What is encouraging in the latest reports (and one such research report was made in 2019 by the British company Arup) is that European cities, such as London, Paris, Milan, Berlin, Madrid, in terms of their historical genesis and morphology, but also organization wise, are already on their way towards 15-minute city [8].

It is evident from this short introduction that this research will try to show that knowledge, opportunities, resources and the realization of them, when it comes to the concept of 15-minute—and therefore healthy cities, absolutely existed in our society before.

Also, it will try to show that the cities planned in the past, were functionally and structurally sufficiently adaptable, sustainable, comprehensible and editable, and were able to respond even in those moments and crisis, for which they had been always preparing, but never wanted to come true.

Sarajevo, as will be shown below, was one of those cities.

2 Literature Review

Concept of the 15-minute city has emerged in public, but also in literature, in the late 2010s.

Connected to the chrono-urbanism, as one of the most influential time policies among several European cities [10], new concept of proximity of additional services and spaces, has taken roots on global level.

According to the Nieuwenhuijsen [11], the 15-minute city can transform a city into one which is experienced as more pleasant, healthy, and flexible, and help the new urban and subrurban sprawl to be less homogenous than previously developed residential districts [10]. The aim is to decrease dependence on private-car use and to make parts of the city more heterogenous, walkable and livable.

It could be achieved by "bringing activities to the neighborhoods rather than people to activities" [12], while providing six essential urban social functions in the space–time frame of 15′, such as: living, working, commerce, healthcare, education and entertainment [10].

Already tested and partially applied in some of the European, Australian and American cities, such as Paris, Munich, Liverpool, Melbourne, Auckland, Portlan, Bogota, etc., the concept has been widely advocated after 2020 and COVID-19 pandemic.

Self-sufficiency of the neighborhoods had been a necessity in the times of crisis, and proximity of the essential functions proved to be imperative.

During the pandemic, resilience of the contemporary cities was tested [13] and results were often disappointing.

While the concept of the 15-minute city is primarily focused on the topology and usage of the living, working and recreational space in a certain spatial context, it is a paramount to acknowledge it's potential health benefits.

As it was pointed out by Moreno and coauthors of the "The 15-minute city offers a new framework for sustainability, liveability, and health" article, these factors will lead to better citizen health, with a reduction in health complications such as premature mortality and diseases of the cardiovascular and respiratory system, obesity, brain diseases (e.g., Alzheimer's disease), and cancer.

According to some authors [14], there is a knowledge-gap in the research on the 15-minute city. But, theory and practice in the former Yugoslavia, specifically in

Bosnia and Herzegovina, proved to be ahead of the concept 40 years before. What is left of the concept, is yet to be seen.

3 Sarajevo Between Planning and Dismantling

The urban plans, specifically, the General Urban Plan of Sarajevo 1965–1986 (GUP) and its subsequent changes, relied on well-known principles, which today are called in recent practice—the 15-minute city concept.

Analysis of the planning documentation was made: The General Urban Plan of Sarajevo 1965–1986, then The Urban Plan of Sarajevo 1986–2015 [3], followed by the comparative analysis of the current situation, based on two examples that most clearly show contradictions of the city planning, in the timeframe of nearly 60 years.

As the first planning document that visualized and envisioned the planning and development of the city, the GUP laid the foundations for the new parts of the city, which, although geomorphologically linear, formed its matrix as a set of urban parts which host up to 5,000 inhabitants. In every of those parts, a radius of 500 m ensured six basic existential functions (see Fig. 1), just as it was suggested by Perry or today, among others, by Moreno.

Housing, work, traffic, health, education and recreation, accessible to pedestrians/cyclists within 15 min, were not utopias in the plans of the mid-twentieth century, but well accepted, applied and adequately used principles of urban planning. It is evident from the network of these function in the urban parts or neighborhoods of the city of Sarajevo.

Fig. 1 Extract from GUP 1965–1986 with a representation of a typical settlement and radius of accessibility; *Source* General Urban plan of Sarajevo 1965–1986, author's drawings over cadastre

Fig. 2 Extract from UP 1986–2015 with a representation of a typical settlement and radius of accessibility; *Source* Urban plan of Sarajevo 1986–2015, author's drawings over cadastre

The urban plan of Sarajevo 1986–2015 continues to cite the previously conceived and established planning tradition, so we can read the same tendency to form residential areas as a functional urban microcosmos in this planning document (see Fig. 2).

It is important to note that this Urban Plan is still in use today.

Overview of the detailed planning documentation up to the 1990s, such as regulation plans, which directly refer to the spatial layout and more detailed specifications of construction and exploitation, proves that they accept and continue to elaborate the elements of the urban plan. It was almost unthinkable that a planning document lacked an elaborate network of support for housing functions or that it treated space in a homogenous and purely zonal way.

If the historical concept of planning is placed within the framework of what Moreno propagates today as the seven dimensions of the 15-minute city, namely (1) proximity, (2) density, (3) diversity, (4) digitalization, (5) human scale urban design, (6) flexibility, (7) connectivity, it is easy to see that, apart from the at the time non-existent digitalization, there was a clear vision and sensibility for the organization and function of the settlement in a sustainable and successful way in the planning discourse.

This kind of relationship of long-term deliberate planning played a role in the periods of greatest devastation and urbicide, which took place in the war years of 1992–1995.

However, the post-war period, which was supposed to bring restoration, revitalization and recovery of society, superstructure and infrastructure, along with the implementation of a new political and social system, failed to continue the cherished and acquired values, when it comes to city development planning.

The urban plan of Sarajevo, which was supposed to be in force until 2015, is today still the only official, valid document, which strategically and developmentally plans and directs the formation of the city fabric and its constituent parts. The plan, itself outdated and inapplicable in a vast number of cases, was not even partially implemented, and almost all post-war, larger spatial interventions were made contrary to the provisions of the plan.

And while the preparation and adoption of the new urban plan has been long overdue, Sarajevo remains in the gap between the new political and social-property arrangement, ambitious investors and the disappearance of support within the nucleus of the neighborhood. This is a result of the 25 years of complete neglection of the practice of forming a network of elements within a radius of 500 m (today's 15-minute city), which was very present until the 90s.

15-Minute City Model in Times of Crisis, Example of Dobrinja

As shown in the Introduction, 15-minute city model can be easily imposed as a good solution, especially for the cities in crisis. It is primarily due to its availability within relatively small distances that can be overcome by pedestrians and cyclists. Knowing how crisis could manifest themselves differently, as manmade, natural or combined factors (wars, riots, climate change, floods, earthquakes, pollution, pandemics and diseases), in each of these crises, the availability of vital functions and the safety of their use prove to be extremely important common ground.

One such crisis emerged on the global level with the COVID-19 pandemic, to which the whole world was simultaneously exposed.

Probably one of the most extreme forms of manmade crisis for the city is war. Bosnian cities experienced this crisis at the end of the last century.

But, if a war is the extreme form of crisis, then the siege of the city is crisis's climax.

If war is an extreme form of crisis in cities, then the siege of the city is the peak of that extreme.

Sarajevo was subjected to the siege in the period 1992–1995. Most likely and partly thanks to its spatial organization based on the concept of proximity in 500 m radius, it managed, as far as possible under such circumstances, to survive functionally.

An example of the Sarajevo neighborhood of Dobrinja, shown in Fig. 2, can testify to an exceptional case of the functioning of the 15-minute city in the period of crisis. The neighborhood was cut off and isolated from the rest of the city and nearly under complete siege for four years.

Way of self-organization and at least a semblance of normal functioning, was a priority for the community.

Housing was reformed and remodelled, trying to respond to the new circumstances and almost permanent loss of electricity, water supply and other infrastructure necessities. At the same time, housing remained the basic function of the settlement (Dobrinja was developed for the Winter Olympic Games less than a decade before), providing the context for approximately 20,000 inhabitants who happened to be in there at the beginning of the War.

At the same time, the educational process had to be continued, and certainly the matrix created within a radius of 500 m played a great role here.

Even though the teaching did not take place in school buildings all the time, but inside business ground floor premises, existence and equipment of the facilities made it much easier to organize educational process.

Healthcare had to be raised from primary to a higher level, but the germ of healthcare, as well as education, was already present and that certainly made the organization easier.

The importance of short distances between the place of residence and the place where other life functions could be satisfied was of unspeakable importance. Security, accessibility and social inclusion are imposed as the most important features of this model. This model contributes to the resilience of the city and speaks of its adaptability, qualities that prove to be crucial in crisis situations.

Degradation of Sarajevo

Overview of the planning tradition in Sarajevo and the obvious benefits of the previous planning rules, which proved themselves in times of crisis, made a solid ground for critical rethinking of the new development practice in Sarajevo after the war.

Tibra settlement, placed on the far west of Sarajevo longitudinal axe and developed in 2000s, is a typical example of contradictory planning practice. It denied all of the norms seen in pre-war planning practice, such as neighborhood organization, proximity, equipment, distances even fire regulations, traffic safety and parking spaces, which are embedded in the legislation. Moreover, Tibra settlement was developed by completely ignoring the Urban plan of Sarajevo 1986–2015, which is still in use (Fig. 3).

Reasons and circumstances of such an ignorant and obviously planned corruption should be studied through the specific research. Focus of this study are functional capacities of settlements like Tibra and proximity of social inclusion facilities within, which are basic elements of the 15-minute city concept.

More than 15,000 inhabitants of Tibra satisfies the basic life functions at significantly greater distances (see Fig. 4), than it is adequate according to the previous models and 15-minute city model as well.

Tibra, sadly, is not an isolated case, as two similar settlements were developed in Alipašin Most V and VI area in 2010s (see Fig. 5).

If it was considered that the Tibra was a reckless incident in the post-war period, the recentness of these two settlements indicates the opposite, and speaks in favor of a new, devastating reality in the urban and architectural public of Sarajevo.

Alipašin Most V and VI are inhabited by new 10,000 people, and besides housing, there are almost no other supporting functions in the expected proximity, making

Fig. 3 Settlement Tibra, Stup, Sarajevo; *Source* author

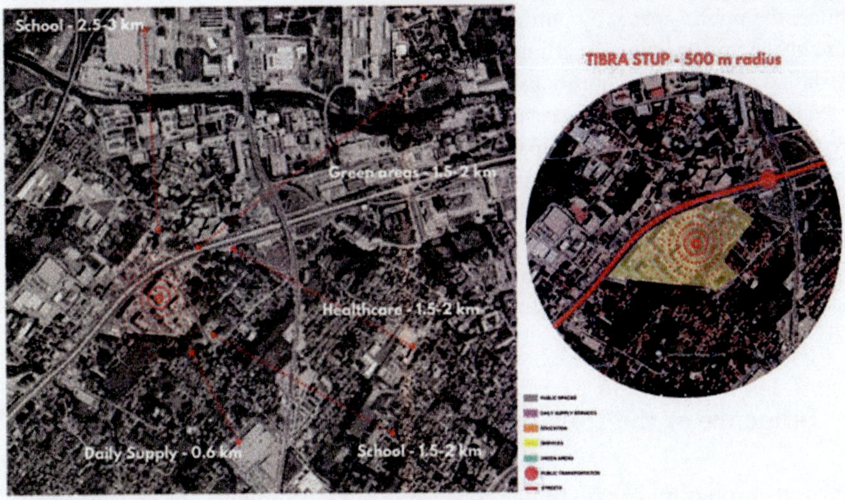

Fig. 4 Tibra neighborhood, Stup, Sarajevo in the context of the radius of availability; *Source* authors over Google maps

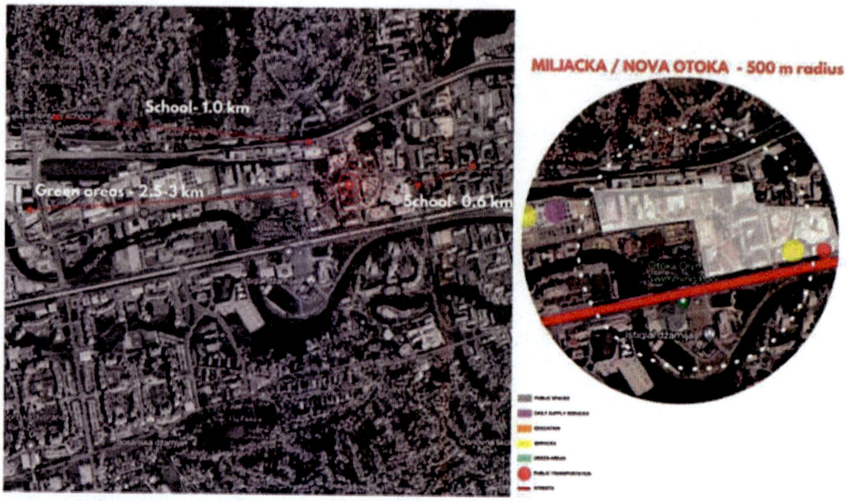

Fig. 5 Miljacka-Nova Otoka neighborhood, Alipašin Most, Sarajevo in the context of the radius of availability; *Source* authors over Google maps

them similar to the sleeping neighborhoods of the 60s and 70s, a model long abandoned and proved to be unsuccessful (see Fig. 5).

It is clear that the new urban policy's [15] narrative, with its critical approach focused on one common goal as "to revive the city as a collective reality" and rehabilitate the public space, by putting up the fight against segregation and monofunctionality and prioritizing the efficient public transport and revival of local democracy, has failed in these neighborhoods.

Measuring the distance of seven basic indicators of the 15-minute city, indicates their absence in Tibra and Alipašin Most V and VI settlements. Impossibility of organizing everyday life, even those basic elements, such as trade, education or healthcare, which for almost half a century were the standard of every Sarajevo settlement, are trademarks of the neighborhoods.

Is there a cure for the neighborhoods built in this way and can the 15-minute model be the path to healing in them?

4 Influence of the 15-Minute Concept on Health of the City

The next chapters make case for the connection of the 15-minute concept and health of the cities. It is explained through the global overview of the meaning of it and case example of Sarajevo, where disappearance of the 500 m radius standards influenced the overall health of the communities, especially in newly built neighborhoods.

(Un)Healthy Cities, (Un)Healthy Communities—Global Overview

The 15-minute city can be a model to apply to cities in crisis. Crises can have different causes, forms and manifestations. What is the common crisis of today's cities, including the settlements built in Sarajevo in the last two decades and which is described in the previous chapters, is the continuous decomposition of the planned development of the neighborhood, which affects the level of health of society in general.

Accepting that today's cities, in their intrinsic and spatial organization, are largely opposed to a healthy way of life is the first step towards understanding the problem and its repercussions for humankind.

The World Health Organization (WHO) started promoting the Healthy Cities pro-gram back in 1986, with the aim of building an integrated approach of health promotion at the local level. If we know that cities are places where local authorities are formed to implement health policies, where citizens identify with their environment and participate in policies that concern their neighborhoods [16], then we can relate the policy of a healthy city to everything mentioned above.

Available research shows a link between the physical characteristics of the neighborhood and the risks of endangering health. Various health factors such as obesity, diabetes, depression are associated with neighborhood characteristics such as population density, presence of local amenities, access to nature, access to work, etc. [17].

Today, some 40 years after the start of the healthy cities movement, we recognize that a healthy city is not defined by outcomes but by processes. The basic model implies a place that encourages the participation of all communities in order to achieve common prosperity and peace [18]. A healthy city was just an idea, started as a promotion, and today the movement, the vision, is more significant than ever.

The 15-minute city as a concept differs from the initial ideas of the 'neighborhood unit' advocated by Perry, and advocates the distribution of functions that are accessible by active transport [19], and through modern understanding and analysis it can be accepted that the idea is not unambiguous and does not suit everyone. That occasionally appears as a criticism of the concept, especially when it comes to the different needs of different people (persons with mobility difficulties or different marginalized groups), and talks about the impossibility of adapting to the "one size fits all" system.

Therefore, it is especially important to observe the concept of the 15-minute city in the context of healthy cities, whose role lies in the cohesion of social, demographic, economic and topographical elements.

If we accept that a city or settlement like those described in the previous chapters, which were used as a example of the lack of a clear development strategy, is in crisis, and that the postulates that we asserted affect resilience, adaptability and functional survival in times of crisis, then it should be easy to accept new ways of planning, which will help them heal.

For healthy cities, there are clearly defined sets of indicators that must be followed for a successful path to a healthy city, a city for citizens. The indicators are organized into four groups: health, health services, environmental indicators and socioeconomic indicators [20]. In addition, the World Health Organization indicates that the health of cities, apart from the mentioned indicators, is measurable and observable through the aspects of social, demographic, economic and geographical factors [21], whose mutual interaction and influence is everyday and is discussed through various aspects of society and speaks the most. The indicators are clear, well explained, and by developing a strategy for their improvement, it seems easy to reach a healthy city.

But the fact is that today we still live in unhealthy cities, and it is obvious that we forget that the connections between our environment and health are very complex, and that we tend to ignore them, which consciously affects the quality of life and the level of health of the residents in the planned settlements.

Healing of the Sarajevo Neighborhood

Neighborhood that is aligned with health planning promotes walkability, builds a complete and compact neighborhood, safe and efficient infrastructure [22]. A neighborhood planned in this way can contribute to the health and well-being of its residents.

The neighborhood concept is extremely important for the city of Sarajevo, and it is also connected with the city's identity. The spirit of the neighborhood is still present, many citizens are proud of the part of the city where they grew up, and it is expected that the development of awareness of such concept could have the active participation and support of citizens.

According to the WHO, awareness of health and the desire to improve it is a process that leads to a healthy city. Sarajevo, as a city that founded its linear development on epochs of development (growth) and stagnation, in accordance with socioeconomic circumstances, does not fall into the trap of suburban planning. Through exciting historical stages, it passes through different social arrangements, and its neighborhoods remain somewhat compact. The actual moment in time can be seen as a crisis in the development of the city. The newly created settlements neither accept the previous approaches in the development of the city, nor position new ones. In order to improve the health and well-being of all residents, including those who live in such settlements, it is necessary to repeat the urban lessons demonstrated during the development of the city and integrate the missing functions.

Planned development has developed areas with appropriate city functions that can bring all functions to 15 min from the place of living with active city transport.

Spatial planning for health can transform Sarajevo's neighborhood units, created during the development of the city in the socialist period, into the modern concept of a 15-minute city [23].

5 Conclusion

The 15-minute city model represents a return to the principles established by urban planners and researchers of the city and city life at the beginning of the twentieth century, its survival and adaptability to change. This model proves to be a good and acceptable solution for cities in crisis situations, regardless the origin of their malice.

At the same time, there are conflicting opinions that connect the 15-minute city with the doctrine of socialism, conspiracy theorists on social networks consider it an attack on personal freedom [7] in the context of COVID-19 and the lockdown, and translate accessibility into restriction of movement.

Criticisms can also be heard in scientific and academic circles, such as the one presented by Jay Pitter at The City Lab Conference 2021 [9], asserting as follows: "It doesn't take into account the histories of urban inequity, intentionally imposed by technocratic and colonial planning approaches, such as segregated neighborhoods, deep amenity inequity and discriminatory policing of our public spaces."

Besides Pitter, who works on public space design in number of U.S cities [14], some economists, such as Edward Glaeser, Professor of Economics at the University of Harvard, suggests that the concept will increase inequalities, especially, the 15-minute city's emphasis on digitalisation. Working in a North American context, Glaeser points to figures that states that only 5% of the American population without a high school degree has been working from home through the pandemic, suggesting that these people will be completely left out of the proximity aspect of the concept [14].

If the model of a 15-minute, and at the same time healthy city, is examined in the elements of urban planning of Sarajevo in the period up to the 1990s, to that extent one can notice both benefits and limitations. Those elements, which are still applicable and measurable, can serve as the basis on which cities of today, who are starting with the application of the model, can learn.

Certainly, a critical review of the model has its basis, but if one takes into account the example of today's Sarajevo settlements, and the historical approach to city planning, which had its 40-year test run, evidence goes in favor of the formation of the city in a concept of the 15-minute model. Proper modifications and adjustments are, of course, inevitable and desirable.

At last, let remind us of what Jane Jacobs thought us [2], when she spoke of the city as a laboratory of constant tries and errors, and not just that, but as a space "in which city planning should have been learning and forming and testing theories. Instead, the practitioners and teachers of this discipline (if such it can be called) have ignored the study of success and failure in real life, have been incurious about the reasons for unexpected success, and are guided instead by principles derived from the behavior and appearance of towns, suburbs, tuberculosis sanatoria, fairs, and imaginary dream cities—from anything but cities themselves."

Bearing in mind the very different and sometimes conflicting thoughts of the thesis discussed in this paper, which examines the indisputable connection of a homogeneous way of planning with the level of the quality of life of the inhabitants, it can be concluded that the 15-minute city, supported by elements that make cities healthy, can contribute to a more functional, safer more healthier and more pleasant, city, the one which is adaptable to crises and more resistant to them, than cities and settlements that are based on monofunctionality and centralization.

Such models can be a good starting point for improving the quality of life in the community and can certainly serve as a lesson for building.

A lot can be learned from the example of Sarajevo, as a city that in the last 40 years went through a planning process in which accessibility, density, diversity, the human scale of urban design, flexibility and connectivity within the nucleus of one neighborhood were the natural environment of urban planners and users.

Especially if you follow the development up to the present day, when the health of the city has been seriously damaged due to non-respect of the basic postulates of the 15-minute city.

Therefore, it can be concluded that the concept may be imperfect, it is not "one size fit all" principle, but it belongs to everyone.

References

1. Perry CA (1929) The neighborhood unit: a scheme of arrangement for the family-life community. In: Neighborhood community Plan. Comm. Reg. Plan N. Y. its Environ, pp 2–140
2. Jacobs J (1974) The death and life of great American cities. Penguin Books Ltd., Harmondsworth, Middlesex, England
3. Generalni urbanistički plan grada Sarajeva 1965–1986 i Urbanistički plan grada Sarajeva 1986–2015.g sa pratećim studijama i programima
4. Kucharek JK Carlos Moreno 15 minutes to save the world, The Riba Jurnal published on https://www.ribaj.com/culture/profile-carlos-moreno-15-minute-city-obel-award-planning. Accessed 23 Feb 2023
5. The City of Paris webpage https://www.paris.fr/dossiers/paris-ville-du-quart-d-heure-ou-le-pari-de-la-proximite-37. Accessed 23 Feb 2023
6. Wilsher K Paris mayor unveils '15-minute city' plan in re-election campaign. The Guardian published on https://www.theguardian.com/world/2020/feb/07/paris-mayor-unveils-15-minute-city-plan-in-re-election-campaign. Accessed 23 Feb 2023
7. The 15-min city: where urban planning meets conspiracy theory. The New York Times. https://www.nytimes.com/2023/03/01/world/europe/15-minute-city-conspiracy.html
8. Green and Thriving Neighbourhoods, ARUP. https://www.arup.com/perspectives/publications/research/section/green-and-thriving-neighbourhoods
9. https://bloombergcities.jhu.edu/citylab. Accessed 25 Mar 2023
10. Di Marino M, Tomaz E, Henriques C, Chavoshi SH (2023) The 15-minute city concept and new working spaces: a planning perspective from Oslo and Lisbon. Eur Plan Stud
11. Nieuwenhuijsen MJ (2021) New urban models for more sustainable, liveable and healthier cities post covid19; reducing air pollution noise and heat island effects and increasing green space and physical activity. Environ Int 157:106850
12. Pozoukidou G, Chatziyiannaki Z (2021) 15-Minute city: decomposing the new urban planning Eutopia. Sustainability 13(2):928
13. Allam Z, Nieuwenhuijsen M, Chabaud D, Moreno C (2022) The 15-minute city offers a new framework for sustainability, liveability, and health. Lancet Planet Health 6:E181-E183. https://doi.org/10.1016/S2542-5196(22)00014-6

14. Zakariasson A (2022) A study of the 15-minute city concept: identifying strengths, risks and challenges through imagining the implementation of the 15-minute city concept in Munich, Germany
15. Supek R (1987) Grad po mjeri čovjeka, s gledišta urbane antropologije. Naprijed, Zagreb, pp 91–91
16. Ashton J, Grey P, Barnard K (1986) Healthy cities—WHO's new public health initiative. Health Promot Int 1(Issue 3):319–324. https://doi.org/10.1093/heapro/1.3.319
17. Grant M (2015) European healthy city network phase V: patterns emerging for healthy urban planning. Health Promot Int 30(suppl_1):i54–i70. https://doi.org/10.1093/heapro/dav033
18. World Health Organization webpage https://www.who.int/europe/groups/who-european-healthy-cities-network/what-is-a-health-city. Accessed 25 Feb 2023
19. Lu M, Diab E (2023) Understanding the determinants of x-minute city policies: a review of the North American and Australian cities' planning documents. J Urban Mobil 3:100040. ISSN 2667-0917. https://doi.org/10.1016/j.urbmob.2022.100040
20. Webster P, Sanderson D (2013) Healthy city indicators—a suitable instrument to measure health? J Urban Health 90(Suppl 1):52–61. Published online 2012 Apr 20. https://doi.org/10.1007/s11524-011-9643-9
21. World Health Organization webpage https://www.who.int/southeastasia/activities/healthy-cities
22. Spatial planning for health: an evidence resource for planning and designing healthier places
23. Kissfazekas K (2022) Circle of paradigms? Or '15-minute' neighborhoods from the 1950s. Cities 123:103587. ISSN 0264-2751. https://doi.org/10.1016/j.cities.2022.103587

Heritage, Preservation and Conservation

Critical Assessment of Industrial Heritage Recording Processes in Bosnia and Herzegovina

Maja Pličanić and Tijana Veljković

Abstract The present methodological approach to developing a database in the field of architectural heritage is based on the establishment of measurable parameters that would ensure the most objective recognition of cultural and historical values. Considering the international guidelines for defining industrial heritage and the limited representation of this category of architectural heritage on the list of protected cultural assets at all administrative levels in Bosnia and Herzegovina, the parameters for identifying industrial heritage were defined. In this regard, this paper examines the characteristics of industrial heritage in the context of Bosnia and Herzegovina by implementing historical and comparative analysis derived from systematic research of industrial heritage in Sarajevo and Tuzla. Two different methodological approaches to documenting individual industrial buildings and complexes, as well as mutually related industrial ensembles, are combined and presented in this article. The standardized form of record sheets intends to separate the parameters that are typical of Bosnia and Herzegovina's industrial heritage so that the method can be implemented in all regions of the country, as well as to provide directions for further valorization and legal protection.

Keywords Industrial heritage · Industrial heritage parameters · Recording methodology

M. Pličanić (✉)
Faculty of Educational Sciences, University of Sarajevo, 71000 Sarajevo, Bosnia and Herzegovina
e-mail: mplicanic@pf.unsa.ba

T. Veljković
Recognized researcher, 75000 Tuzla, Bosnia and Herzegovina

M. Arslanagić-Kalajdžić et al. (eds.), *Interdisciplinary Advances in Sustainable Development II*, Lecture Notes in Networks and Systems 804,
https://doi.org/10.1007/978-3-031-46692-2_11

1 Introduction

The need to develop a comprehensive inventory for industrial heritage arose from extensive research conducted in the authors' doctoral theses, in which systematic studies of industrial heritage of Sarajevo and Tuzla aim to identify its cultural significance in the context of Bosnia and Herzegovina (in continuation: BH) [1, 2]. Regardless of the different recording approaches, their comparative analysis will define common identification parameters within this paper in order to ultimately contribute to industrial heritage valorization criteria.

The scope of industrial heritage at national, regional and individual levels should be clearly defined and documented. Inventories play a key role in integrating industrial heritage assets into development policies. However, without the engagement of the professional and public realms in decision-making, it is difficult to develop a sustainable model of heritage-led regeneration, and inventories can be confined to a mere collection of fundamental information [3].

The Issue of Industrial Heritage Protection in Bosnia and Herzegovina

Numerous restoration and reconstruction projects of damaged cultural assets associated with ethnic groups of Bosnia and Herzegovina, or damaged sites of historic and cultural importance at national levels have been initiated since the war ended in 1995 with the objective of restoring peaceful coexistence. At the time, industrial sites were not considered relevant for post-conflict rehabilitation. Moreover, the transition from a socialist to a capitalist economic model, the loss of foreign markets and capital, and the consequential high unemployment rate contributed to the negative perception of industrial enterprises. Thus, neglected and degraded industrial buildings have long been linked with places of struggle between public and private interests, rather than places of cultural and historical significance. Considering the aforementioned historical events, it is easier to understand why Bosnia and Herzegovina's industrial heritage has been neglected for many years, both by responsible institutions and by the local population whose identity is directly linked to a specific area's industrial activity.

BH's heritage is characterized by cultural diversity and multiple historical layers. Its preservation and protection at the institutional level is currently managed by sixteen bodies[1] with corresponding regulatory legal acts.

The lack of a single register of architectural and archaeological heritage across the entire territory of BH[2] renders impossible assessing the real state of BH's heritage, including industrial heritage. The Federal Institute for the Protection of Monuments provides a list of "*registered and protected industrial and commercial facilities*" (Bosnian: *evidentiranih i zaštićenih industrijskih i privrednih objekata*) which includes 48 buildings. According to the Institute for the Protection of Cultural, Historical and Natural Heritage at the Ministry of Education and Culture of the Republic of Srpska, the list contains only 20 buildings or sites.[3] Since the registry lists within BH's entities overlap, a total of 65 buildings or sites belonging to the current monument classification of "*industrial and commercial buildings*" (Bosnian: *industrijskih i privrednih objekata*) [3] were recorded, of which only five[4] are legally protected (that is, registered in the Federal Institute for the Protection of Monuments in BH) on the basis of historical events from World War II.

The Commission to Preserve National Monuments, as the only institution at the national level of BH, declared a total of 920 properties as national monuments of BH from 2002 to January 17th 2023. Of these, only three monuments are classified as industrial heritage, namely:

- industrial heritage—Salt production in Tuzla, which consists of three sites, declared a national monuments in November 2007;
- industrial buildings ensemble—Electric Power Station on Hrid in Sarajevo, declared a national monument in December 2009;
- industrial buildings ensemble—Hydroelectric power plant in Jarak, declared a national monument in March 2011;

[1] The intricacy of the protective services structure stems from the complexity of the country's administrative system. The Commission to Preserve National Monuments is responsible for cultural heritage protection at the national level, which is followed by two institutes at the entity level: the Institute for the Protection of Cultural, Historical, and Natural Heritage at the Ministry of Education and Culture of the Republic of Srpska, and the Institute for the Protection of Monuments at the Federal Ministry of Culture and Sports. Protection of historical buildings at cantonal levels is administered by institutes for the conservation of cultural, historical, and natural heritage (Sarajevo Canton, Tuzla Canton, Central Bosnia Canton, Herzegovina-Neretva Canton, and Una Sana Canton), while special services at departments related to the Ministry of Culture are provided in the remaining five cantons, which do not have established institutes [4].

[2] It is certainly the result of the absence of a single Law on the Protection of Cultural Heritage at the national level.

[3] The list provided by the Institute from RS partially overlaps (three facilities from Banja Luka) with the records of the Institute of the Federation of BH, because the latter has a database made on the basis of the inventory from 1986 of the then Republic Institute (SRBiH).

[4] The building of the Elektrobosna factory in Jajce, the warehouse of the Elektrobosna factory in Jajce, the foundry building in Drvar, the production hall of the former Drvar Pulp Factory and the machine shop with the foundry of the Ljubija mine in Prijedor. These five buildings are "protected on the basis of significant events from the national liberation struggle (NOB)—registration decision of 1950", and not on the basis of heritage values that testify about their industrial purpose.

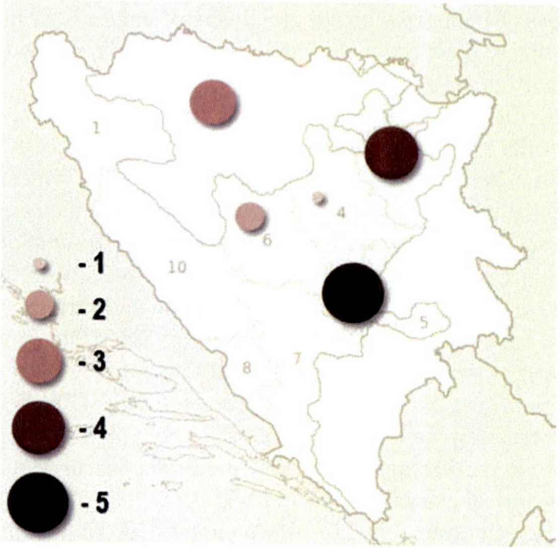

Fig. 1 Depicts the number of industrial heritage sites that have been designated as national monuments but do not have the prefix "industrial": 1—a paper factory and a workers' settlement in Zenica-Doboj Canton; 2—a steam locomotive and a train station in Central Bosnia Canton; 3—three train stations in Srpska Republic; 4—a train station, a villa built for the caustic soda factory, and a miners' memorial in Tuzla Canton; 5—an electric power plant, two train stations, and an old train workshop and warehouse in Sarajevo Canton. (the original image was modified by the authors, attribution: TUBS, CC BY-SA 3.0 <https://creativecommons.org/licenses/by-sa/3.0> , via Wikimedia Commons)

Furthermore, 14 decisions from the list of national monuments were recognized as historical buildings or monuments, architectural ensembles, and sites that do not bear the industrial prefix (Fig. 1), but, by all accounts, belong to this category, according to the adopted charters and decisions of the international bodies: The International Committee for the Conservation of the Industrial Heritage (TICCIH) [5], the International Council on Monuments and Sites (ICOMOS) [6], and Council of Europe (CoE) [7].

Based on the present fund of BH's industrial heritage, it is difficult to read, namely, to understand the country's industrial and socioeconomic past, its connections with other areas of the world or the region, [6] which should be of essential importance.

Materials and Methods

A methodological approach was devised through comprehensive research in the Sarajevo and Tuzla areas, which will be presented in the following study through the basic parameters for identifying Bosnia and Herzegovina's industrial heritage. A typology of industrial heritage was defined based on production branches, in accordance with the specifications of the American classification of industrial heritage [8] and the existing classification of activities in Bosnia and Herzegovina [9]. The architectural-morphological typology's parameters were determined based on the defined identity of this particular type of structure during key periods of the country's industrialization. On the basis of a relevant sample, analysis of archival material, and field research on the spatial coverage of Sarajevo, the main sets of features of this style of architecture were defined.

Workers' housing, cultural, sports, educational, and health facilities, which originated from the industrial culture of defined historical periods, were integrated into the documentation process on selected examples in Tuzla in order to ultimately determine the social value of industrial heritage in the context of BH. Salt production in Tuzla was chosen as a representative example for determining technical–technological parameters since it includes both pre-industrial and industrial production processes. A comprehensive study of original archival resources, monographs of industrial companies, and field research were employed for this aim.

In order to ultimately determine the cultural significance of industrial heritage, based on UNESCO's Operational Guidelines for the Implementation of the World Heritage Convention [10] and Principles and Guidelines for the Preservation of National Monuments [11], authenticity and integrity, which are defined as a basic qualification criteria, are included in the record sheets.

2 A Brief Review of BH's Industrial Development

The information presented in the histrical review is derived from secondary literature. Mainly, the work of the industrial geographer Đ. Marić [12], economic historian K. Hrelja [13], and extensive research by historians who dealt with the industrialization of BH in the key periods of its development [14, 15]. BH's industrial development may be separated into two distinct phases. The capitalist socioeconomic system characterizes the first industrialization, which spans the Austro-Hungarian period and the period until the end of World War II (1878–1945). The socialist organization of the state marks the second industrialization wave, which begins after WWII and lasts until 1992.

Until the end of the nineteenth century, BH was predominantly an agrarian country, with agriculture being the main occupation of 90% of the population, and its traditional economic and social development, based on trade and craftsmanship, did not provide the prerequisites for modernization [15]. Capitalist social relations in

BH developed under the Austro-Hungarian administration in the context of the state's dominant role as a first-rate political and economic element in the growth of commerce, industry, and other economic branches other than agriculture. The state had the majority of the province's forests and mineral richness, but lacked the necessary financial resources for exploitation and was compelled to seek capital for initial investments. The first acquired capital, raised on the Austrian and European financial markets, is used by the state to establish railway and road transit, allowing for the exploitation of forests and ores, i.e. the natural resources based on which the first industrial facilities were built [15]. Thus, the wood industry was the prevailing industrial sector in BH, not only in terms of the number of employees and enterprises, but also in terms of obtained revenue. The iron ore mines in Vareš and Ljubija are particularly important for mineral reserves exploitation, and two iron-works (in Vareš and Zenica) were developed during this period, i.e. industries that exploit raw materials for the metal processing industry. In addition to iron ore mines, coal mines (Zenica, Kreka, Kakanj, and Breza) were opened, as well as mines of manganese, chromium, and other metal ores, on the basis of which a solid base and prerequisites for the further development of ferrous metallurgy in BH were created. The chemical industry, which developed on the basis of its own raw materials (mostly rock salt from the Tuzla basin),[5] was no less substantial. The geographical distribution of industry during this time period, which was oriented towards the exploitation of natural resources for export, resulted from the principle of concentrating industrial production in the most economically beneficial locations. The Sarajevo-Zenica basin developed an industrial core with about half of the total number of industrial facilities in Bosnia and Herzegovina [13]. Heavy industry was predominantly focused along the Sarajevo-Doboj railway line, relying on iron ore in Vareš and coal mines at Kakanj and Zenica. The wood industry grew in the Bosna river basin, with industrial centers developed around Zavidovići and Sarajevo, then in Drvar, Tuzla (Živinice and Kladanj), and the upper course of the Drina River [14].

The extractive nature of BH's first industrialisation wave was retained throughout the next historical period. Specifically, the industrial sectors developed during the Austro-Hungarian period served as the foundation of BH's industrial development in the time between the two wars. During the Kingdom of Serbs, Croats and Slovenes, the main industrial enterprises in BH were founded before 1919. Following a comparison of data on the distribution of industrial businesses by branches in 1938 with data from 1918, it is concluded that the expansion of industrial stock (129 new industrial enterprises) had no effect on the industry's structure [14].[6] In this period, the

[5] Thus ten buildings were built, the most important of which were: the Soda Factory in Luka-vac, the Electrochemical Industry in Jajce, Wood Distillation facility in Teslić, the Oil Refinery in Bosanski Brod, the Match Factory in Dolac near Travnik and others.

[6] The largest number of companies built in this period belonged to the wood industry, then in the food and construction materials industry, and then in the textile, graphics, chemical and metal processing industries. Although several factories from the group of propulsive in-dustrial branches such as metal processing and chemical were built, the structure of BH's industry retained the extractive character it had before 1918 and at a very low technical level [12].

inherent unevenness in the geographical distribution of industry from the Austro-Hungarian period became even more pronounced. New industrial capacities were mostly constructed along the railway line Bosanski Brod-Sarajevo and in larger urban areas [12].

BH had 43 mining companies in 1939, 112 sawmills that were considered industrial, 11 metal processing companies, 13 chemical industry companies, 33 construction (mostly brick) enterprises, 26 food industry companies, 19 textile, 8 leather industries, and 46 electrical power stations [14].

By the end of WWII BH had 45% of industry and mining still in operation [14]. The finding from Marić's study [12] that industrial production in 1947 was 13% higher than in 1939 appears incredible. One of the primary goals of economic policy at the period was to overcome the legacy of backwardness, i.e. to accelerate overall growth in order to join the circle of developed industrial-agrarian countries. To complete the mission, the industrialization path chosen had to imply quicker development of industry than agriculture, which could only be accomplished by faster development of basic industry (energy, mining, ferrous metallurgy and mechanical engineering). With this, BH gained a forefront primary role in the first five-year plan, when constructing new, as well as expansion and modernisation of existing industrial facilities, particularly mines and hydroelectric plants, had begun. However, the true process of industrialization (the so-called second industrialization of Bosnia and Herzegovina), which means the construction of that industry that absorbs labor force from the countryside to a significant extent and thus changes the structure of the entire society, is thought to have started in 1948, when the center of the military industry was relocated to Bosnian territory.[7] During this period, the military industry became the leading force of industrialization.[8] Because of extensive military investments, BH was viewed for a long time as a "developed republic," and it did not receive civil investments that were transmitted to other republics, staying objectively one of Yugoslavia's poorest sections [16]. It is also noted that one of the most significant changes throughout these decades was the increase in industrial workers from the countryside, who resided in rural areas and frequently traveled long distances to work in factories.[9]

---

[7] The newly created political situation (1948—the exit of Yugoslavia from the Cominform) and a high rise in tensions with the Soviet Bloc. "Fearing a simultaneous attack from the north (Hungary) and the east (Bulgaria) and guided by the success of the Partisans in World War II, Tito turned the remote region of central BH, with its wild mountains, densely forested areas and natural caves, into a Yugoslav fortress and center its military industries [16]."

[8] The most important companies of the military industry were: Bratstvo (howitzers and rocket launchers) in Novi Travnik, Slobodan Princip Seljo/Vitezit (explosives) in Vitez, Slavko Rodić (fuses for mines and grenades) in Bugojno, Pobjeda (detonators) and Istra in Goražde, Soko (jets) in Mostar, Famos (engines for tanks and armored vehicles) in Sarajevo, Rudi Čajevac (radar and electronics) in Banja Luka and others [12, 14, 16].

[9] In the 1950s, a study of one of's BH industrial power Zenica Ironworks revealed that workers from rural areas traveled eight hours a day from their villages. Internal migration was present within the majority of the workforce working in Yugoslavia's textile, timber, and construction industries, as well as 60% of miners in Serbia and Bosnia [16].

3 Identification of Parameters for Documenting Industrial Heritage in Bosnia and Herzegovina

The guidelines of the methodological framework for the systematic assessing, i.e. documentation of the industrial heritage, are based on the fundamental principles of the English methodology, since the majority of European countries evaluated them as the most effective.[10] One of the initial challenges England faced in valorizing its industrial past was a lack of a scientifically based information base. A method for systematically examining the complete typology of industrial architecture had to be chosen first. There were two options considered: division based on historical periods of creation, or division based on technological process. The majority of industrial archaeologists agreed with the latter, believing that the architectural form of industrial buildings is a product of spatial requirements imposed by technological processes that took place in buildings, and that all successive modifications are the result of changes in the technological processes. Moreover, the technological process is the cause of all construction features in the industrial complex, and all of its buildings are equally vital in comprehending it. Finally, it was agreed to accept a "thematic approach" to evaluation based on the classification of industrial architecture into functional groups and the assessment of individual groups, the key benefit of which is the possibility of comparing buildings and complexes belonging to similar industrial branches.

As a result, analogy becomes the primary analytical method for analyzing industrial complexes. The overall outcome of the selected industry's thematic research would be: defining the time period in which production took place (or is still taking place), determining the main historical periods in the development of industrial complexes of the associated industry, and analyzing the basic spatial-functional, constructive, and architectural characteristics of complexes as well as individual buildings.

The parameters described in the remainder of the article are the results of adapting the British model of industrial heritage documentation to the Bosnian context. The Association for Industrial Archeology (AIA) and the Royal Commission on the Historical Monuments of England (RCHME) [18] methodological approaches were assessed in order to design the record sheets that are presented in the Appendix of this article.

[10] At the "Recording the Industrial Heritage" meeting held in London in 1989, expert conservation circles agreed that the English system is the most developed at the European level, and as such, it is recommended as a model for inventorying industrial heritage [17].

Functional Typology

The first step, according to the English methodical approach, is to generate a list of existing industries. Thus, the structure of BH's current classification of activities (KD BiH 2010) [9] served as the starting point for developing a list that resulted in functional types of industrial complexes and buildings on BH's territory. The classification is concluded at the level of activity areas (Bosnian: *područja*) and fields (Bosnian: *oblasti*) since further segmentation into activity branches (Bosnian: *grane*) and classes (Bosnian: *razredi*) is difficult due to major changes in technological advancement that have accompanied industrial growth over the last 140 years (such as life's modern demands, altered production techniques, development of new materials, etc.).

Based on the HAER classification[11] of industry used by Falser [8] in the UNESCO report, a list of functional typologies tailored to the presented historical context was compiled, resulting from systematic research of relevant spatial units within Bosnia and Herzegovina. As an outcome, the following activity areas, which are recognized in accordance with current industrial heritage definitions, were included on the list: B—Ore and stone extraction, C—Manufacturing industry, D—Electricity production and supply with gas, steam, and air conditioning, E—Water supply, waste water disposal, waste management and rehabilitation activities, H—Transport and storage, J—Information and communications.

To incorporate the social dimension into the identification and valorization of industrial heritage, an additional field of buildings classification in the domain of housing, education, culture, sports, and service activities is required. These functional typologies are included in the category of "Specialized structures and objects" in the aforementioned American classification system, which is also applied to the model of documenting industrial heritage in BH.

Architectural-Morphological Typology

The architectural-morphological typology of industrial structures was defined based on a comprehensive analysis of industrial buildings and complexes in Sarajevo [1] using all available sources, including collected archival material, historical plans and maps, as well as analysis of the current state and photo documentation made during field research. The distinctiveness of this analytical approach was reflected in the

[11] After attending some of the early industrial archeology conferences in England, Robert M. Vogel arranged a similar seminar at the Smithsonian on April 11, 1967, to initiate the American version of the movement, largely based on British practice. For instance, the British industrial classification system was introduced back to the United States to serve as the foundation for HAER's Inventory Program [18].

analysis of the original state as well as those industrial sites that no longer exist "in situ", providing the most accurate definition of this architectural typology.[12]

According to modern architetural heritage conservation methodology, industrial heritage is analyzed on three different levels: as a unit, in detail, and in relation to the environment, which Ifko [19] adapts to the aspect of the considered complex construction of industrial units and interprets at the levels of: a complex (micro-urban units), individual building within the complex (detail), and environment/landscape, i.e. the placement of the considered industrial building or complex within the wider environment. The selected methodological process allowed for the classification of the architectural-formal, micro-urban, and environmental-landscape aspects that constitute this architectural typology, i.e. its development on the examined spatial coverage across all key chronological sections. The important periods of industrial architecture development are defined in continuation.

In the **first industrial period of development, from 1878 to 1900**, the basic production-activity took place in buildings that can be divided into two types based on morphology: production buildings (one-story and two-story) formed in the spatial composition of the pavilion system and ground-floor factory halls that responded to the functional requirements of production technology for flexible and open work spaces. The structural and architectural features of these buildings adhered to previously established principles of technical architecture construction. They were simple in design, massive brick buildings with a rectangular ground plan and gable pitched roofs, whose façade bore the modest aesthetic elements of historicism, divided by pilasters and segmentally arched window openings. Building spans were solved by implementing a skeleton frame, consisting of initially wooden and then cast iron columns supported by steel beams. The factory hall spans were also solved, first with wooden and afterwards with steel lattice frames. The morphology of the complex, that is, its functional content and organization, was directly dependent on the basic production process of individual industries. Along with the main production facilities, there are power stations for production, boiler rooms or water-powered turbines, storage and workshop facilities, and associated residential facilities for owners of production facilities, as well as special buildings for workers' accomodation. Spatial relations within the first industrial complexes were adapted to the functional scheme of production technology, which would have been developed in the composition of the pavilion system (that is primarily influenced by the spatial possibilities of the region). The formal characteristics of the construction structures derived from the technological requirements of specific production units (warehouses, dryers, malt houses, kitchens, ovens, factory halls, boiler rooms, etc.[13]) indicated the utilitarian

[12] The choice of spatial coverage as a relevant sample for this systematic investigation was supported by the number of registered industrial facilities on the list of the Institute for Protection of the Federation of BH, which accounts for 27% of the total fund. Preliminary studies of BH's industrial development confirms that the area of Sarajevo as an administrative center was also the country's economic center through almost all historical periods (the Austro-Hungarian period and the period from 1945 to 1992), inheriting the typologically most diverse industrial building remains.

[13] In certain industrial complexes in Tuzla, goods were transported by horse-drawn carts, which required the construction of stables.

nature of these structures. The factory chimneys of the boiler rooms and the sawtooth roofs above the manufacturing halls are the most notable semantic aspects of industrial units during this period. When it comes to placing industry in the environment, the basic location conditions imposed during this first period of development were raw material sources (construction industry),[14] watercourses (food industry), railway infrastructure (tobacco, metal, and electricity production industries), and workforce proximity (textile, graphics industries). Industrial development has created a substantial influence on the city's urban spatial development.

The second stage of development, from 1900 to 1945, witnessed the introduction of a new form of production facility, with the creation of single factories that unified the entire technological process of production. In most cases, these industrial structures have the appearance of a more modest urban residential palace created primarily with historicism and art nouveau style aspects.[15] This time period also witnessed a major expansion of existing industrial sites, as well as the construction of new production facilities in established industrial zones located outside the defined city limits. This period of growth is distinguished by the use of reinforced concrete in the construction of industrial structures. With the introduction of electricity into these complexes, new functional building typologies emerged (substations and engine rooms for housing dynamo machines).

The third period of development, from 1945 to 1975, was characterised by increased industrialization, with new factories created in all previously established industrial areas and existing ones substantially expanded. Production facilities are designed using the functionalist design principle, which follows the technological process, with essential stylistic qualities of modern, mass construction after the Second World War. The production facilities' structures are made of a mix of masonry walls and frame systems, monolithic, as well as prefabricated, reinforced concrete structures. During this time, typical production halls consisting of prefabricated steel structures started to be built. The new industrial complexes built during this time period are situated near industrial railways. The composition of these complexes is primarily formed by a mixed pavilion-block spatial system, which is frequently conditioned by the plot's location and space limitations. The administrative and main production facilities are the key components of the building block (most often connected by a single entrance, and the entrance to production is achieved through the changing room block for workers). The complex's morphology, or functional units, contains, in addition to the previously stated ones: production, energy, storage, workshop, administration, social facilities for workers, laboratories, and garage facilities.

[14] Also applicable to extractive industries developed in the city of Tuzla.

[15] Trending modernist design with large window openings and simple geometrical compositions were also documented in this period, such as the engine house of the Spirit factory in Tuzla, designed in 1922.

Workers' apartments could still be found within industrial complexes and in their immediate vicinity. High factory chimneys and functionalist design are semantic features in the third period of development, which with their forms, such as shed roofs, the structure of the facade on which the load-bearing structural elements of the building are readable, and the arrangement and size of the openings, clearly hint at the utilitarian nature of these spatial entities.

The fourth period of development, from 1975 to 1992, is also the final phase defined by the time frame of this research, and it is marked by major modernization and expansion of practically all industrial facilities. The building structures' decorative characteristics also adhere to a purely utilitarian design. There are often production facilities reduced to simple volumes, with only the exterior membrane visible from all building elements. The development of industrial architectural style design based on postmodernist principles as well as brutalism principles is evident.

Social Infrastructure/Workers' Housing, Cultural, Sports, Healthcare and Educational Facilities

The social significance of architectural heritage is revealed in the material resources inherited from the past that reflect values, beliefs, knowledge, and customs of a certain group of people [20]. Given that the aforementioned features are constantly evolving and adapting to the historical context, the transformational character of architectural heritage's social dimension is apparent. As a result, integrating communities in the research, who have a direct or indirect relationship with an industrial building complex, or area is critical for analyzing the context in which the industrial heritage is currently located. Because the material dimension is measurable during field study, social infrastructure objects designed to meet the demands of the working community will be incorporated into the documentation process.

The transition of workforce from agriculture and crafts to the factory system had a direct impact on the formation of new types of social relations during industrial development, as a broad socioeconomic phenomenon. Both functional diversification and spatial distribution of amenities inside industrial zones can be studied to identify the earliest changes caused by industrializatio in BH's society. Thus, using Kreka as an example, the first industrial zone in Tuzla, an urban pattern was identified that clearly demonstrates the hierarchical relationship between the ruling and working classes of the population.

During the Austro-Hungarian period, the housing typology that emerged within industrial complexes and in the immediate vicinity of the production zone reflects the paternalistic attitude of the capitalist society, which was maintained until the development of workers' consciousness and the introduction of the socialist social order. During the socialist era, the participation of industrial enterprises in the construction of collective housing resulted in a rigorous separation of residential and production zones. Participation of industrial businesses in the construction of collective housing was advantageous in actualizing the vision of a modern city during the socialist era, whereas all public works were geared towards serving the requirements of society.[16]

Given that the early industrial facilities' production methods necessitated a staff trained to work on imported technology, the first wave of skilled workers' migration to these areas also led to cultural diversification in the first industrial zones. As a result, the first facilities for sports and cultural activities were constructed in the immediate vicinity of factories. The ideological shift in the next phase of industrialization of the socialist period created the need to build the first educational institutions with the aim of generating a professional workforce in both production and administrative branches of industry.

Defined social parameters will help to determine the interaction between industrial sites and the environment, as well as the extent to which industrial culture influences a region's social development.

Production Technology and Technical Equipment

Documenting industrial production equipment provides information on the technological level of the analyzed production plant as well as the major manufacturers of technological equipment at a certain historical period. This approach documents technological transfer, which was accelerated during Europe's second wave of industrialization, and depicts the process of developing, applying, and disseminating technological innovations [21].

Based on previous study, it is possible to conclude that BH's industrial history differs from that of other countries particularly in terms of imported production technology and machinery. Examples of industrial plants where knowledge and technology have been transferred demonstrate the impact of global industrialization trends in BH. Their technological value can be assessed via the perspective of technical and technological advances that transcend local and national levels.

---

[16] "In the socialist system, land management, urban planning, infrastructure and housing policy were focused on the needs of the social sector… Planning was a top-down process, where central planners determined the location of housing blocks and socially owned enterprises." [16]

Traditional craft skills that existed prior to the introduction of the factory system, that is, the division of labor within the factory system, can be identified as indigenous expression in domestic industrial output. As a result, the material relics of the economy's pre-industrial development can be a significant testament to sustaining the continuity of production and the development of skills in order to adapt to the challenges faced throughout history.[17]

Salt production in Tuzla is a representative example of industrial heritage in terms of progress in technology. The preservation of salt production material remains, ranging from Neolithic clay pot fragments for primitive evaporation to more complex vacuum evaporation technology, helps to understand the evolution of the salt industry and its impact on production capacity and local socioeconomic development.

Given that the early period of industrialization's reliance on imported technology conditioned the development of the initiative for a self-sustaining production chain during the socialist period, it is possible to find the first innovative technical and technological solutions created on BH's territory (coal industry—Kreka, Energoinvest, etc.).

Qualification Criteria

Although this article does not discuss the valorization of the industrial heritage itself, the following text will present the basic parameters of evaluation based on which it is possible to systematize their basic features, leading to the determination of micro-urban (those related to the industrial complex level), architectural-formal (those related to the individual objects level), and environmental (complex in relation to the environment) [1].

In this section, it is required to give the basic definitions of authenticity and integrity, which are recgnized as a basic criterion for the qualification of a property in UNESCO's Operational Guidelines for the Implementation of the World Heritage Convention.[18] [10]

Authenticity, sometimes known as an "authenticity test",[19] is related to the concept of truthfulness in the sense of providing true proof of something. Authenticity in the context of heritage is the truth of ideals determined by specific criteria. As a result,

---

[17] Example of Vareš Majdans—iron production, and salt production in Tuzla, protected as national monuments of BH.

[18] "Authenticity, integrity, and originality are not values in and of themselves, but rather qualifiers for all other particular values identified by individual authors. They are separated from the values in the UNESCO Operational Guidelines and related to the good as a state of authenticity and integrity." [10]

[19] Cultural property is considered authentic if its values have been explored truthfully and convincingly through various attributes (parameters) such as shape and design; materials and construction; use and function; traditions, techniques, and management systems, location and spatial context, languages and other forms of intangible heritage; spirit and feeling; and other internal and external factors [11].

specific features are associated to the authenticity test, which would accurately and convincingly convey the intended cultural values of the heritage asset. Specifically, the historical review of development at the national, regional, and individual levels will allow for the recognition of the authenticity of the potential heritage asset, as expressed through the attribute of the spatial context that attests to the material evidence of the history of a place, which is one of the qualifiers of the value of the cultural asset.

The measure of completeness and intactness of the natural and/or cultural landscape, i.e. the preservation of the property, its values, and its environment, is referred to as integrity.[20] Integrity is also connected with ascribed values, which are qualities that are valued on a specific heritage asset and assessed using specific attributes. In the case of industrial buildings and units, physical integrity is considered the most important consideration for selecting an asset worthy of future protection, according to the English conservation practice. This methodology defines physical integrity as the quantity of authentic building and technological structure preserved until the initiation of the valorization process [23].

4 Proposed Method of Documenting Industrial Heritage in BH

Based on the outcomes of the authors' research, two complementary record sheets will be presented in this section of the work, which are integrated in a unique methodological approach for a holistic consideration of all considered levels. The method of cross-references, proposed by the "Guidance on inventory and documentation of the cultural heritage"[21] [22], was developed to facilitate the establishment of connections between a group of objects and the environment/landscape.

The first example refers to comprehensive records of industrial systems, developed on the example of the salt industry in Tuzla. The second example refers to the record sheet for individual buildings and complexes prepared as part of the doctoral dissertation on the research of Sarajevo's industrial architectural heritage.

[20] The following characteristics are implied by the assessment of integrity: completeness (physical completeness and compactness); continuity of use; complete representation of features and processes that convey the meaning of the asset; intactness of condition; social-functional integrity; and visual integrity [11].

[21] The provided guidelines were adopted by the Commission for the Preservation of National Monuments of Bosnia and Herzegovina to create a database compatible with European standards during the Integrated Rehabilitation Project Plan/Survey of the architectural and archaeological heritage (IRPP/SAAH) project [4].

Record Sheet for Industrial Systems

The complexity of the industrial heritage will be considered as a type of industrial system consisting of various site activities incorporating mutually connected technological and historical components [6]. The purpose of the **record sheets for industrial systems**, adapted from the case study of industrial heritage in Tuzla, is to determine the extent and boundaries of the protected area based on established values, as well as to define the relationship of protected buildings with the environment and related industrial sites, to facilitate heritage resource management.

The record sheet is organized into nine thematic fields that provide information about: name and associated buildings/sites/fixtures/machines, location, functional typology, historical names, key historical figures and organizations, materials and types of construction systems, general physical and legal preservation state, and additional observations (Appendix—Table 1).

The first thematic field aims to connect individual buildings located within the studied area (individual building-complex-environment). To enable precise identification of territorial units, the proposed coding consists of the postal code, the name of the settlement, and an abbreviation of the name of the industrial plant undergoing evaluation. The applied coding system also serves to establish a link with the record sheets of individual buildings, which will include the code, allocated to the associated industrial system, in the reference number.

Given the uniqueness of BH's historical context and the frequent alteration of names of industrial enterprises resulting from ideological shifts, the original names identified through a preceding historical study are listed in the field of historical periodization. As a result, a starting point for additional research of archival resources will be provided.

Historical figures and organizations associated with the development of the industrial heritage asset, depicted in the following field, will be applied to assess historical and technological relevance. The proposed field is based on the research of relevant sources such as monographs, books, articles, and archival documentation in which the manufacturers of technological equipment, the associated industrial enterprises, and the architect, in case the facility has a distinctive architectural expression, are listed.

General information on the main building materials and types of structures provides an understanding of the wide range of buildings that existed in the studied area. This field, however, is elaborated in detail in the record sheets for individual buildings and complexes.

Record Sheet for Individual Industrial Buildings and Complexes

Individual building and complex record sheets should give the basic data required for understanding and final evaluation of this specific typology of architectural heritage. The record sheet itself is divided into numerous sections in order to carry out an analysis of the main characteristics of industrial buildings and complexes, from which it will be possible to determine their architectural-formal, micro-urban, and environmental-landscape values.

The first section contains general information about the industry in consideration (name and location), as well as its original and current purpose. Data linked to the recorded industry's original or current purpose are required to determine the authenticity attribute connected to the use and function of the observed entity. To determine the attribute of authenticity, which relates to the location and geographical context, as well as the attribute of integrity, which refers to functional continuity, the period of construction and key historical periods of development of the observed site (building or complex) are required. Defining the time period during which a specific industry existed (or continues to exist) might, on the other hand, provide evidence regarding the historical, social, and economic aspects of a given geographical location, which surely contributes to the relevance of the site in consideration.

The second section discusses the recorded industry's spatial and functional aspects. There is a clear distinction in this section of the record sheet between Table 2 (Appendix) and Table 3 (Appendix), and it depends on whether it is an individual industrial object or the entire industrial complex. The spatial system of the complex's construction, the number of buildings in the complex, and the basic purpose (and preservation) of individual buildings, observed according to the functional typology (divided into: production, administration, storage, service, energy, residential, and so on), are data that should enable a functional-morphological analysis of the observed spatial objects and complexes, on the basis of which the authenicity of the site can be assessed, expressed by attributes related to use and function, as well as form and design. Following an analysis of the studied industry's location in the urban landscape, depending on whether the building or complex is located in the narrower urban zone or within the formed industrial zone, as well as an analysis of its (original and current) transport connectivity (railroad, road, or both), it will be possible to determine specific characteristics related to the micro-urban and landscape values of the industry.

The following two sections discuss the property's building construction and architectural properties, that is, data on each individual object within the considered complexes that will allow functional, constructive (structural), and stylistic analysis in relation to the architectural-formal values of the industrial heritage.

From the viewpoint of authenticity and preservation, the basic structural qualities exhibited through vertical, horizontal, and roof construction, i.e. their structure and material, are examined. It is required to establish the extent to which the original structures have been preserved (in their original state, partially modified, completely modified, or not possible to determine), as well as the degree of preservation (good, bad, dilapidated and not possible to determine).

In the case of architectural elements, this record sheet determines how much of their original features have been retained by assessing the authenticity of the floor plan layout, volume, and (stylistic) treatment of the facade.

Given that authenticity is a sum of significant historically determined characteristics from the original to the current state, created as a result of various transformations that occurred over time [23], the values of these changes (which can be: carried out in accordance with the original state, neutral, incongruous, and finally they can add a certain value to the original state) in the phase of recording the encountered state will only be documented in additional notes. A column for supplementary available information (descriptions and historical references) as well as images from field visits (exterior and interior) and a map of the place are also included within the sheet.

5 Conclusion

Considering that the record sheets provided in this paper are the result of a detailed analysis of industrial heritage parameters in the Sarajevo and Tuzla areas, further analytical studies covering other parts of the country may disclose new information that could be beneficial in the data collection process. Therefore, it is important to note that the industrial heritage documentation process is not limited to the five identification parameters discussed in this study. However, with the proposed functional classification, identification of architectural-morphological typology, and definition of the scale and scope of industrial heritage, a significant step has been taken toward recognizing industrialization's unique contribution to Bosnia and Herzegovina's socio-cultural and historical development. The proposed documentation technique aims to aid in the integrated conservation of industrial heritage by holistically documenting individual industrial buildings, complexes, and systems that are intrinsically related to the country's modern communities and cities. As a result, the proposed record sheets were designed to serve as tools for further assessing the historical, socio-cultural, and scientific-technological value of the material remains of industrial development in the context of Bosnia and Herzegovina.

Appendix

See Tables 1, 2 and 3.

Table 1 Industrial heritage system record sheet

<table>
<tr><th colspan="12">Names and CR (Cross-References)</th></tr>
<tr><td colspan="4">Reference Code:</td><td colspan="8">e.g. 75000SOLANA-S</td></tr>
<tr><td colspan="4">Current name:</td><td colspan="8">e.g. Solana</td></tr>
<tr><td colspan="4">CR to individual buidlings:</td><td colspan="8"></td></tr>
<tr><td colspan="4">CR to fixtures/machines:</td><td colspan="8"></td></tr>
<tr><td colspan="4">CR to archeological sites:</td><td colspan="8"></td></tr>
<tr><td colspan="4">CR to associated industrial sites:</td><td colspan="8"></td></tr>
<tr><th colspan="12">Location</th></tr>
<tr><td colspan="2">Address:</td><td colspan="10"></td></tr>
<tr><th colspan="12">Functional typology</th></tr>
<tr><td colspan="2">B</td><td colspan="2">C</td><td colspan="2">D</td><td colspan="2">E</td><td colspan="2">H</td><td>J</td><td>SB</td></tr>
<tr><td colspan="12">B - Ore and stone extraction, C - Manufacturing industry, D - Electricity production and supply with gas, steam, and air conditioning, E - Water supply, waste water disposal, waste management and rehabilitation activities, H -Transport and storage, J -Information and communications, SB – Special buildings for the industrial community</td></tr>
<tr><th colspan="12">Historical names of the industrial site</th></tr>
<tr><td colspan="3">Before 1878</td><td colspan="3">1878-1918</td><td colspan="3">1918-1945</td><td colspan="3">195-1992</td></tr>
<tr><td colspan="3"></td><td colspan="3"></td><td colspan="3"></td><td colspan="3"></td></tr>
<tr><th colspan="12">Key historical figures and organizations</th></tr>
<tr><td colspan="4">Name(s)</td><td colspan="4">Role</td><td colspan="4">Year/period</td></tr>
<tr><td colspan="4"></td><td colspan="4"></td><td colspan="4"></td></tr>
<tr><td colspan="4"></td><td colspan="4"></td><td colspan="4"></td></tr>
<tr><td colspan="4"></td><td colspan="4"></td><td colspan="4"></td></tr>
<tr><td colspan="4"></td><td colspan="4"></td><td colspan="4"></td></tr>
<tr><th colspan="12">Materials and construction systems</th></tr>
<tr><td colspan="2">Materials:</td><td colspan="2">brick</td><td colspan="3">cast iron</td><td colspan="3">steel</td><td colspan="2">reinforced concrete</td></tr>
<tr><td colspan="2">Construction systems:</td><td colspan="2">load bearing masonry wall</td><td colspan="3">frame construction</td><td colspan="3">mixed</td><td colspan="2">prefabricated</td></tr>
<tr><th colspan="12">Current physical and legal state of preservation</th></tr>
<tr><td colspan="3">Physical state:</td><td colspan="3"></td><td colspan="3">Legal status:</td><td colspan="3"></td></tr>
<tr><td colspan="12">A - Good, B - Bad, C - Dilapidated, D - Not possible to determine
I – National monument, II - Listed at Federal/RS level, III – Listed at Cantonal level, 4 - Not listed</td></tr>
<tr><td colspan="2">Notes:</td><td colspan="10"></td></tr>
</table>

Table 2 Industrial heritage complex record sheet

Record sheet **Ref.code:**	*Name:*			
	Address:			
Current function:		*Original function:*		
Year or period of construction:		*Site development periods:*		
I – Ottoman period, II – Austro-Hungarian period, III – period between WW1 and WW2, IV - period between 1945 and 1992				
Spatio-functional properties				
Total number of buildings in the complex:		*Morphological type:*		
(map)		*Purpose of the building:*		
		current	original	
Location:				
Transport connectivity:				
Notes:				

Ref.number:								*Year of construction:*															
Building construction properties																							
Vertical construction								*Horizontal construction*								*Roof*							
preservation				authenticity				preservation				authenticity				preservation				authenticity			
A	B	C	D	1	2	3	4	A	B	C	D	1	2	3	4	A	B	C	D	1	2	3	4
A - Good, B - Bad, C - Dilapidated, D - Not possible to determine *1 - Original state, 2 - Partially modified, 3 – Completely modified, 4 - Not possible to determine*																							
Architectural properties																							
Floor plan layout								*Volume*								*Façade*							
preservation				authenticity				preservation				authenticity				preservation				authenticity			
A	B	C	D	1	2	3	4	A	B	C	D	1	2	3	4	A	B	C	D	1	2	3	4
1 - Original state, 2 - Partially modified, 3 – Completely modified, 4 - Not possible to determine																							
Description: Photographs (map, interior i exterior, machines, and similar):																							
Notes:																							

Table 3 Industrial heritage building record sheet

Record sheet **Ref.code:**		*Name:*	
		Address:	
Current function:		*Original function:*	
Year or period of construction:		*Site development periods:*	
I – Ottoman period, II – Austro-Hungarian period, III – period between WW1 and WW2, IV - period between 1945 and 1992			

Spatio-functional properties					
Individual building:					
Functional preservation:	A	B	C	D	E
A – Original function,, B - Partially modified, C – Different function, D – Abandoned, E- Not possible to determine					
Position of the building:	*e.g. Urban area*				
Traffic connectivity:	*e.g. City roadway*				

Building construction properties																							
Vertical construction								*Horizontal construction*								*Roof*							
preservation				authenticity				preservation				authenticity				preservation				authenticity			
A	B	C	D	1	2	3	4	A	B	C	D	1	2	3	4	A	B	C	D	1	2	3	4
A - Good, B - Bad, C - Dilapidated, D - Not possible to determine *1 - Original state, 2 - Partially modified, 3 – Completely modified, 4 - Not possible to determine*																							

Architectural properties																							
Floor plan layout								*Volume*								*Façade*							
preservation				authenticity				preservation				authenticity				preservation				authenticity			
A	B	C	D	1	2	3	4	A	B	C	D	1	2	3	4	A	B	C	D	1	2	3	4
1 - Original state, 2 - Partially modified, 3 – Completely modified, 4 - Not possible to determine																							

Description: Photographs (map, interior i exterior, machines, and similar):
Notes:

References

1. Pličanić M (2021) Prilozi za metodološki postupak zaštite graditeljskog naslijeđa kroz aktivnu zaštitu industrijskog naslijeđa grada Sarajeva (PhD thesis). University of Sarajevo—Faculty of Architecture
2. Veljković T (2022) Endüstri miras alanlarının bütünleşik korunması için bir öneri: Tuzla (Bosna Hersek) (PhD thesis). Mimar Sinan Fine Arts University—Department of Restoration and Conservation
3. Krstić B (2010) Spomenička Baština svjedočanstvo i budućnost prošlosti—tematska enciklopedija. Synopsis, Sarajevo
4. Commission to Preserve National Monuments of BH: Report on assessment of the architectural and archaeological heritage-Project coordinator: Mirela Mulalić Handan. shorturl.at/AMOZ6, last accessed 2012/02/22
5. TICCIH: The Nizhny Tagil Charter for the industrial heritage. https://ticcih.org/about/charter/, last accessed 2023/01/26
6. TICCIH: Joint ICOMOS—TICCIH Principles for the Conservation of Industrial Heritage Sites, Structures, Areas and Landscapes. https://ticcih.org/about/about-ticcih/dublin-principles/, last accessed 2023/01/26
7. CoE: Rec(90)20 on the protection and conservation of the industrial, technical and civil engineering heritage in Europe. shorturl.at/BM458, last accessed 2023/01/26
8. Falser M, Yang M (2023) Industrial heritage analysis. World Heritage List and Tentative List, Global Strategy Studies, UNESCO, World Heritage Center, Asia-Pacific Region. https://whc.unesco.org/archive/ind-study01.pdf, last accessed 2023/01/26
9. Agencija za statistiku Bosne i Hercegovine, Klasifikacija djelatnosti BiH (2010). https://fzs.ba/wp-content/uploads/2016/06/KDBiH2010-bosanski.pdf, last accessed 2023/01/26
10. UNESCO: The Operational Guidelines for the Implementation of the World Heritage Convention, https://whc.unesco.org/en/guidelines/, last accessed 2023/01/26
11. Commission to Preserve National Monuments of BH: Principi i smjernice za očuvanje nacionalnih spomenika. http://kons.gov.ba/Publication/Read/principi-i-smjernice-za-ocuvanje-nacionalnih-spomenika?lang=bs, last accessed 2023/01/26
12. Marić Đ (1991) Industrijalizacija Bosne i Hercegovine—geografska studija savremenog razvoja. Geografsko društvo Bosne i Hercegovine, Sarajevo
13. Hrelja K (1974) Razvoj industrije u Bosni i Hercegovini do drugog svjetskog rata. Acta historico-oeconomica: časopis za ekonomsku povijest 1(1):17–35
14. Kamberović H (2007) Karakteristike društva u Bosni i Hercegovini neposredno nakon Drugog svjetskog rata. In: Editor, Filipović, M.: Naučni skup—Bosna i Hercegovina prije i nakon ZAVNOBIH-a, Knjiga 37, pp 214–226. Akademija nauka i umjetnosti Bosne i Hercegovine, Sarajevo
15. Hadžibegović I (2004) Bosanskohercegovački gradovi na razmeđu 19. i 20. stoljeća. In: Editor, Kamberović, H.: Institut za istoriju, Contributions No.33, Sarajevo
16. Evropska inicijativa za stabilnost—ESI. Uprava i demokratija u BiH: Post-industrijsko društvo i autoritarno iskušenje [1. dio—Procjena upravljanja u BiH]. ESI, Berlin—Sarajevo (2004). https://www.esiweb.org/publications/post-industrijsko-drustvo-i-autoritarno-iskusenje, last accessed 2023/03/29
17. Rogić T (2005) K cjelovitom pristupu valorizaciji industrijskih zgrada: funkcija, forma, prostor. In: Editor, Smokvina, M.: I. međunarodna konferencija u povodu 150. obljetnice tvornice torpeda u Rijeci i očuvanja riječke industrijske baštine. Pro Torpedo, Rijeka
18. Palmer M (1998) Industrial archaeology: principles and practice. 1st edn. Routledge, London. https://doi.org/10.4324/9780203022993
19. DeLony E (1999) HAER and the recording of technological heritage: reflections on 30 Years' Work. IA. The Journal of the Society for Industrial Archeology, pp 5–28
20. Ifko S (1999) Industrial architectural heritage—scheme for a methodology of evaluation. Varstvo spomenikov 38(1):123–151

21. CoE: Council of Europe Framework Convention on the Value of Cultural Heritage for Society (Faro Convention). https://rm.coe.int/1680083746 last accessed 2023/01/26
22. Douet J (ed) (2016) Industrial heritage re-tooled: the TICCIH guide to industrial heritage conservation, 1st edn. Routledge, New York
23. Rogić T (2007) Converted industrial buildings: where past and present live in formal unity (PhD thesis), Technische Universiteit Delft
24. Bold JA (Ed.) (2009) Guidance on inventory and documentation of the cultural heritage. Council of Europe
25. Chabbouh—Akšamija L (2009) Autentičnost ruralnog graditeljstva kao preduvjet aktivne zaštite: doktorska (PhD thesis). University of Sarajevo—Faculty of Architecture

Neglecting the Protection of Urban Cultural Heritage; Examples of Jajce Military Barracks and Residence Konak

Dizdarević Lejla

Abstract The study presents the current situation of urban cultural heritage of the City of Sarajevo. The cultural and historical heritage of the urban core of Sarajevo is a fundamental element of the attractiveness of the city and the reason for the visit of numerous tourists from all over the world. The main goal of this research is to present the identified problems and omissions in the processes of cultural heritage restoration and protection. Thus, the main research question was: Has the cultural and historical heritage been sufficiently restored and protected? In order to get an answer to the research question, the in-depth interview method was used, and the interviewees were representatives of institutions responsible for issues, regulation and implementation of the process of protection and restoration of cultural and historical heritage. In order to complete the research, an analysis of the institutional and legal framework for the issues of cultural and historical heritage was carried out. The collected data as well as the analysis of the answers made it possible to identify the problems that lead to the negligence of the cultural heritage of the urban core of the city. In accordance with the identified problems, the necessary recommendations were proposed.

Keywords Urban cultural heritage · Sarajevo city · Protection · Restoration · Institution · Legal framework

1 Introduction

Urban regions represent regions of great importance, both from the economic, social and cultural aspects. Cities are considered urban regions together with their surroundings [1]. The greatest significance of these regions, economically speaking, is that they greatly contribute to the national economy through the concentration and representation of various activities. Urban regions with a wide range of offers create opportunities for a large number of economic activities, and therefore a greater

D. Lejla (✉)
University of Sarajevo, 71000 Sarajevo, Bosnia and Herzegovina
e-mail: lejla.dizdarevic@unsa.ba

M. Arslanagić-Kalajdžić et al. (eds.), *Interdisciplinary Advances in Sustainable Development II*, Lecture Notes in Networks and Systems 804,
https://doi.org/10.1007/978-3-031-46692-2_12

opportunity for employment and an increase in income. Looking from a geographical point of view, urban regions can be considered to represent a source of both problems and opportunities. This is mainly reflected in higher incomes, access to technologies, infrastructure, better social conditions, etc. On the other hand, urban regions around the world, especially in underdeveloped countries, are an example of concentrated poverty, inadequate infrastructure elements, environmental problems and various sociological problems and social differences [2]. Culture and cultural-historical heritage is a factor of tremendous importance for these regions that makes them different from each other. The culture of urban regions is the culture of cities, formed by the population of those cities, firmly devoted to the specific lifestyle [1].

The main goal of this research is to present the identified problems and omissions in the processes of cultural heritage restoration and protection. For the purposes of this research, the main research question was formed: Has the cultural and historical heritage been sufficiently restored and protected? A qualitative method was used, more precisely, an in-depth interview was applied, which enabled the identification of the source of the problem. In addition, an analysis of legal regulations was applied and two cases of objects of cultural importance were presented, during which it was necessary to analyse archival material, make observations, and photograph the objects.

The City of Sarajevo, as the capital of Bosnia and Herzegovina, is one of the cities that occupy an important place on the cultural map of Europe. The cultural heritage, especially the architectural one, is a reliable guide through the history of Sarajevo and in addition to the natural heritage it is one of the main assets of the tourist offer. The cultural and historical heritage of the city dates back to ancient times. The heritage located at the urban core of the city is the result of the influence of different cultures, where one can find architectural works from the Middle Ages, Ottoman reign and the Austro-Hungarian period, which eventually created unique multinational and multi-religious environment [3]. Saved and preserved cultural heritage of Sarajevo is an important factor in its development, establishing the attractiveness and competitiveness of the entire region and potential economic growth. Certainly, this is not possible if there are no instruments and measures to encourage the process.

Number of historical monuments located at the urban core of the city have been neglected to a large extend in terms of restoration and protection, although having the status of monuments of national importance. For the purpose of this study two examples of historical monuments located at the urban centre of Sarajevo City will be presented in terms of their importance, restoration, protection and their current state. Those two monuments are Konak (residential villa) and Jajce kasarna (military barracks). Furthermore, for the purpose of a comprehensive analysis of the situation, the institutional and legal framework from the domain of cultural and historical heritage will be presented, as well as the results of in-depth interviews with representatives of official institutions in respect to the research question.

2 Meaning and Importance of Urban Cultural Heritage

Cultural historical heritage is considered a link to the past [4]. This connection is reflected in efforts to pass on all that has been inherited to the current and future generations, and includes all forms of cultural historical heritage [5]. Thus, for example, Tweed and Sutherland [6] describe architectural heritage as a very important part of the overall cultural heritage of cities. They view cultural heritage as a very broad category that contains a very diverse collection of phenomena, mainly because it is recognized as an important factor in the sense of belonging and cultural identity of the population, which is linked by locus to the geographical area where the heritage artefacts are found. Other authors argue that the preservation of cultural heritage contributes to the creation of different values which in turn, significantly influence the creation, revival and preservation of national identity, the importance of space, collective memory and social unity at all levels [7–9]. In fact, cultural heritage is a complex concept that is changing in the context of time, and which includes a combination of cultural, aesthetic, symbolic, spiritual, historical and economic values.

Cultural heritage is an indispensable part of urbanization process, which is a process whose beginning is linked to the appearance of urban centers, mainly cities, as early as the thirteenth century [10]. This process, which started a long time ago, is still popular, so some call it today's trend. As the term "urban" is closely associated with cities, so everything that bears this prefix is brought into the same relationship. Whether it is an urban region, urban population, urban culture or even urban tourism, in each of these situation it pertains to the city, or what is happening in the city. Urbanization is an unstoppable process, which has affected the big cities of developed countries, but it is also present in the cities of developing countries, which are largely unprepared to deal with the consequences of urbanization. Quality management of urbanization may bring prosperity to millions of people or result in the creation of poverty-stricken urban regions. In the process of urbanization, cultural heritage is often found between different actors and not so rarely under threat of damage or even destruction. Such situations are mostly related to the urban development of cities or regions [11].

Cities may be considered as to represent the mirror of the society within which they are located, while at the same time they reflect the image of the society that is located in them. Today, cities are considered places of cultural exchange, mostly due to the presence of tourism industry. Also they are considered a source of education and knowledge with a great influence on the creation or change of ideas, values, beliefs and ways of life. They provide great opportunities for participation in social and political spheres to all groups of population [12]. According to Hewison [13], cultural and historical heritage is an important tool in all development processes, so its promotion is an important part of urban development. And Al-Hagla [14] agrees with the previous views, claiming that culture in all its forms represents one of the most important elements in the process of urban development and that the importance of cultural historical heritage exceeds the value of monuments and handicrafts.

In the context of urban spatial planning and design, cultural and historical heritage is a very important element, which is explained by certain characteristics and features it possesses. Thus, Vasi [15] claims that each individual form of architectural heritage is built to respond to different needs of people while at the same time providing a sense of belonging and identity and having different architectural values that should be preserved because ignoring them is contrary to the principles of urban development planning. Furthermore, he states that as time passes, the value of cultural and historical heritage increases, and its protection and care is one of the tasks of urban development. Architectural heritage as well as other forms of heritage represent, as he states, part of the public values of the city, and in that case it is not only important for the organization and establishment of urban places, but also in the context of the public life of the city. Then, he emphasizes that the cultural heritage of cities is not only those objects located in city centers or those monuments that are linked to certain events, but that it includes all places that generally have their own story and value, which contributes to the diversity of cities. As the last characteristic of cultural heritage, he lists cultural values that have become an important source in the creation of economic values in the modern world [15].

In addition to the fact that cultural heritage is widely recognized as an important factor in creating the uniqueness of cities and its role in improving their competitiveness [6, 16, 17], it is also considered an important element of sustainable development. According to Nijkamp and Riganti [18] cultural heritage is a social, economic and cultural resource, and it also has its role in politics. They emphasize the necessity of adequate management of cultural and historical heritage and its valorization in the process of urban development, while claiming that urban development in which special attention is not paid to adequate management of cultural assets cannot be sustainable in the economic, cultural and social sense. Special attention is paid to the evaluation of the cultural heritage in order to achieve its valorization, the final result of which is the improvement of the existing ones and the creation of new values.

Cultural heritage is significant asset of the community, society, and the world, and represents its cultural capital. The theory of "*cultural capital*" was introduced by Bourdieu [19] in his work in 1986, explaining it through the prism of sociological and ideological settings, while the importance and its application to culture and everything it includes, including architectural heritage, is presented by Throsby [20]. He presents the importance of the theory of cultural capital, explaining its application to culture and cultural assets. Among other things, he also explains that the theory of social capital, which is described in Bourdie's work, in certain aspects, is identified with the theory of human capital, as one of the most important economic theories. This is supported by the statements that in the economic framework of defining human capital, culture is counted as its component.

Cultural heritage is one of the most important topics of today's world institutions, e.g. Council of the European Union; International Centre for the Study of the Preservation and Restoration of Cultural Property; United Nations Educational, Scientific and Cultural Organization; International Council on Monuments and Sites; International Council of Museums, etc. considering it a strategic choice of 21 century. These institutions consider that cultural heritage has an important economic impact, as part of culture, creative sectors and especially through cultural tourism, which, among other things, contributes to urban development. Based on this, the Council calls for the promotion of the importance of cultural heritage, increasing its role in sustainable development with a focus on urban and rural planning.

3 Methodology

The main aim of this research was to obtain data and information that will enable the identification of the source of the problem related to the protection of the cultural heritage of the City of Sarajevo. The aim of the research defined the research approach and the application of the qualitative method was chosen. For the purposes of this work, various methods were used like in-depth interview, observation method, archival research, content analysis, case study and photography.

The presentation of relevant literature required a detailed analysis of the content, that is, an analysis of the literature that is relevant and in the context of the research problem. This was preceded by the collection and analysis of literature related to the significance of cultural heritage in the context of urban development, as well as the analysis of the content of the legal framework regulating issues of cultural heritage. In addition, a review of archival documentation related to selected objects of cultural importance presented in the work was carried out. The choice of a qualitative approach, which was caused by the very nature of the research problem, required the application of the in-depth interview method.

An in-depth interview was conducted among respondents who were chosen to represent institutions that are most responsible in the field of cultural heritage, and as well urban development. When choosing respondents, special attention was given to the function they perform in the institution where they are employed, so that their views and opinions represent the views of the institution. In this sense, leading representatives were elected (see Table 1).

The in-depth interview consisted of an introductory part where the aim and purpose of the research was presented, and the main part which included the main research question, namely: Is the cultural-historical heritage sufficiently restored and protected? As it was an in-depth interview, the interviewees were free to express their views and opinions in relation to the question and the topic being discussed.

After the information have been collected through an in-depth interview, a transcription, coding and mapping was made with focus on the research question. In the part of the text where results are presented, the answers or entire quotes are shown.

As already stated, the legal framework related to the regulation of the area of cultural heritage was analyzed (e.g. Law on the Protection and Use of the Cultural and Historical Heritage of SRBiH; Law on the Implementation of Decisions of the Commission for the Protection of National Monuments; Law on Spatial Planning and Land Utilization in the Federation of Bosnia and Herzegovina; Law on Spatial Planning of the Canton of Sarajevo; Law on the Protection of Cultural Heritage of the Canto of Sarajevo).

Table 1 Institution, field, and job position of participants

	Institution	Field	Job position of participants
1	Commission to Preserve National Monuments	CH	President
2	Federation Ministry of Culture and Sports	CH	President of the institute for the protection of monuments
3	Cantonal Institute for the Protection of Cultural, Historical and Natural Heritage Sarajevo	CH	Senior expert for cultural heritage
4	Cantonal Institute for the Protection of Cultural, Historical and Natural Heritage Sarajevo	CH	Expert for the protection of architectural heritage
5	Sarajevo Canton Development Planning Institute	CDP	Coordinator of the social development and infrastructure sector
6	City of Sarajevo—City administration	CDP	Expert in the urban planning service
7	Cultural Heritage Without Borders—NGO	CH	President
8	Association of Architects of Bosnia and Herzegovina	CDP	Member

Note CH—Cultural heritage; CDP—City development planning

In the research process, the case study method was also applied, and presentation of two objects of cultural importance situated in the urban core of the City of Sarajevo was chosen. The case study included the analysis of archival materials, that is, the acquisition of literature related to selected objects of cultural heritage (historical data), and data that record activities and works on objects from the earliest period. These two examples were chosen for several reasons, in particular because they represent two significant objects in the process of urban development of the City of Sarajevo, they are located in the very urban core of the city, they have great potential to be used for tourist purposes and although they are objects of national importance, nothing was done to protect and restore these objects for a long period of time. In addition to these examples, in the urban core of the city there are other buildings that share the same fate, as one may say, for example the medieval fortress of Bijela Tabija (white bastion), or the archaeological site where the foundation of Isa-bey Ishaković's tekke is located, then the Goat's bridge and other objects. Nevertheless, for the purposes of this research, it was considered appropriate to choose two examples that will best show the attitude towards objects that are not only of cultural and historical importance, but are objects of national importance, as well as.

The process ended by presenting the state in which the selected objects are today. In addition, exhaustive observation of the state of cultural heritage was applied over a longer period of time with a focus on selected heritage objects with the aim of gathering information about changes, works or similar activities on the objects. That activity was monitored using the photography method.

4 City of Sarajevo; Presentation of Two Urban Cultural Monuments

Sarajevo is the capital of Bosnia and Herzegovina and the largest urban, economic, cultural and traffic center of the country. The history of this city goes back to ancient times, which is confirmed by the sites of the first human settlements just outside the city. The Slavs came to this area in the seventh century after the period of Neolithic, Illyrian and Roman rule. In the thirteenth century, Sarajevo was part of an independent Bosnian state called Vrhbosna, as evidenced by a charter from this period and unique in the world tombstones known as "stećci". After the arrival of the Ottomans in the middle of the fifteenth century, the formation of the City of Sarajevo on the banks of the Miljacka river began, whose founder is considered to be Isa-beg Ishaković, who issued the order to establish the city in 1462. The current name of the city was mentioned for the first time in 1455. Sarajevo got its name from the Slavicized abbreviation of the Turkish words "saraj" (courtyard) and "ovasi" (field). Shortly after the arrival of the Ottomans and the founding of Sarajevo, the period of construction, development of crafts and trade begins, which soon turns Sarajevo into one of the richest cities in Europe at the time. This period records the construction of numerous religious institutions of all denominations and the arrival of Sephardic Jews, who contribute to the already present multi-religious community. The oriental style of construction from this period is evident even today, which significantly contributes to the authenticity and attractiveness of the city [21].

The modern era begins with the arrival of Austro-Hungarian rule in 1878. Forty years of Austro-Hungarian rule brought significant changes. A lot of attention was given to the construction of numerous educational and cultural institutions as well as infrastructure, which complemented the existing oriental building style with the European one. This modern European influence also left its mark on the customs, manners of business operations and other aspects of civilizations, which is evident even today. As a result of these different influences, in Sarajevo, rarely or not found anywhere else in the world, there are religious buildings of all four major religions in just a few hundred square meters. That is why Sarajevo is often called the European Jerusalem [22].

Sarajevo City acquired its first urban functions even before the fall of the medieval Bosnian kingdom [3], however official development of Sarajevo began in the fifteenth century, while the city experienced the greatest territorial and economic development in the sixteenth century, with the arrival of the Ottomans. In the first years of the sixteenth century, significant progress was recorded in Sarajevo, in the growth of the population, followed by economic and urban development. At that time, Sarajevo was the largest city in Bosnia and one of the most developed in the Balkans [23]. The most significant period in the urban development of Sarajevo is the period when the Bosnian Sandžakbeg Gazi Husrev-beg (1521–1541) became the governor of Sarajevo, which resulted in the raising of numerous endowments and Sarajevo grew from a village to a city. This contributed to Sarajevo becoming the most important economic and trade center in Bosnia and beyond [24]. Numerous buildings built in that period have retained their basic function for a long time, and some have become museums, and today they represent the indispensable content of tourist tours, especially because they are located in the urban core of the city.

With the arrival of Austro-Hungarian rule in 1878, major changes took place in the urban context. Instead of a "mahala", as an urban unit in the Turkish period, in the Austro-Hungarian period it is a street. Sarajevo continues to develop in an urban sense and the city has undergone a transformation from traditional through modern and until western style. A particular role in this process was played by Benjamin Kállay von Nagy-Kálló, who held the position of joint minister of finance between 1882 and 1903. It is recorded that he had organized the remodeling of the urban core and that by his order the largest religious, cultural and educational institutions of all religious communities were located in the immediate vicinity, around the central square, which makes Sarajevo a particularly attractive city even today [21]. The end of Austro-Hungarian rule marked the end of the city's development period, but that period resulted in a rich legacy in terms of education, culture and science.

The assassination of the Austrian heir to the throne Franz Ferdinand and his wife Sofia marked the beginning of the First World War, which resulted in mass killings and causing immeasurable damage to the city [21]. The development and urbanization of Sarajevo also declined significantly in the period between the two world wars. The period after 1945 was marked by renovation, and the first Urban Plan of Sarajevo was presented in 1947. The development of the city in the period from 1945 to 1953 was described in the work of Taubman [25].

The big construction project of Sarajevo followed the selection of the host of the XIV Winter Olympic Games by the International Olympic Committee in 1978. By choosing Sarajevo to host the Olympics, an ambitious construction campaign was launched. Based on the Long-term Social Plan of the City of Sarajevo for the period 1986–2000 and Spatial Plan of the City of Sarajevo for the period 1986–2000/2015 the Urban Plan of the City of Sarajevo for 1986–2000/2015 was drawn up. In the years that followed, major socio-economic, political, and administrative changes took place.

Sarajevo suffered tremendous destruction in the period 1992–1995. As a result of the war, a great deal of buildings, residential, religious, administrative, educational, cultural and other, suffered great damage, and some were completely destroyed, so it was necessary to proceed with their reconstruction. During that period, there was also a large expansion of illegal construction, which needed to be regulated. In 2006, the Canton Sarajevo Assembly passed the Decision on joining the drafting of the Sarajevo Canton Spatial Plan for the period 2003–2023. In the Plan, the cultural and historical heritage was recognized as the tourist potential of the Sarajevo Canton. A special aspect is on the urban core of the city where the largest number of cultural heritage buildings are located including the two monuments which will be presented in the following text.

Residential Villa "Konak"

Sarajevo got its name from the palace, which was built there by the first governor in Bosnia, Isa-beg Ishaković, which was recorded in his census in 1455. The Turkish governors had their headquarters in Sarajevo on three occasions, and the residences where they stayed were called palaces or lodgings [26]. One such palace was built near the Emperor's Mosque by Isa-bey Ishaković, and later there were more of them. According to the 1455 census, it can be concluded that the manor already existed. The residence discussed here, as an architectural entity, stands out for its historical significance, but also for its aesthetic and ambient values.

The original palaces, residences and palaces changed their purpose to a large extent later on, and a huge part of them perished in a fire in 1697. New ones were built, but as Kreševljaković [26] writes, they lagged behind those from the seventeenth century in terms of architecture. Fadil Pasha's lodging, which was located in the place where the Franciscan monastery of St. Ante is, was used for residential purposes, and Omer Pasha Latas stayed there for less than two years. That building was used until the construction of a new one in 1869, which was built by Topal Sherif Osman Pasha. With the arrival of Topal Sherif Osman Pasha in 1861, the Constitutional Law was established, and there were significant positive developments at the cultural and spiritual level. Thus, the construction of the government building, which the people call the residence, comes to the fore. This building served as the place of the Austro-Hungarian government until 1918. From 1918 to 1941, the building was used to house generals. During the period of the Kingdom of Yugoslavia, it served as the king's residence.

The residence Konak is also known for the fact that after the assassination on July 28, 1914, the bodies of the Austro-Hungarian Archduke, heir to the throne His Highness Franz Ferdinand and his wife Sofia, Duchess of Hohenberg, were exhibited there, as well as their coffins. In 1944, the first Faculty of Medicine in Bosnia and Herzegovina was opened there, and after the Second World War, this facility returned to its residential function. The then president of the Socialist Federal Republic of Yugoslavia, Josip Broz Tito, stayed in it during his visit to Bosnia and Herzegovina.

Since the establishment of this building, various changes have taken place, both in appearance and in the purpose of the building. Some of the works on this building include those undertaken in 1906, 1926, 1929, which included adaptation and improvement, then masonry and painting works, and the replacement of the damaged cover made of Eternité plates. Until the period of 1990, Konak had a residential character and one may say that it was in good condition (Picture 1).

Although the building of the Konak residence suffered minor damage during the war devastations of 1992–1995, the condition indicated the need to undertake various works for the purpose of rehabilitating the internal and external parts. Since 1992, Konak has served as a representative facility of Bosnia and Herzegovina where high-ranking state officials stay, and the facility serves the requirements of the Presidency of Bosnia and Herzegovina. Immediately before the visit of the high delegation of Qatar (the visit of His Highness Shaikh Hamad bin Khalifa Al-Thani, Emir of

Picture 1 Residential villa "Konak". *Source* Author/2020

the State of Qatar), the repair and commissioning of internal and external hydrant equipment was carried out, and certain interventions took place in the bathrooms, the purchase of beds and club tables, and lamps for nightstands and several smaller repairs were made.

Constructions on the restoration of other parts of the building began in 1999, when the worn vertical gutters, horizontal gutters, lightning protection installations and the replacement of tin gutters were replaced. In the same year, measures were taken to revive the fire and hydrant installations, clean and repair gutters, and repair the building's roof. In 2000, rehabilitation construction works were undertaken on the entrance plateau and staircase. According to the revised report of the Sarajevo Cantonal Institute for the Protection of Cultural, Historical and Natural Heritage, the state of the Konak facility is described as "restored". For this building, the Elaborate of the project task for the renovation of the Konak residential villa in Sarajevo was prepared [27]. In 2007, the Commission for the Preservation of National Monuments issued a Decision declaring the Konak building a national monument.[1]

The results of the 2010–2020 field research showed that access to the facility is highly restricted. On the external walls, damage is visible on the facade and on the external protective parts of the windows. The building has a residential function and serves the requirements of the BiH Presidency. According to the assessment of the Commission for the Preservation of National Monuments of Bosnia and Herzegovina, the building is on the list of endangered monuments, so we conclude that the building

[1] The decision on declaring the property a national monument is available at: http://aplikacija.kons.gov.ba/kons/public/uploads/odluke_bos/Sarajevo_Konak%20kompl%20BOS.pdf.

is in bad condition. According to the Institute for the Protection of Monuments, the project documentation for rehabilitation, restoration and conservation is currently in the drafting phase, managed by the Service for Joint Affairs of the Institutions of Bosnia and Herzegovina, while the drafting of the technical documentation is led by a consortium of private companies.

Military Barracks "Jajce Kasarna"

Jajce barracks are one of the most beautiful architectural units within the Old Town of Vratnik. It was built on the plateau of the southern part of the walls of the medieval old town, with a visible strategic position. As such, it is visible from almost all points in the city. It is a building of exceptional value, whose construction began in 1886. For unknown reasons, the original building was removed, and the construction of the barracks in their current form, which was built in 1914 during the Austro-Hungarian period and for the needs of the Austro-Hungarian army, began at the same place [26]. Originally, this building was named after Prince Eugene of Savoy, the Austrian general and statesman, who burned down the City of Sarajevo in 1697, in his campaign against the Ottomans. In 1915, it was renamed to Jajce barracks, after the military hospital from the town of Jajce, which moved to this facility that year. For many years after the World War II, the barracks housed the Yugoslav People's Army (abbreviated as JNA, the military of the Socialist Federal Republic of Yugoslavia).

According to the available data, restoration and construction on the Jajce barracks building were carried out in: 1948, 1955, 1956, 1967 and 1968 respectively. As part of these constructions, the barracks building was adapted, certain parts of the building were reconstructed, the covering was changed, the exterior joinery was changed, the main entrance to the building was moved, and certain changes were made in the interior. The reconstruction of the water supply and sewage system and the adaptation of the bathrooms and wetrooms were carried out, as well as regular maintenance constructions on the roof, facade and sheet metal (Picture 2).

Certain reconstructions of the actual barracks building were undertaken by the JNA in 1975, 1983, 1988 and 1989 respectively, which did not significantly affect the change in the character of this building. Until the beginning of the war in 1992–1995, the building was used by the JNA, and then by the Army of Bosnia and Herzegovina. According to the available data, it is considered that the condition of this facility in 1990 was solid considering that it was fully used and maintained by the JNA.

In the period of war events, 1992–1995, the Army of Bosnia and Herzegovina was stationed in the barracks complex. During that period, the building was severely damaged and destroyed. Since 2002, when the building was abandoned by the Army of Bosnia and Herzegovina, the building has not been used. Due to various factors as well as inadequate control of access to the object, additional damage occurred. Although there were indications that something would be done regarding its rehabilitation, conservation, restoration and adaptation of this building, nothing concrete has been done until the moment of writing this paper. It is considered that the biggest

Picture 2 Military barracks "Jajce kasarna". *Source* Author/2020

obstacle in the planning of construction works on this building is created by unresolved property rights. In 2009, the Commission for the Preservation of National Monuments made a decision to declare the Jajce barracks a national monument.[2]

Numerous damages are visible on the building, which will eventually result in the need for greater allocation of financial resources for the restoration of the building. The Commission for the Protection of National Monuments, in its protection measures, gave the possibility of using the facility for catering, educational and cultural purposes. While the issue of its ownership, possibilities of restoration and the manner of its use is still unclear, additional damage to the building is evident on the ground (Picture 3).

Analyzing the relevant literature and conducting field research in the period 2010 through 2020, information was obtained that gives an insight into the condition of this facility. So, since the situation in 2009 was already worse compared to the situation in 2004, and no construction was undertaken on the building per se, it can be concluded that the situation has now further deteriorated. Entry to the Jajce barracks facility is prohibited and the facility has not been used since 2002. The condition of the building is quite bad, the building is destroyed, and it is threatened with collapse. In 2019, the Institute for the Protection of Monuments issued a positive Expert Opinion for the possibility of cutting wild vegetation on the site, at the request of the City's Urban Planning, Investments, Housing and Utility Services Sarajevo. According to the assessment of the Commission for the Preservation of National Monuments of Bosnia and Herzegovina, the building is on the list of endangered monuments.

---

[2] The decision on declaring the property a national monument is available at: http://aplikacija.kons.gov.ba/kons/public/uploads/odluke_bos/Sarajevo_Jajce%20kasarna%20kompl%20BOS.pdf.

Picture 3 Military barracks "Jajce kasarna". *Source* Author/2020

5 Institutional and Legal Frame Work: Analysis of Protection of Cultural Heritage

Commission for the Preservation of National Monuments of Bosnia and Herzegovina

The Commission for the Preservation of National Monuments of Bosnia and Herzegovina[3] is an institution formed in 2001 by the Presidency of Bosnia and Herzegovina, which operates throughout the country and is headquartered is in Sarajevo. This institution was established by Annex VIII of the General Framework Agreement in Bosnia and Herzegovina, with the aim of protecting and preserving the cultural and historical heritage in the entire territory of Bosnia and Herzegovina.

The task of the Commission is to determine all the characteristics and values and, according to the degree of meeting the criteria, to make a final decision on the declaration of the property as a national monument. The Law on the Implementation of the Decisions of the Commission for the Preservation of National Monuments comes into force upon its promulgation and inclusion on the list of national monuments. This Law regulates the required cooperation of all institutions of the Federation of Bosnia and Herzegovina, cantonal and municipal levels, especially in situations where the protection of national monuments is an issue.

[3] www.kons.gov.ba.

Federation Ministry of Spatial Planning

The Federation Ministry of Spatial Planning[4] operates in the field of spatial planning, preparation of planning documents and their implementation. Planning documents, issued by this Ministry, determine the manner of the use and purpose of the land, as well as measures for the protection of the area.

This Ministry is responsible for issuing approvals for construction works on the protection of national monuments, which were previously declared national monuments by the Commission for the Preservation of National Monuments. All actors in society, especially the legal authorities of the FBiH, cantons, city and municipal services respectively, have the obligation to refrain from actions that would result in damage to national monuments or endanger their protection and rehabilitation. Accordingly, the task of the Government of the Federation is to provide funds for the protection of national monuments every year, and to co-finance protection projects.

Federation Ministry of Culture and Sports

The Federation Ministry of Culture and Sports[5] is responsible in the field of cultural heritage and the culture, sports, youth, economic, financial, legal and general affairs, and the protection of monuments. The Department for Cultural-Historical Heritage and Culture deals with administrative, professional and other tasks related to the protection and use of cultural heritage as well as other numerous cultural and artistic activities. It deals with the drafting of laws and regulations in the field of cultural heritage and culture, keeps records of cultural manifestations in the territory of FBiH, maintains a database of legal entities in culture, prepares proposals and plans for the distribution of funds, finds methods of improving the distribution of budget funds for financing the protection of cultural heritage and culture in the FBiH and performs other tasks in the aforementioned field.

The Ministry of Culture and Sports of the Canton of Sarajevo

The Ministry of Culture and Sports of the Canton of Sarajevo[6] performs administrative and professional tasks determined by the Constitution, the law and other regulations, which pertain to the performance of the Canton's duties in the field of culture and sports tasks. As part of this ministry, there are sectors for culture, sports and economic and legal affairs. The sector for culture deals with administrative and professional work, which includes the work of determining and proposing policies

[4] https://fmpu.gov.ba/.

[5] https://www.fmks.gov.ba/.

[6] https://mks.ks.gov.ba/.

for the development of culture and cultural activities, preparing, implementing and monitoring laws and regulations in the field of culture, establishing and supervising the work of public institutions of culture within the responsibilities of the Ministry, organizing and participating in the organization of cultural manifestations of importance for the Canton, regulating and providing conditions for the protection, revitalization, restoration of cultural and historical heritage, museums, monuments and memorial ensembles in the area of the Canton of Sarajevo and other professional tasks related to the competence of the Ministry in the field of culture.

Cantonal Institute for the Protection of Cultural, Historical and Natural Heritage of Sarajevo

The Sarajevo Cantonal Institute for the Protection of Cultural, Historical and Natural Heritage is the oldest institution dealing with the protection and preservation of natural and cultural heritage in Sarajevo. It was formed in 1963. This institution is responsible for the entire cultural and natural heritage, located in the territory of the Sarajevo Canton. The work performed by this Institution is aimed at achieving the following goals:

- protection and preservation of monuments,
- recording of movable and immovable heritage,
- collection of documentation,
- valorization of heritage,
- preparation of heritage protection documents in the field of spatial planning,
- creation of projects, reports and studies,
- works on monuments and
- works on the popularization of cultural, historical and natural heritage.

The Institute performs a number of other tasks, which are important for museum operations, protection of movable heritage, documentation of cultural and historical heritage, popularization of protection of natural and cultural heritage, implementation of laws and regulations in the sphere of heritage protection.

Bearing in mind the above, it is clear that there are institutional capacities that deal with the protection of cultural heritage. However, the situation on the ground shows that this is not the case. In addition to the listed examples of buildings of cultural importance, there is a number of other buildings that are endangered and, in most cases, neglected.

6 Legal Framework: Documents in the Service of Cultural Heritage Protection

Law on the Protection and Use of the Cultural-Historical and Natural Heritage of SRBiH

This Law was published in 1985 in the Official Gazette of the Socialist Republic of Bosnia and Herzegovina.[7] The Law was adopted by the then Assembly of the Socialist Republic of Bosnia and Herzegovina (SRBiH) at a session of the Council of Joint Work, a session of the Social-Political Council, a session of the Council of Municipalities and a session of the Assembly of the Federation of Cultural Communities of Bosnia and Herzegovina. Due to the lack of the most recent laws on the protection and use of cultural and historical and natural heritage, the mentioned Law is considered valid in the territory of the entity of the Federation of Bosnia and Herzegovina. Within the general provisions of the Law from 1985, the categories belonging to the cultural and historical heritage are clearly regulated.

For example, Article 5 of this Law lists goods of general interest, which are treated with special protection and are used under the conditions and in the manner prescribed by the law. Article 7 regulates that working people, organizations of joint work, socio-political communities, socio-political organizations, local communities and other self-governing organizations and communities are obliged to protect, but also have the right to use cultural, historical and natural assets inheritance. When talking about the protection of natural and cultural heritage, Article 11 stipulates numerous activities and measures that can or should be taken in order to protect heritage, including recording, research, evaluation, conservation, restoration, revitalization, adaptation, preservation, maintenance, proper use, declaration of goods as protected, prohibition of permanent removal from the country, etc.

According to Article 12, the use of cultural goods implies the creation of conditions that enable protected cultural goods to be available to the public in order to meet the educational and scientific needs of working people and citizens. It is necessary to provide such use of cultural and natural heritage that will not cause any damage to the heritage itself, but also to socio-economic and cultural development, which is regulated in Article 13.

Article 55 of the Law is particularly noteworthy, which regulates the necessity of preparing an expert report in cases of conservation, restoration, adaptation and revitalization by a potential contractor, which must be approved by the proficient institution. In this part of the Law, the area of use of cultural goods is also dealt with in more detail, like it is stated in Article 67 where it is written that these goods can be adapted to contemporary needs and purposes like artistic or scientific activities, educational activities and hospitality-tourism and other activities. As already mentioned, this law is still in force until the adoption of a new one that will be in line with the Constitution of Bosnia and Herzegovina and all its structures.

---

[7] Available at: https://www.fmks.gov.ba/kultura/legislativa/bih/58.pdf.

Law on the Protection of the Cultural Heritage of Sarajevo Canton

The parts of this Law, which are most closely related to the topic of this paper, are the protection and use of cultural heritage. Therefore, from the protection section, we highlight Article 8, which, *inter alia*, regulates the actions aimed at protecting assets and the obligation to valorize localities, real estate and movable property, determine the properties of heritage assets, and revitalize and create conditions under which cultural assets would serve the needs of society and individuals in accordance with the goal of their preservation.

With the adoption of this Law and its publication in the Official Gazette of the Canton of Sarajevo,[8] the application of the Law on the Protection and Use of the Cultural, Historical and Natural Heritage of the SRBiH in the area of the Canton of Sarajevo ends, but remains at the level of the Federation of BiH. In the comparison of these two laws, one can point out a great similarity in many positions, especially in the part regulating the use of cultural and historical heritage.

Law on the Implementation of the Decisions of the Commission for the Protection of National Monuments in Bosnia and Herzegovina

The Law on the Implementation of the Decisions of the Commission for the Protection of National Monuments established under Annex VIII of the General Framework Agreement for Peace in Bosnia and Herzegovina establishes measures for the protection and rehabilitation of assets that have been determined as national monuments of BiH by the decision of the Commission for the Protection of National Monuments. In the section on the protection of national monuments of this law, Article 4 obliges everyone, especially the responsible authorities of the Federation, Canton, city and municipal services, to refrain from any actions that may damage national monuments or call into question their protection and rehabilitation.

Furthermore, and what is considered significant for this research, is that the Federation Ministry of Spatial Planning is responsible for issuing approvals for the protection, conservation, presentation and rehabilitation of national monuments. This Law gives mandates to the level of the Federation of BiH to ensure legal, scientific, technical, administrative and financial measures for the protection, conservation, presentation and rehabilitation of national monuments. This Law specifically emphasizes the obligation of cooperation between institutions responsible for national monuments in the Federation of BiH.

---

[8] Available at: https://mks.ks.gov.ba/sites/mks.ks.gov.ba/files/zakon_o_zastiti_kulturne_bastine_2_00.pdf.

It is obvious that the legal framework and all documents that are current and concern the protection of cultural heritage have not been implemented, because otherwise the situation on the ground would not be as it is. Enforcement of the Law apparently is not a priority, and the most frequently used excuse is jurisdiction and lack of financial resources.

7 Results and Discussion

In this part of the paper, the focus is on the presentation of the research results, more precisely on the data collected through the in-depth interview method. The presentation of the results of the data collected is organized in accordance with the subject and problem of the research, in order to get a clearer picture of the attitudes and opinions of the selected institutional representatives. The collected information provides insight into:

- attitudes and opinions about the state of cultural heritage of Sarajevo City in terms of restoration and protection;
- causes and sources of identified problems;
- possible solutions to identified problems.

The representatives of the institutions gave their views and opinions in relation to the posed research question, which is presented in Table 2.

Based on the above as well as other collected data, the results indicate the overlapping of responsibilities and non-cooperation of the state institutions responsible in the field of cultural heritage in Sarajevo, and as such are considered the main problems that directly contribute to the obviously problematic state and inadequate protection of cultural heritage. There are also other problems, namely: lack of financial resources, (in)adequate attitude of the population towards heritage, inadequate legal regulations, and problems related to the high level of corruption in state institutions.

The prevailing opinion is that cultural heritage is not adequately protected and that additional efforts should be made in terms of its restoration and usage. It is noted that the views of some respondents partially deviate from the others, and to a lesser extent they are of the opinion that the state and restoration of heritage is good compared to the state it was in 1995. The fact that the cultural and historical heritage of Sarajevo suffered the great damage during the period of siege 1992–1995 was certainly taken into account.

The attitude of all involved institutions regarding the condition, restoration and protection of the cultural and historical heritage of Sarajevo City can be characterized as negative. The analysis showed that in the background of the current situation, there are much more in-depth problems that need to be solved at the level of state institutions, but certainly in cooperation with lower levels that have jurisdiction over the sector of cultural heritage as well as sector of urban development and spatial planning. The analysis of the collected data allowed insight into the problems of this sector, which is of great importance for understanding the current situation.

Table 2 Responses (opinion and attitudes) of institutional representatives to the research question

Number	Codes assigned to research participants	Responses of institutional representatives to the research question
1	K1-KONS	• Not enough has been done; • A systematic planning approach should be made; • There is no control; • Inspection for urban construction is completely incompetent for cultural heritage; • Follow-up is the weakest link, when it comes to inspection; • In Sarajevo, heritage is highly compromised
2	K2-FMKSZZS	• The inspection is not doing its job; • Buildings are legalized even though they contradict the protection of cultural and historical heritage; • Politics is everywhere
3	K3-KZZKHPNS	• Not destroyed by the war but time; • Problem of restoration and investment in cultural and historical heritage is only piling up and becoming more and more complex
4	K4-KZZKHPNS	• It could be better; • Lack of vision, strategy, thoughtfulness, professional approach and thinking about how this cultural and historical heritage can be used as a potential for tourism development
5	K5-ZPRKS	• The situation is not satisfactory; • A lot of work has been done considering the situation in 1995; • We have to work a lot more on it
6	K6-KNBG	• After the war, a lot of work was done on reconstruction in Sarajevo; • It is an area that should be invested in; • It should be integrated into other areas; • Think about their future function
7	K7-AABH	• Right treatment of cultural and historical assets is necessary; • Interventions should somehow raise the value of these cultural monuments; • Monuments in urban spatial plans are important
8	K8-GSUPISKP	• Very slow process; • The situation with the cultural and historical heritage should be better

The results of the analysis of the collected data through in-depth interviews are in accordance with research in the field, and with the analysis of archival materials based on which the recorded activities and constructions on the selected objects from the period of their creation are presented. The results indicate that the activities and work on the selected facilities after 1996 were more prevalent than in the period after 2010 until today.

By analyzing the institutional and legal framework of the cultural and historical heritage sector, it was determined that there are certain problems. Although there are institutional capacities for carrying out all procedures in this sectors the existence of several different documents regulating the protection and use of cultural heritage cause problems. It was observed that the methods of securing financial resources for the protection of cultural heritage were not determined in detail, and the process of economic valorization was completely omitted. In addition, there is also the problem of transferring responsibility from one level to another, where lower levels of government are again conditioned by the decisions and consents of higher levels, while no one was held responsible for securing funds for the protection of the culture heritage in reality. Complex and lengthy procedures of administration, management, use of cultural heritage, and coordination with institutions at various levels seriously complicate the regulation and any attempt to solve problems which means that immediate improvement is necessary.

The situation on the ground, through case studies, supports this view. More precisely, the analysis of the state of the selected cultural heritage objects found that there was a significant deterioration of both objects in 1996 compared to the state in 1990. The reason for this is the damage caused by the war. In the period from 1996–2000, there were no significant changes and constructions on the buildings. Until 2000, minor works and restoration of the fire and hydrant installations were recorded in the Konak residence. In the period 2000–2010 both buildings were declared national monuments. In this period, for the Jajce barracks, a study was prepared on the assessment of the construction value of the building and a preliminary report was prepared on the condition in which the building was found, and for Konak, a study was prepared with defined objectives of the restoration program. An analysis of the state of the selected buildings in 2020 found that there are no significant changes to the Konak residence building, while the Jajce barracks was quite destroyed.

Therefore, the results indicate the need to urgently take all the necessary steps that will ensure that the legal institutions take their respective responsibility, which in this case is cultural heritage. An important step in that process is the establishment of cooperation among sectors for the reason that cultural heritage does not exist independently, but in symbiosis with other elements of society and the environment.

8 Conclusion

The purpose of this research was to determine the state of the cultural heritage of the City of Sarajevo. For that purpose, the collection of relevant literature and data that will provide an answer to the research question have been undertaken. Archival literature related to the objects of historical importance that were selected for this research was analyzed in order to reach an answer to the research question through the application of the case study method. The selection of the buildings of the Konak residence and the military building of the Jajce barracks was purposeful because they represent buildings of national importance, they are located in the central urban part of the City of Sarajevo, they played a significant role in the urban development of the town, and their values represent unique examples of cultural heritage.

In order to identify the sources and reasons of the problems that are evident when it comes to the protection and restoration of the city's cultural heritage, the institutional and legal framework was analyzed. All the collected data and information were analyzed, and then the collection of primary data was started through an in-depth interview method. Respondents were carefully selected so that the data would be relevant, current, objective and accurate. Eight individual interviews were conducted with the representatives of the institutions that are responsible not only for the cultural heritage sector but also for city development planning. A representative of a non-governmental organization also participated in the survey.

The analysis of all collected data using different research methods enabled an insight into the state of the cultural heritage of the City of Sarajevo. Namely, the conclusions of this research can be presented on the basis of analyzed collection of literature, archival material, an insight into the state of the selected objects, and information collected by the in-depth interview method. Thus, it can be concluded that although there are institutional capacities for the protection and restoration of cultural heritage, the legal regulations have not been aligned, and there are numerous overlaps in the competences and responsibilities of the institutions. Such a situation complicates the processes of restoration and protection of cultural heritage as an important aspect of the city's urban development. This is supported by the results of a case study of two selected examples of cultural heritage, which confirm that cultural heritage is largely neglected and abandoned. The lack of financial support from institutions also contributes to this, which was confirmed in the conversation with the representatives of those institutions followed by the lack of cooperation among different sectors.

In order to solve the mentioned problems, they need to be approached in a systemic way. First of all, it is necessary to align the legal regulations, which implies the revision of the current documents, competences, but also the manner the institutions operate. It is also necessary to review the manner of valuing the cultural and historical heritage, define priorities and ways of using and protecting cultural heritage. Protecting cultural heritage does not mean restoring and limiting access to it, but on the contrary, restoring and ensuring adequate valorization of that cultural capital. An active approach to the protection of cultural heritage requires creativity and the development of ideas that fit into the framework of the contemporary and sustainable development, ensuring that cultural heritage does not lose its values, but expands and strengthens them. Also it is recommended to ensure activities that will ensure the establishment of cooperation between sectors, for example with the sector for planning and development of the city, the sector for spatial planning, the sector for tourism and other sectors that are in any way connected with cultural heritage.

According to the above, we conclude there is a lack of a strategic and holistically coordinated approach to cultural heritage, actually the approach which would accept the cultural heritage as a factor of urban development and many other processes of society followed by improvement of human capital in terms of education that encourages creativity, innovation, leadership, cooperation and the ability to perceive and solve identified and all other problems that may appear in future.

Future research should integrate other areas related to cultural heritage, such as the aspect of their use and valorization. In addition, this research is limited to two examples of objects of cultural importance, and that number could be larger in future research. Also, a number of other strategic documents, initiatives, as well as international agreements should be the subject of future analyses.

References

1. Forman TTR (2008) Urban regions: ecology and planning beyond the city. Cambridge University Press
2. Polyzos S (2012) Urban development. InTech. Zagreb
3. Kurto N (1997) Sarajevo: MCDLXII-MCMXCII. OKO. Sarajevo
4. Timothy DJ, Boyd SW (2003) Heritage tourism. Pearson Education
5. Hardy D (1988) Historical geography and heritage studies. Area 20:333–338
6. Tweed C, Sutherland M (2007) Built cultural heritage and sustainable urban development. Landsc Urban Plan 83(1):62–69
7. Lowenthal D (1997) History and memory. Public Hist 19(2):30–39
8. Hubbard P (1993) The value of conservation: a critical review of behavioural research. Town Plan Rev 64(4):359
9. Rowntree L (1981) Creating a sense of place: the evolution of historic preservation of Salzburg, Austria. J Urban Hist 8(1):61–76
10. Serageldin I, Shluger E (2001) Historic Cities and Sacred Sites: Cultural roots for urban futures. The World Bank
11. Karlström A (2014) Urban heritage. In: Smith C (eds) Encyclopedia of global archaeology. Springer
12. Martine G, McGranhan G, Montgomery M, Fernandez-Castilla R (2008) The new global frontier; urbanization, poverty and environment in the 21st century. Earthscan
13. Hewison R (1987) The heritage industry. Methuen, London
14. Al-Hagla K (2010) Sustainable urban development in historical areas using the tourist trail approach: a case study of the cultural heritage and urban development (CHUD) project in Saida, Lebanon. Cities 27:234–248
15. Vasi IB (2009) New heroes, old theories? Toward a sociological perspective on social entrepreneurship. In: Ziegler R (ed) An introduction to social entrepreneurship. Edward Elgar Pub., pp 155–173
16. Nurse K (2006) Culture as the fourth pillar of sustainable development. Small States Econ Rev Basic Stat 11:28–40

17. Đukić MA, Vukmirović M (2010) Creative cultural tourism as a function of competitiveness of cities. Multiscience 2(1):27–35
18. Nijkamp P, Riganti P (2008) Assessing cultural heritage benefits for urban sustainable development. Int J Serv Technol Manag 10(1):29–38
19. Bourdieu P (1986) The forms of capital. In: Richardson J (ed) Handbook of theory and research for the sociology of education. Greenwood, Westport, CT, pp 241–258
20. Throsby D (1999) Cultural capital. J Cult Econ 23(1–2):3–12
21. Donia RJ (2006) Sarajevo: bografija grada. Institut za historiju. Sarajevo
22. Sparks M (2014) The development of Austro-Hungarian Sarajevo, 1878–1918: an urban history. Bloomsbury Publishing
23. Filipović N (1952) Pogled na osmanski feudalizam. Godišnjak Društva istoričara Bosne i Hercegovine, IX/1952, 146
24. Zlatar B (1996) Zlatno doba Sarajeva: XVI stoljeće. Svjetlost. Sarajevo
25. Taubman I (1954) Razvoj grada Sarajeva. Sarajevo
26. Kreševljaković H (1969) Sarajevo za vrijeme austrougarske uprave (1878–1918). Sarajevo
27. Wiener B (2006) Elaborat projektnog zadatka za sanaciju rezidencijalne vile Konak u Sarajevu. Sarajevo

Implementing the Global Approaches in Local Context: Case of 'Conservation Works' at Nineteenth Century 'Epiphany Church', Gurugram, India

Mohit Dhingra and Surbhi Anand Roy

Abstract Heritage conservation ideologies have been under constant change; from 'antiquity to modernity', from material based to value-based, and finally turning its course towards sustainability-oriented approaches since the turn of the millennium. Through Convention for the Safeguarding the Intangible Cultural Heritage (2003) followed by the Convention on the Protection and Promotion of the Diversity of Cultural Expressions (2005) and finally, the 'historic urban landscapes' (HUL) approach adopted worldwide in 2011, the paradigm shift can be seen. These discourses, on one hand, enabled the inclusion of long-left communal value not only in the heritage conservation domain but also in the urban development domain, thus paving the way for the notion of "heritage as a tool for sustainable development". In reference to the above, this research adopts an evidence-based methodology in examining the case of 'Conservation of Epiphany Church' in Gurugram, India in the year 2021. The church which was once the symbol of association for its parish was in the early stage of decay until November 2020. The church has now been conserved through the efforts of the Epiphany Church Committee, community, and other concerned stakeholders under the supervision of Delhi based conservation team 'adapt'. The second author has been instrumental in leading the initiative as project manager whereas the first author was the faculty mentor from the Jindal School of Art and Architecture who led the group of students to critically analyze the ongoing conservation works and formulate a conceptual revitalization program for the church complex. The conclusions drawn from this research are recorded, considering the detailed step-by-step process undertaken for conservation, the negotiations involved, and if these methods can work as a model for revitalization of the local heritage in peri-urban areas of Northern India.

Keywords Sustainable development · Local heritage · Conservation · India

M. Dhingra (✉)
Jindal School of Art and Architecture, OP Jindal Global University, Sonipat, Haryana, India
e-mail: mdhingra@jgu.edu.in

S. A. Roy
Adapt, New Delhi, India

M. Arslanagić-Kalajdžić et al. (eds.), *Interdisciplinary Advances in Sustainable Development II*, Lecture Notes in Networks and Systems 804,
https://doi.org/10.1007/978-3-031-46692-2_13

1 Introduction

Heritage and Sustainable Development: The Global Context

Heritage is not a thing, site, or place, nor is it 'found', rather heritage is the multiple processes of meaning-making that occur as material heritage places or intangible heritage events are identified, defined, managed, exhibited, and visited [1].

Heritage conservation from 'antiquity to modernity' has always been an intrinsic part of these multiple processes. The discourse of heritage and sustainable development is not new and can be traced back to antiquity with the onset of the Athenian oath [2] which reads as below:

"We will transmit the city, not only not less but greater and more beautiful than it was transmitted to us".

Though the idea of heritage led sustainable development which lost its way during the course, pertaining to diverse pressures like ever-increasing demands of urbanization, armed conflicts, and various socio-economic and socio-cultural challenges. This notion found its way into the developing world at the onset of new millennia through various strategic approaches developed during various conventions.

The Convention for the Safeguarding the Intangible Cultural Heritage (2003) aimed at safeguarding the expressions of intangible heritage that are endangered by the processes of globalization and presses to raise awareness at the local, national, and international levels. The convention respected the communities, groups, and individuals thus bringing them into the mainstream of development planning. This was followed by the Convention on the Protection and Promotion of the Diversity of Cultural Expressions (2005) which had the core aim to support cooperation for the international development through human capacity building in developing countries as stated in Article 14 of the convention. Whereas Article 16 clearly states the facilitation and prioritizing of artists and other cultural professionals and practitioners, as well as cultural goods and services from developing countries [3].

The historic urban landscape approach (UNESCO 2011) further strengthened the perspective of seeing the heritage in totality as a social, cultural, and economic asset for the development of cities rather than an isolated set piece. The Historic Urban Landscape Approach moves beyond the preservation of the physical environment and focuses on the entire human environment with all its tangible and intangible qualities. It seeks to achieve sustainability of planning and design interventions by considering the existing built environment, intangible heritage, cultural diversity, socio-economic and environmental factors, and local community values.

The year 2015 marked a significant change with the inclusion of culture including heritage as a part of the international framework also called Sustainable Development Goals [3]. Out of 17 SDGs and 169 targets, SDG 11, target 11.4 identifies conservation and management of historic assets as an important tool for sustainable development, moreover, New Urban Agenda developed in the year 2016 explicitly calls out both the intangible and tangible heritage as important attributes in sustaining and supporting urban economies to progressively transition towards higher productivity [4]. These discourses enabled the inclusion of long-left communal value not only in the heritage conservation domain but also in the urban development domain and paved the way for the notion of "heritage as a tool for sustainable development". The reflection of this changing paradigm can also be seen in some of the recent conservation projects in India, undertaken by some of the non-profit organizations, trusts, and community groups.

Heritage and Sustainable Development: Challenges in the Local Context with an Emphasis on the Case of the Millennial City: Gurugram, India

In the developing world, the meaning of heritage is different from that of many developed countries. With rapid demographic growth, as in the case of the Indian subcontinent, heritage is disappearing at an alarming rate [4]. Moreover, the issue has a multiplier effect while dealing with peri-urban areas as they are usually subjected to unplanned and rapid growth. Pertaining to their porous boundaries, planning ambiguity and sudden changes in land use from rural to urban puts tremendous pressure on these areas.

Though there are multiple studies focusing on these effects from the point of view of environment and infrastructure, but they say little or nothing at times about the state of heritage. In these socially polarized regions, not only the tangible architectural or material values are lost but also the associational value of the heritage is lost under the burden of real estate prices and the growing demands of the city. Heritage is only seen as a burden or a commodity to be erased or insensitively used.

This research will focus on Gurgaon town, now known as Gurugram, which is in the northern Indian state of Haryana. It is about 19 miles southwest of the national capital New Delhi. It has the third highest per capita income among other Indian cities. It has rapidly urbanized in the past two decades. Gurugram has experienced phenomenal growth in terms of population which was 870,539 in 2001 to 1,514,085 in 2011 [5].

The current population of Gurugram is estimated to be close to 1,726,452. Such explosive population growth has adversely affected not only the physical infrastructure of the city but also the land value. The prices of real estate have put pressure on the available social spaces in the city. These transformations have led to various decay patterns pertaining to change in land use leading to altered patterns of resources and associated values, creating socio-cultural and economic changes. The community, which was primarily agrarian, has now lost the sense of its own identity and the town as well.

The city, which was thriving with step wells, forts, mosques, colonial structures, colonial railway lines, and many other archaeological sites (refer Table 1) is now inundated with glass skyscrapers, insensitive urban infrastructure, and a migrant population with no sense of belonging and many dilapidated heritage sites waiting to be erased.

In order to understand the underlying cause of the situation, the first author's hands-on experience while working with historic cities both at the project as well as policy level with multiple state bodies, communities, and multilateral organizations, and the second author's extensive experience of working on various conservation projects in peri-urban and rural areas across India, coupled with various research papers and development reports the following issues can be elucidated below:

Absence of local heritage in peri-urban areas in development planning

The idea of local heritage is missing from various development schemes such as Jawaharlal Nehru National Urban Renewal Mission (JNNURM), Heritage City Development and Augmentation Yojana (HRIDAY), and Pilgrimage Rejuvenation and Spirituality Augmentation Drive [7] that eventually initiated looking heritage as a focus area also tend to skip these peri-urban areas and their associated heritage.

Complex multi-decision-making process [8]

Peri-urban transformations in Gurugram indicate the complex bureaucratic setup. At the macro level, the planning structure is divided between the central government, state government, and the local bodies (urban or rural), supported by various regional bodies and other technical institutions. Though at the local level, such urban and rural governance systems are distinct, these urban–rural interactions are leading to transformation issues that are urban in nature but governed by rural bodies. These are often ill-equipped to deal with the attendant complexities of urban densities and land uses and to provide the required urban infrastructure and amenities in these areas.

Table 1 Built heritage of Gurugram district [6]

Forts and Palaces	Fort remains in Sohna, Sheesh Mahal in Farrukhngar, Summer Palace in Pataudi
Religious structures	Aliwardi Masjid and Masani temple in Gurgaon, Church in Gurgaon, Jama Masjid at Farrukhngar, Shahi Jama Masjid in Sohna
Havelis and bungalows	Aman Nath Haveli near Badshapur, Seekhonki Haveli in Farrukhngar, many havelis in the towns of Farrukhngar and Sohna, a few British Bungalows in Gurgaon
Water bodies	Ali Ghosh Khan baoli in Farrukhngar, Sakhanjati Kund in Sohna, Stepwells in Badshapur
Public buildings	Purani kachehri in Jharsa, Kaman Sarai in Gurgaon, Akbar mandi in Pataudi, and few more British buildings in Sohna
Gates	Delhi gate, Jhajjar Gate in Farrukhngar
Tombs and Memorials	Qutub Khan's tomb, Hazrat Shah Najm al Haq Mosque, Lal Kalal tomb at Sohna, Sethaniki chattri in Farrukhngar

Lack of capacity among decision-makers

Sectoral thought processes and compartmentalization of heritage in peri-urban areas have resulted in the neglect of heritage assets and surrounding historic areas. There are often weak implementation, monitoring, and penalizing mechanisms in the absence of definite heritage byelaws, especially dealing with a heritage of local importance. This has led to massive encroachments of many historic structures and their surroundings.

Loss of identity and spirit of a place

The sudden densification of settlements in Gurugram due to the overflow of the new working population to provide residential and commercial needs has led to conflicting land use, extreme spatial fragmentation, the emergence of new and complex forms and urban conditions, and varying infrastructural demands and conditions of access.

Commoditization of heritage

The majority of the land in peri-urban areas like Gurugram is privately owned, these in turn lead to high land prices due to an increase in demand. Thus, heritage precincts are seen as commodities to be erased in lieu of high prices, therefore many historic structures are deliberately left in shambles and some like the 'Epiphany church' are constantly persuaded to give away their land for road widening projects.

In reference to the above, the research aims to analyze a bottom-up approach that was undertaken by various stakeholders to preserve the lost identity of ' the Epiphany Church' and its surroundings but also worked as a demonstration project which urges the need for more such localized approaches in the heritage sector.

2 Research Methodology

The research comprises three approaches. Through a theoretical approach, the literature search on the subject was performed which focused on various charters, international conventions, and research papers focusing on the status of peri-urban areas. To comprehend the current status of Gurugram the study by the National Institute of Urban Affairs [8] and the Centre for Science and Environment (CSE) [9] and was referred.

The practical or evidence-based approach comprised semi-structured interviews and focus group discussions with various stakeholders. Moreover, the hands-on experience of the second author helped in bringing forth the issues, challenges, and aspirations related to the project from its inception till its handing over. The first author critically analyzed the project as an outsider with exposure in handling similar sites in India from the perceptive of international trends and protocols. The conclusions drawn from this approach are recorded, considering the detailed step-by-step process undertaken for conservation, the negotiations involved, and if these methods can work as a model for the revitalization of the local heritage in peri-urban areas of Northern India.

Together these approaches led to a theme-based approach through which the project was analyzed. Further, these themes are discussed in detail in the subsequent section (Fig. 1).

3 About the Project—'Conservation of Epiphany Church, Gurugram, India

Background

Sustainable development goal (SDG-11) identifies the conservation and management of historic assets as an important tool. This is particularly applicable to cities like Gurugram which are rapidly urbanizing and are losing their cultural and social identity. Inclusive Heritage conservation can help in building up a "sense of place" which can unify both the resident as well as the migrant population. One such effort was the conservation project of the Epiphany Church in old Gurugram. Built in the year 1862 [10], in a quiet neighborhood of Civil lines, it served as a Garrison Church visited by a few British residents and military officers stationed on the outskirts of Delhi. The architecture of the church strongly resembles the Parish churches which were built in rural England in the nineteenth century.

The parish of the Church grew by the 1940s and thus the original building was extended on the western side. The extension was meticulously designed and executed to blend with the historic chapel. The parish grew further and thus in 2009 a new church building was built next to the historic chapel.

The historic church has continued to be used since its inception. The chapel was in fair condition as it was regularly maintained by the Church Committee (Diocese of Delhi). It had some seepage issues in the roof and all along the base of the exterior walls. But in 2018 a small portion of the south section of the roof tiles slumped, due to which the church committee decided to close the church for the public. The conservation work was taken up in November 2020 and it was completed in October 2021.

When the church committee met the conservation team in November 2020, they had only one thing on their mind, "fix the roof!!" With a small amount of liquidated funds and no expertise in the conservation process, they felt that roof was all that was needed to be done. The conservation process diligently explained by the conservation team enabled the stakeholders to look at the church as a part of the larger cultural urban landscape. Based on the comprehensive consultations with various stakeholders, coupled with best practice approaches a detailed statement of significance was formulated which is elucidated below (Fig. 2):

Historical Value: The Epiphany Church is the oldest church in Gurugram and has a strong local and regional significance. It forms part of a larger narrative and stands as a testimony to the colonial rule in India. Similar churches were built during the same time in many other British cantonments. It also contributes to the history of Gurugram and its changing political position during the nineteenth century.

Architectural Value: The spatial layout and the architectural vocabulary of the church has a strong resemblance to the nineteenth century Parish churches built in rural England. It has Gothic style influences in architecture, much of which are still intact. Some of the original building embellishments are also retained in the form of brass hardware, mosaic tiles, casement windows, ornate lime molding, and metal bell.

Cultural Value: The church has continued to be used for various religious, social, and cultural events since its conception with the active support of the church committee. Christmas and Epiphany Day are important events that are celebrated with huge zeal by the church and its parish. The 150 years of the establishment of the historic church was celebrated in the year 2016 with fervor by people from various walks of life. Many have fond memories of their marriage, child's baptism, and other religious events which have commenced in the historic church.

Since this project was completely funded by the church committee it was envisaged into three stages (Immediate, Urgent, and Desirable) based on the necessity of the intervention. The conservation process spanning almost a year was designed around various thematic approaches:

Place-based Approach

The church is a living heritage site that has continued to be used for various religious and cultural events. It has a strong historic and religious significance for its parishioners. There are approximately 200 families and about 700 individuals who are part of the parish. For any place-making intervention, it was important to understand the various stakeholders, their needs, aspirations, and visions.

The church sits on 1.6 acres (0.65 hectares) of land in a prime real estate area. The sanctity of the place had to be maintained so the interventions in the open area were minimalistic and low in maintenance. The church hosts various community functions like Annual Fete, Christmas Feast, Epiphany Day, Christmas Day celebrations, etc. which need large open space, so the rear side of the complex is used for this purpose. The front lawns are used as a buffer space between the main road and the church building. Stone paving is provided as a building apron around the historic church. The surface water is drained towards the green patches on the north and south of the church.

The historic church had no foyer before entering the nave. After much deliberation, the team designed a glass porch on the western side which served the utilitarian purpose without overpowering the historic façade (Figs. 3, 4).

Crafts-based Approach: Revival of Traditional Crafts

The Epiphany Church that we see today is primarily built in two stages. The extension was carefully designed and executed to blend seamlessly with the historic chapel. Both phases are different layers of history that indicate the growing needs of the parish. Minor difference in the size of bricks, detail of the scissor truss, and door window hardware, clearly indicates the different construction periods.

Phase I: The first phase of construction was around 1862 when a humble Chapel was built for a small European population that came to the newly developed district headquarters. The Church had a typical Latin cross plan. It was east–west aligned with the north and south vestibule perpendicular to the nave. The altar was visually segregated from the nave by a huge, pointed arch and a raised floor. The nave and altar floor was provided with beautiful Victorian mosaic tiles, the remains of which can be seen even today on the altar. Maybe there was an entrance vestibule on the western side which was later demolished in phase II.

The Gothic features that adorned the 19th-century church were a steep gable roof, a front gable crowned with the bell tower like a pinnacle, lancet windows, and maybe stained glass along the altar. The bell tower has Celtic Cross that has a nimbus or a ring around the Christian cross. This style emerged in Ireland, France, and Britain in the Early Middle Ages.

The fenestrations had beautiful lime molding detail along the profile and brass ornate window handle cum latch. The casement windows have typical Victorian diamond-shaped mullions to hold small pieces of glass. The highly steep pitch roof similar to Welsh Churches is supported by the timber scissor truss, wooden purlins, and planks. Slate tiles were provided as a waterproofing layer (Fig. 5).

Phase II: In 1942 an extension was planned on the western side of the historic church. The entire new roof was covered using the original detail. The buttress was built to strengthen the masonry wall. The original door was relocated to the west side of the extension. The bell tower on the west side probably houses the original bell from the nineteenth century.

Material conservation was based on international charters and conservation protocols. Burra Charter (2013) was followed for the conservation process. The conservation interventions are tabulated below (Table 2).

Table 2 List of priority interventions

Conservation Interventions:
• Repair of the pitched roof
• Removal of incompatible renders and additions
• Removal of Reinforced Cement Concrete (RCC) roof and relaying of the roof in the north and south transept as per original detail
• Masonry stabilization, crack stitching
• Conservation of original doors, windows, and furniture
• Lime plastering and repair to the original moldings
• Internal flooring- conservation of the exposed mosaic tiles and laying of new flooring in missing areas
• Conservation of the painted glass along the altar
• Perimeter surface drainage

Revival of traditional crafts was at the heart of the 'Epiphany church conservation project'. From setting up the lime mill on the site to sourcing the matching materials in case of replacement, to carefully documenting the original details for reproduction, authenticity was always the priority. Local masons with no experience in handling lime were trained by expert lime masons providing them not only a new skill set but a new way of enhancing their livelihoods. Similarly, the senior carpenter with experience in historic carpentry headed the team of local carpenters. The conservation architects and architects worked closely with craftsperson to produce minimalistic and cost-effective solutions. In-situ repairs were preferred above complete replacement (Fig. 6).

Community-Led Approach: Sensitization and Demonstration

The conservation initiative was an effort of a sensitive and expert team that comprised of the Church committee, parish, community representatives, trained historic masons and carpenters, conservation architects, and senior advisors. The entire team worked towards safeguarding the significance of the building and its setting, preventing further decay, and prolonging the life of the structure so that it can be used for several years.

As mentioned in the Burra Charter (2013), understanding of cultural significance became a vital tool in the formulation of a sensitive and holistic conservation methodology. Understanding the values, user needs, available resources, and the building condition helped the team to develop mitigation measures in case of conflicts.

The community was involved in every decision, and fortnightly focused group discussions were organized with the committee representatives and the conservation team where the important milestones, conservation techniques, and fund requirements were discussed. Not all the interventions got an instant nod from the committee. The conflicts were mitigated based on facts, sampling, and constant persuasion. For example, the committee refrained from redoing the RCC roofs of the transepts in the original timber detail considering it as an unnecessary expense. The conservation team sensitized them that the reinforcements have decayed and the masonry arches supporting the roof have developed cracks which will further aggravate and would need interventions in the near future. The committee understood that is economically and logistically more viable to intervene now.

Labor-intensive works were first demonstrated on a small sample to sensitize the committee before extensive application. The active involvement of the church committee and the parish in the conservation interventions further strengthened the sense of ownership within the community. What started as a 'roof repair project' ended up as a historical and architectural revival project. Today the committee is very keen on nominating their historic church for the UNESCO Asia Pacific Awards.

After almost 3 years, in December 2021 the church opened its doors to its people on the auspicious day of Christmas. Decked up in lights and a regenerated soul, the historic chapel welcomed hundreds of people from various walks of life who had gathered to relive the memories of the bygone era (Fig. 7).

4 Discussion

This research started with the premise that heritage has now been identified as a tool for sustainable development globally and reflections of the same can be seen in developing countries like India as well, where local heritage has for long remained the responsibility of the government, but now nonprofit organizations and trusts are coming forward to preserve their heritage in order to reconnect it with the city. To test the above argument this research adopted an evidence-based methodology in critically examining a community-led conservation project 'Conservation of Epiphany Church' located in the Peri-urban area, of Gurugram. In doing so, some interesting and key findings come to the fore, which are discussed below.

It must start from the bottom, but is it enough?

The conservation of Epiphany church though demonstrated a perfect example of a community-led conservation project where the complete process demonstrated best conservation practices. It also created a ripple effect with the Bishop of the Church of North India who came for the inauguration of this project in December 2021 is now willing to revive more historic church buildings in the state through a similar process.

But it is important to understand that the maximum impact of any demonstration project can only be reflected if all the stakeholders are involved in the process therefore involving local municipal stakeholders would have ensured in scaling up of the project at the state level.

On the other hand, government bodies at the local and state level should also come out of their complex administrative structure (refer Sect. 1.2) and become enablers in facilitating these kinds of projects to rebuild the lost image of the city because the process of inclusive revitalization is neither bottom-up nor top down, it's always both ways.

Decoding the Theme-Based Approach

Conservation of Epiphany church followed a comprehensive consultative approach which not only ensured stakeholder involvement with the interventions and its management thereafter but also enabled the expert team to come up with various tangible and intangible themes. Based on which interventions (urgent/ necessary/ desirable) were carried out within the given budget. The theme-based approach also ensured to carry forward the project which originally started with an intent to just consolidate the roof to its present state (Table 3).

Understanding the significance of financing heritage management.

It is important to understand that economic viability is imperative to sustainable management. Though the church committee along with other stakeholders managed to fund the conservation works in the church but were hesitant to accept the various revenue models proposed for its upkeep by the students of Jindal School of Art and Architecture. Some of the proposals suggested including opening of a souvenir shop, and cafeteria, running lime training workshops in the rear area, and opening of tuition center in the basement of the adjoining new church. The committee has

Table 3 Themes derived post consultations with various stakeholders

Summary and outputs	
Stakeholders with which discussions were undertaken	Pastor, treasurer, community group, families associated with the church, maintenance staff, conservation architect, craftsmen, contractor, supervisor, and senior advisors
Tangible themes derived from the discussions	Place-making, reconnecting with the street, conservation material sensitivity, structural stability, quality assurance, imageability, and proactive maintenance
Intangible themes derived from the discussions	Quietness, calmness, pride, memories (emotions/ events associated with the Church)

always been economically dependent on donations. This is probably the reason why they were resistant to adopting a revenue generation strategy. They were hesitant to comprehend that in the long run, a self-sustainable revenue model will ensure the longevity of not only the tangible assets but will also give them the freedom to undertake various philanthropic and socio-cultural initiatives at a much larger scale.

It is equally significant for the church committee to collaborate with the state and local bodies to explore convergence opportunities within various state schemes.

5 Conclusion

This paper has examined the context of the challenges involved in the revitalization of heritage in peri-urban areas. In doing so, this paper highlighted how grass root level initiatives can be seen as a way forward in reconnecting the city with its heritage.

Some of the theoretical insights that emerged from this research are as follows:

Firstly, a bottom-up approach ensures flexible, tailor-made, cost-effective solutions along with the revival or strengthening of intangible values. An inclusive urban heritage revitalization initiative involving all members of society revives the lost associations.

Secondly, converging the funds to larger state and national schemes will embed heritage conservation principles within the local planning instruments. It also ensures that if these grass root initiatives are dovetailed with other urban-level initiatives, then it will have a larger impact on the urban heritage of the city. Most importantly for any project to be sustainable it should also have financial viability; a self-sustainable revenue model not only ensures longevity and maintenance of tangible assets but also provides an opportunity to scale up the socio-cultural initiatives.

Finally, a thematic approach presents a holistic perspective towards the heritage which is usually looked at from a singular standpoint. It helps in adapting to new meanings that the living heritage may have with changing times and aspirations. Therefore, it is essential for all stakeholders to work towards the common goal and

Fig. 1 Methodology of the research

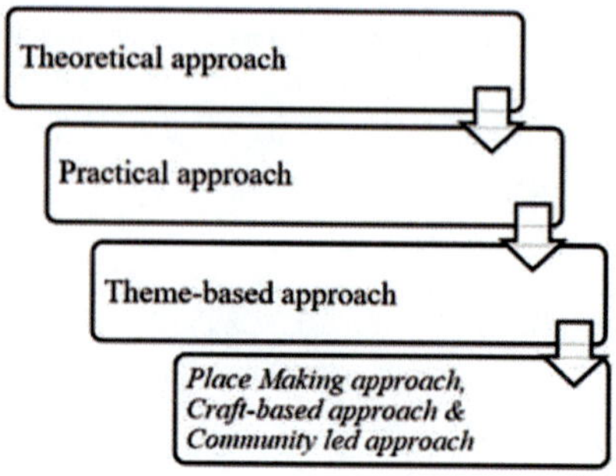

Fig. 2 Church of Epiphany post conservation. *Source* second author, adapt

ensure the heritage places of the past are not lost in the race of urbanization but rather work as a catalyst in making our built environment inclusive, resilient, and sustainable.

Fig. 3 Annual fete in November 2022 at Epiphany Church. *Source* second author, adapt

Fig. 4 Christmas celebrations in the historic church post conservation works. *Source* second author, adapt

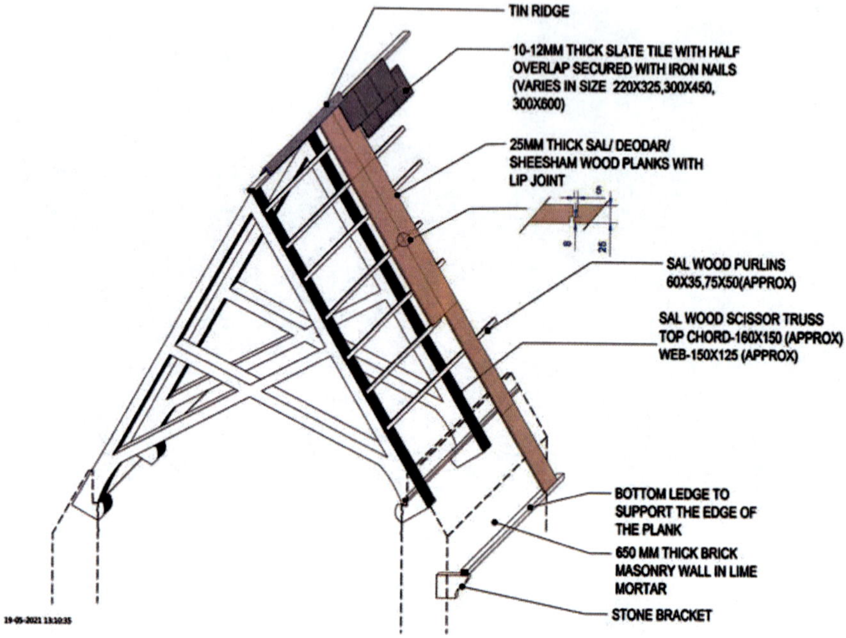

Fig. 5 Original roof detail. *Source* second author, adapt

Fig. 6 Original 19th-century tile flooring revealed when the later additions were carefully removed during the conservation process. *Source* Second author, adapt

Fig. 7 Kids engaged in a painting competition in the historic Church. *Source* second author, adapt

Acknowledgements The authors would like to acknowledge the Epiphany Church committee and all the associated stakeholders for providing valuable contributions toward the realization of this research. The authors would also like to thank the team of experts, craftsperson, and Mr. Sidharth Roy, senior conservation architect at adapt for providing us with project insights.

References

1. Smith L (2011) El 'espejo patrimonial'. ¿ilusión narcisista o reflexiones múltiples ?. Antípoda Rev. Antropol. Arqueol (12), 39–63
2. Glendinning M (2013) The conservation movement: a history of architectural preservation. Routledge, Great Britain
3. Labadi S (2022) Rethinking heritage for sustainable development. UCL Press, London
4. Udeaja C, Trillo C, Kwasi C, Patel DA, Ncube D, Mansuri LE, Jha KN (2020) Urban heritage conservation and rapid urbanization: insights from Surat. India. Sustainability 12(6):2172. https://doi.org/10.3390/su12062172
5. Commissioner, Registrar General, and Census. Census Report. Census Report, New Delhi: Ministry of Home Affairs, Government of India (2011).
6. Jain,S, Dandona B (2012) Haryana: Cultural Heritage Guid. New Delhi: INTACH
7. Munjal PG (2020) Integrating built heritage in development planning: small and medium towns of gurgaon district. Int J Innov Technol Explor Eng (IJITEE) 9 (S), 78
8. Affairs, National Institute of Urban. Managing Peri-Urban Expansion—A Study of Planning Approaches adopted in Delhi, Beijing, and New York to Manage Regional Growth. Research report, New Delhi: National Institute of Urban Affairs, Cities Alliance (2015)
9. Environment, centre for science and gurugram, a framework for sustainable development. New Delhi: Centre for Science and Environment (2017)
10. Epiphany, The church of the heritage church—the church of the epiphany (2022). https://churchoftheepiphany.org.in/, Accessed 5th February, 2023

How to Revive Antic Monument: A Case Study on the Late-Antique Basilica at Crkvina in Breza

Nerma Smajlović Orman, Irhad Mrkonja, and Ahmed El Sayed

Abstract 18 years after being declared a national monument, the Late-antique Basilica at Crkvina, Breza is in an unenviable state, in terms of the condition of the walls, but also in terms of interpretation and visits to the monument. This research paper will provide insight into the current state of the stone remains of the Basilica and all its parts on the site. The existing condition of the stone material at the site is documented, as well as the existing degree of deterioration concerning the last documented condition. Following the qualitative comparative methodology, after the determination of the current condition and the degree of stone decay on the archeological remains, a comparative analysis of the Basilica at Crkvina in Breza with other referent examples is given. Other historically significant buildings for this analysis are selected based on their spatial characteristics and the degree of preservation of archeological excavations that are very reminiscent of the Basilica in Breza. In addition, answers to research questions about importance of the site in this area and comparison with other facilities in the region and beyond resulted in a proposal for future conservation work on the same. So, it is expected that this paper will contribute to the archeological site in Breza coming out of the shadows, long forgotten by scientific circles dealing with architecture, archeology, and conservation.

Keywords Archaeological site · Late antique basilica at crkvina · Breza · Stone · Conservation · Rehabilitation

1 Introduction

Breza is known as a mining town in which there are almost no other economic activities, except for trade and service activities that in the post-war period, after 1995, experienced an expansion in almost all cities in Bosnia and Herzegovina. Back in history in 1907, mining in Breza began to develop, when was opened the

N. Smajlović Orman (✉) · I. Mrkonja · A. El Sayed
International Burch University, 71000 Sarajevo, Bosnia and Herzegovina
e-mail: nerma.smajlovic.orman@ibu.edu.ba

M. Arslanagić-Kalajdžić et al. (eds.), *Interdisciplinary Advances in Sustainable Development II*, Lecture Notes in Networks and Systems 804,
https://doi.org/10.1007/978-3-031-46692-2_14

first mining area and exploitation began [1]. In that period, two late antique basilicas were discovered in Breza, about 25 km north of Sarajevo: one on the left bank of the river Stavnja, in Srđ, and the other on the right bank of the Stavnja, at Crkvina [2]. The Basilica at Crkvina is excavated and conserved until today [3]. The archeological site of the Late-antique Basilica at Crkvina in Breza has been declared in 2004 as a national monument of Bosnia and Herzegovina in a Decision No. 05.1–02-31/04/3 (in later text Decision of designating the site as a national monument) at a session of the Commission to Preserve National Monuments of Bosnia and Herzegovina held in 6th to 10th July 2004 [3].

In October 2020, a private company has done the Main Project of the exterior decoration which includes the first and second protective zones of the Late-antique Basilica in Breza. This project treated mentioned area as three specific parts: a pedestrian street (Desitijatska street), existing remains of the Basilica with an information table, and a park with different features such as pedestrian paths, a green oasis, an open stage, an open exhibition space, children's playground, and sports terrain. This project is presenting an urbanistic solution for the whole area where Basilica is located. The project envisages the protection of the Basilica by forming a glass structure—a cube that will frame it and thereby preserve it from further deterioration and destruction by humans, but also partly by natural disasters. It is planned that the Basilica will be illuminated with reflectors with LED lighting, which can also be seen in other parts of the site in the project. A walkway around the Basilica paved with granite slabs, as well as other pedestrian roads on the site, is planned [4].

Preserving cultural heritage contributes to environmental sustainability, cultural sustainability, and economic sustainability, which are also recognized as main principles in sustainability [5]. But economic globalization poses a major risk of cultural globalization, endangering a place's distinctive identity and built environment, and hindering sustainable development. However, research from around the world demonstrates that preserving cultural heritage can be a key factor in achieving sustainable development goals [6]. This paper will predominantly center on the economic dimension of sustaining the Basilica's site, which will certainly contribute to highlighting these ancient remains for local identities.

This research paper provides insight into the current state of the Basilica's stone remains and all its parts on the site. The existing condition of the stone material at the site is documented, as well as the existing degree of deterioration concerning the last documented condition. In this paper, it will be applied qualitative comparative methodology, i.e., after determining the current condition and the degree of decay of the stone of the archeological excavations, a comparative analysis of the Basilica at the Crkvina in Breza with other relevant examples will be given.

Furthermore, this paper will provide an overview of the Basilica's site significance in this area, and through comparisons with referent examples. The examples of good practice were selected based on an analysis of the condition of the Basilica site, as well as the needs that emerged through observing the potential of this historical asset. Furthermore, in searching for reference examples, those with similar spatial characteristics and a degree of preservation of archaeological excavations similar to the Late Antique Basilica at Crkvina in Breza were identified.

2 Literature Review

The area where Breza is located is historically always a good place to live because of its position. Breza is located on the northeastern edge of the valley (today known as Sarajevo-Zenica valley), at the end of the long, narrow, and gorge valley of the river Stavnja [1]. In a prehistoric and antic period, the area where today Breza is located was the central Illyrian region and ethnic land of the Daesitiati tribe [3, 7]. This tribe lived in the area that included the upper valley of the river Bosna, west to the upper valley of the Vrbas, and east to Rogatica. The Desitiati is one of the most important and numerous Illyrian tribes. They had 103 decuries (fraternities) and belonged to the Naron Conventa (judicial district). They are often mentioned in ancient sources, especially after the rebellion raised against the Romans in the sixth century AD [7].

In written sources, the area of Breza where the Deasitiati tribe developed one of the most important centers, is mentioned for the first time at the beginning of this era. This is most possibly noted in sources as a Hedum Castellum Desitiatium [8]. The rise of settlements on the plains, along with larger and smaller waterways, began in the third and second centuries BC following the breakdown of old tribal structures among the Deasitiati. These settlements were still protected by fortifications. Such a settlement developed in Breza, where a Deasiti necropolis was discovered [9]. One of the most significant researchers on the site of Basilica at Crkvina while excavating discovered the roman inscription of Ulpije Prokule, installed as a spolium. On this spolium are also mentioned Valens, son of Varon (Princeps Daesitiatium) i.e., the prominent Illyrian tribe of the Daesitiati, and a castellum (castellum Daesitiatium) [10]. Spolium is a Latin term that is used for the description of an old building part that is used for building material for a new design. Most often these are stone blocks decorated with fragments of reliefs or capitals, and the like [11].

According to Bojkanovski and Ćelić (1969), the Basilica on Crkvina in Breza is dated from the sixth century. More precisely, from the year 536 AC when Pannonia and Dalmatia belonged to the eastern Roman Empire, and no later than 614 AC and the fall of Salonika, when Byzantine rule ended. There is a possibility that this structure stands from the reign of Justinian in the period from 527 to 565 AC, or even one of his immediate successors [12]. Bojkanovski and Čelić (1969) provide a detailed description of the Basilica and its location. By following this guideline, the decision to declare area a national monument also provides details on the structure and surrounding area.

The building's uniqueness stemmed from its elaborate stone sculptures and distinctive architectural concept. Columns were ornately designed, with grooves and a central entasis, and lacked a clearly defined plinth. The capitals featured two plinths with pointed arches resembling imposts. The frontispiece's trifora included more compact columns [3, 12]. The Basilica has inscriptions on two columns—one in Latin alphabet, the other with a futhorc (futhark) of the older kind with 24 characters whose meaning is unknown. Runic symbols for "d" and "th" were observed on one capital. The graffiti includes the Latin alphabet, the term "misericordia," the letter "P," and an outline of a horse. Runic symbols' height ranges from 0.5 to 2.6 cm, and the "h" symbol suggests a western Germanic origin [3, 13]. The Basilica's history is not without turmoil—it was destroyed by fire, possibly a result of warfare. However, a fascinating discovery was made within its ruins: a tomb holding a Byzantine-made shield's bronze umbo dating back to the sixth century. Despite the damage, remnants of the Basilica's walls endured and were uncovered through the discovery of Early Slav pottery fragments adorned with unique comb-like ornamentation (Kammstrich-muster) from the seventh century. These fragments were found in the narthex region, buried beneath layers of soot [3, 12].

Table 1 provides an overview of the conservation and restoration works that have been carried out on the site and the Basilica since its discovery by Ćorović in 1913 until present day [3].

During a site inspection in May 2004, it was noted that the Basilica is situated in an area surrounded by private properties and houses. Despite some recent physical damage and two fallen columns, the walls are mostly well-preserved. Fragments of the building can be found both inside and outside, particularly in the north colonnade and entrance. The narthex and nave show bare patches where the grass has been worn away, and it is reported that children use the area to play football [3].

To gain insights and inspiration that could be applied to the restoration of the Basilica in Breza, the authors visited the Ostrožac fortress in Cazin, Bosnia and Herzegovina. This impressive medieval stone structure was officially recognized as a national monument in 2013 [14]. Two extraordinary things were noticed: a naturally drained pathway that blends with the surroundings and artistic stone sculptures lying beside a walkway, showing respect for the site's history and modern aspirations. The second example is an archeological site Baelo Claudia which is located on the south coast of Spain, excavated between 1917 and 1921. Reconstruction works carried out on the Basilica of Baelo Claudia were in 1971. Alfonso Jimenez made documentation of all architectural features prior to the removal. Jose Mese Alanis worked with his team on the trials in the treatment of stuccos in the building. Reconstruction works that have been done in the following years included connecting a great number of blocks and architectural elements that had fallen in situ [15]. A suggested visit plan for the archaeological site is available online, featuring a site tour plan with explanations of zone locations and features for optimal presentation and interpretation [16].

3 Methodology

The methodology used in this paper will be qualitative comparative methodology. Qualitative comparative analysis (QCA) acts toward social science categories as sets and views cases in terms of their multiple memberships [17]. Charles Ragin created it in the 1970s intending to use it as a strict procedure. As a result, the various phases are quite well-defined and ought to be used consistently throughout all QCA investigations [18].

A comparative analysis of the Basilica at the Crkvina in Breza with other relevant examples will be given. Examples are selected which, with their spatial characteristics

Table 1 Summary of conservation and restoration works done on the Basilica in Crkvina Breza since its discovery in 1913

Years	Works
1930–1931	Dr. Gregor Čremošnik, the curator of the National Museum in Sarajevo, discovered remains of a Late-Antique Basilica and published a preliminary report [10]. The artifacts were later transferred to the museum
1958–1961	Excavations led by Dimitri Sergejevski did not result in published findings
1961–1968	The Institute for the Protection of the Cultural, Historical, and Natural Heritage of BiH finished conservation and restoration works, under the division into four phases: ***I phase*** *(1961–1963):* conservation works by Mato Biško, a technician from the Institute ***II phase*** *(1965):* a design project for partial restoration of the Basilica was done, led by Ivo Bojanovski and with input from Professor Sergejvski. The restoration works were led by Ivica Tometinović and E. Dimitrijević. The project considered minimal and maximal variation for the elevation of walls and reconstruction of sculptural features. Ultimately, restoration was carried out with minimal variation ***III phase*** *(1967–1968):* finishing works in a frame of minimal variation of restoration and presentation of the building, under the leadership of Ivo Bojanovski, following additional program dating from 1967. Finalization of the upper line of the walls, the portal on the east side, part of the columns and capitals, and were drawn up by Đuro Basler ***IV phase*** *(1968):* works are finalized and presented, based on the project made by A. Ninković, a conservator working for the Institute for the Protection of Monuments of Bosnia and Herzegovina, I. Bojanovski, who managed the works, and Dž. Čelić, consultant
1990	The Institute for Architecture in Sarajevo drew up a Regulatory Plan named "6th April" at the request of Municipality Breza. The plan involves extending the fence, adding an entrance gate, and rest area, landscaping, and designing new roads. However, these actions have not been carried out yet
2000 *latest works*	The Institute for the Protection of the Cultural, Historical, and Natural Heritage of Bosnia and Herzegovina recently carried out conservation and restoration work on the badly damaged walls and columns of the Basilica, which suffered during the 1992–1995 war. As part of the project, a new 1.20 m high metal fence was installed three meters from the walls of the building
Note	*The rest of the movable archaeological material is located and kept in the National Museum in Sarajevo*

and the degree of preservation of archeological excavations, are very reminiscent of the Basilica in Breza. Steps in this methodology are graphically shown in Fig. 1.

Research questions and hypotheses are set after determining the methodology.

RQ1: What role does the Basilica at Crkvina in Breza play in history?

RQ2: What is the condition of the Monument today concerning the last documented condition from the Decision on the Designation as a National Monument?

RQ3: Taking into account the current state of the Monument, what are the possible interventions and measures that would bring this Monument into an enviable state?

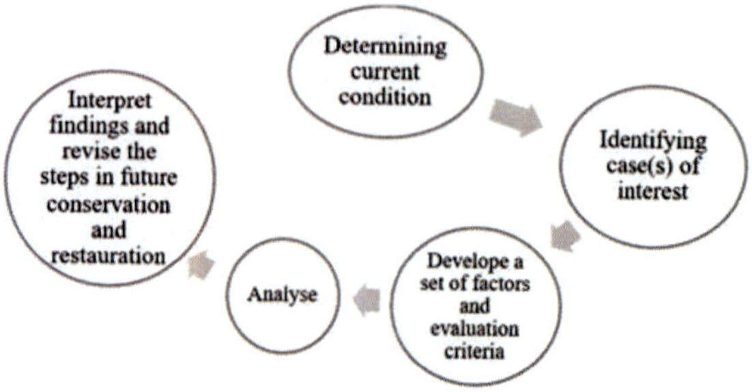

Fig. 1 Methodology flowchart made by authors

H1: The basilica at Crkvina in Breza has significantly deteriorated compared to 2004 when it was declared a national monument.

This paper aims to bring the archaeological site in Breza into the spotlight and prompt experts in architecture, archaeology, and conservation to consider the next steps in preserving this valuable piece of history, so it is not forgotten again.

4 Results and Discussion

On June 29, 2022, the authors of this paper conducted a physical inspection of the Basilica in Breza and identified numerous problems facing this archaeological site. The examination was based on the remaining stone walls and pillars, which revealed a clear trend of deterioration due to neglect and inadequate preservation works. This inspection occurred 18 years after the site was declared a protected cultural asset by the Commission, and the situation had worsened since the last recorded state. The fence that was installed in 2000 is now damaged, without an entrance gate, leaving the Basilica vulnerable to human and animal interference. The grass inside and outside the fence has been mowed, but the hay remains inside the complex, making it a potential hazard for visitors due to the possibility of snake infestations. Figure 2 depicts the current state of the archaeological site of the Basilica in Breza.

Furthermore, weeds are growing in between the stones of the remaining walls of the Basilica, and other bushy vegetation has been found nearby, along, and on

Fig. 2 Archeological site's current condition on June 29th, 2022 (*Source* Photo is taken by Authors)

the walls. This is clearly visible in Figs. 3 and 4. The main issue with the surviving walls of the Basilica, which have previously been conserved with unsuitable mortar, is shown in Figs. 5 and 6. This unsuitable mortar has caused the decay of the stones due to prolonged exposure to external conditions such as precipitation, winter frost, and summer heat. In all current photos taken by authors on June 29th, 2022, vandalism in the form of paint on the walls is also evident.

One of the most notable findings from the physical inspection of the site during the summer of 2022 was the significant difference in the condition of the walls compared to the photos taken in 1968 and 2004. The north porch, in particular, showed the greatest difference, with the remains of the walls and steps of the Basilica being damaged. This damage was likely caused by human negligence, as this area is most exposed. Additionally, the fence in the same area had been damaged and removed, leaving the archaeological site completely open to the street. While it is unclear to what extent the walls of the northern porch had changed between 1968 and 2004, changes were clearly visible from 2004 onwards. The inspection conducted in 2022 has revealed that a number of stones have become dislodged from both the northern and southern walls of the porch, resulting in a significant modification to the final

Fig. 3 Bushy vegetation is deteriorating stone walls on June 29th, 2022 (*Source* Photo is taken by Authors)

Fig. 4 Weeds grew from south wall joints on June 29th, 2022 (*Source* Photo is taken by Authors)

Fig. 5 (left) Stone decay on the south wall due to unsuitable mortar, documented June 29th, 2022

aesthetic of the walls. Table 2. is a summary of detected damages and decay of the Basilica's walls compared to conditions from the photos in 1968 and 2004. It is important to mention that the photographs from 2004, referred to in this analysis, are accompanying documentation of the Decision to designate the site as a national monument.

Figure 7 presents a comprehensive list of the physical damages and deterioration detected on the surviving walls of the Basilica. These issues can be ascribed to two primary causes, namely, anthropogenic factors described in chart as a man-made and a range of climatic and environmental variables. After scrutinizing the chart, it can be deduced that one of the most crucial missteps was the incorrect selection of conservation materials, which were presumably installed in the year 2000. This category of mortar is not suitable for historical stone structures and was the initial factor that led to the observed decay in the Basilica. Nevertheless, this mortar was employed in conservation work that was carried out in the past 22 years, with no

Fig. 6 (right) The external south wall of the Basilica and the remaining column base, revealing previous unsuitable mortar conservation leading to stone decay, documented on June 29th, 2022 (*Source* Photos are taken by Authors)

evidence indicating any other restoration or conservation endeavors being performed from 2000 to the present date.

The exemplary cases cited at the outset of this paper have demonstrated the crucial role that smart management plays in the interpretation, conservation, restoration, and maintenance of historically significant edifices. Foremost on this list is the development of a comprehensive plan for preserving the extant walls of the Basilica, which involves arresting the decay of the stones and subsequently rehabilitating this cultural legacy that bears witness to the period when Breza served as a pivotal gathering place for residents, travelers, traders, and merchants. The Basilica itself must become the site for the telling of this story. Finally, an essential aspect of the economic sustainability of the site will be achieved through regular maintenance and conservation efforts.

The project for the exterior decoration of this archaeological site created by a private company which is mentioned in this paper proposing a form of intervention

Table 2 Summary of detected damages and decay of the Basilica's walls, done by Authors

Parts of Basilica	Condition in 2004 compared to 1968	Condition in 2022 compared to 1968 and 2004
North porch	Unknown	The western external wall of the porch is severely dam-aged: a photo from 1968 shows a wall intact to the right of the entrance; the wall's condition in 2004 is uncertain, but as of the summer of 2022, the porch wall has been extensively damaged, extending to the northwest corner of the Basilica The bordering wall between the narthex and north porch is also impaired, with missing stones and a change in shape in the middle Weeds and bushes are growing on the top and along the entire length of both internal and external walls
Remains of stairs	Unknown	The stairs are overgrown with weeds
West porch	The structure of the wall and columns remained the same	The external wall's top retains its shape from 2004 bushes grow at the top of the northern part of the external wall, and weeds grow all along the wall The remains of the pillars were overgrown with grass, damaged and destroyed by vandalism: stone engraving
Deaconion	The structure of the wall remained the same	Top damaged: stones missing, but shape same as 2004 weeds grow on the walls' top and both external and internal sides The outer north corner of the deaconion is overgrown with weeds and bushes
Apse	The structure of the wall remained the same	Top damaged: stones missing, but shape same as 2004 Weeds growing on the top and along the entire length of both internal and external walls
Martyrdom	The structure of the wall remained the same	The external south-east corner is damaged: missing stones created a hole in the wall from the external side Top of the walls: missing stones even shape remained from the condition in 2004 Weeds growing on the top and along the entire length of both internal and external walls
East porch	The external wall is damaged: missing stones northern from the columns but the wall has remained in continuity	The external wall's top retains its shape from 2004 bushes grow at the top of the northern part of the external wall, and weeds grow all along the wall The remains of the pillars were damaged and destroyed by vandalism: stone engraving The outer north corner of the porch is overgrown with weeds and bushes
Nave and Narthex	The border wall between the nave and narthex is damaged: Missing stones at the top of the west side wall	Top on the walls: missing stones in places even shape remained from the condition in 2004 Weeds growing on the top and along the entire length of both internal and external walls

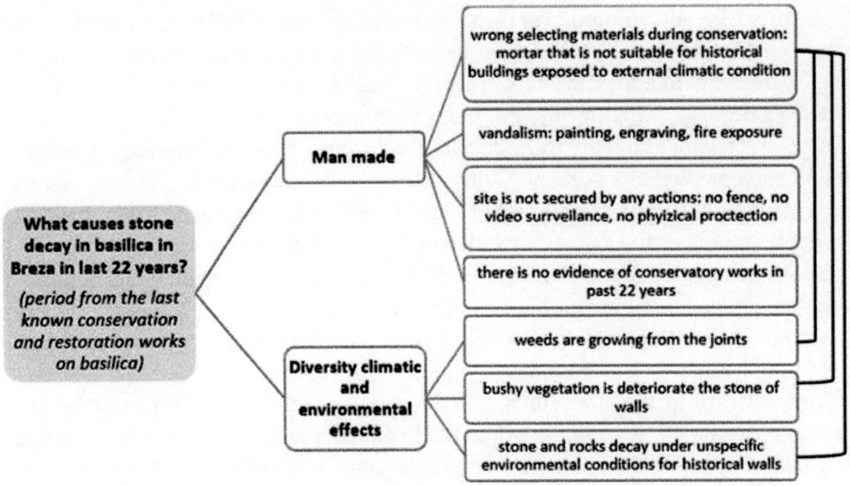

Fig. 7 Stone decay caused on the Basilica remains from the last conservation and restoration works, done by Authors

on the archaeological remains of the Basilica such as the design of a glass cube used for protection of the structure. It is imperative to consider the potential production of greenhouse gases within the confines of this glass cube. As greenhouse gases are known to trap heat and elevate the planet's temperature, the absence of proper ventilation within such structures can cause the temperature and air humidity to rise by an additional 30 degrees Celsius [19, 20]. The presence of greenhouse gases within such structures can accelerate the deterioration of stone structures that are connected with mortars, as the excess moisture caused by these gases can lead to a more rapid decay of the walls in question.

When it comes to interpreting the site, there is currently no means of conveying the story of the Basilica on location, such as through signage, plaques, or illustrated information board about the site name and basic heritage information. This could potentially be addressed through the implementation of a static board, such as the one located on the front wall of Ostrožac Castle in Cazin, or an interactive exhibit with video storytelling and a 3D model presenting the assumed appearance of the Basilica.

Following the examples set by referent cases, it is imperative that the Basilica begins to incorporate commercial features linked to Breza, thoughtfully planned as daily, monthly, seasonal, or yearly events. In consultation with the town's residents, the authors have learned of a recent promotion by the library located near the Basilica, which showcased writers and books within the Basilica's remains. While this is but one of many methods for promoting this cultural heritage, it should also be developed into an income of Basilica's site. The municipality must establish a fund to collect money from such events for the maintenance of the Basilica's area, as well as for ongoing conservation efforts. Inhabitants have also noted that Breza becomes a hub for social and cultural events in the Sarajevo-Zenica Valley for a week in the summer. There is ample potential for including the area around the Basilica in this event, perhaps with a transformable stage for cultural events located adjacent to the Basilica's remains, which would not harm the archeological site.

One crucial measure involves devising a cultural heritage interpretation plan, ideally implemented as a workshop with students specializing in architecture, history, and archaeology, demonstrating a deep respect for sites like the Basilica. Not only will this approach foster fresh and innovative ideas, it will also encourage future professionals to acknowledge the significance of these remains. However, it is essential to note that detailed documentation, including drawings, 3D models, and comprehensive condition assessments, is currently absent. A workshop presents an affordable solution to address these challenges.

It is not rare to find museums or exhibition spaces situated adjacent to such historical remains. The example of the Baelo Claudia archaeological site is particularly inspiring due to the enclosed museum built in close proximity to the Basilica's ruins, providing additional facilities. Alongside a museum space, the location could also feature multipurpose spaces for exhibitions or workshops. This presents a great opportunity for Breza, where there are no other museums despite the rich history from the Basilica's earliest construction to the present. The proposed building should be contemporary yet constructed with materials that respect the site's cultural heritage, such as concrete, steel, or glass. Such a project would not only provide work opportunities for the inhabitants of Breza but also contribute to the fund for regular maintenance of the Basilica's remains and the area that the municipality designates to the site. It is undeniable that the museum would support the cultural identity of the place, but it is not necessary for the exhibition artifacts to be exclusively part of this cultural heritage, but also of the identity that Breza has built throughout the centuries that have followed.

Table 3. is showing proposed steps in future conservatory and rehabilitation works at the archeological site of the Basilica at Crkvina Breza. These steps must be done following the existing document Decision for designation as a national monument.

5 Conclusion

Breza has always had an important place in history, from the Illyrian state, through the Byzantine Empire, when the Basilica was built. There is a break in history, when it was a simple inhabited place, with reduced potential for growth and development, which was interrupted by the opening of the Breza Mine at the very beginning of the twentieth century when the present-day city core of this place began to develop. During that period, many workers came from other parts of Bosnia. Later in 1930 and 1931, according to the writings of Dr. Gregor Čremošnik, the Basilica was discovered, the excavation of which began then. The Late Antique Basilica in Breza, dating back 1500 years, serves as a significant archaeological site that echoes the stories of rituals, conversations, events, and people from the past. However, the site's

Table 3 Summary of proposed steps in future conservatory and rehabilitation works at the archeological site of the Basilica at Crkvina Breza done by Authors

#	Steps in Progress	Benefits
	Creating a fund managed by the Municipality of Breza that will collect assets for later conservation, rehabilitation, or maintenance of the archeological site	Regular conservation and maintenance of the archaeological site, including job creation for residents
	Creation of project documentation that will be proof of existing condition before any steps in a matter of conservation or rehabilitation of archeological site and Basilica	Monitoring of the current situation, which will determine the further intensity of work on the remains of the Basilica
	Examine the stone remains and determine the degree of damage and other significant elements related to the deterioration of the stone	
#	**Conservation Measures**	
	The most important step is to stop the deterioration of the archeological site, i.e., stone decay!	Extending the life of the remaining walls of the Basilica
1	Remove unfavorable mortar used for previous conservation works on the Basilica, that caused stone decay by catching the water and moisture inside	Steps will make the area of the Basilica more accessible to visitors and give importance to the building itself Note: proposed walking pathway must not allow rainwater to collect in place or destroy underground part of stone remains. Also, walking paths must not make pressure on the walls of the basi
2	Clean low vegetation and weeds on all remaining walls of the Basilica	
3	Clean the damage on the Basilica's remains that is caused by fire, graphite, and other vandalism traces	
4	Clean the site around and inside the Basilica that could in the future cause damage or deterioration of stone remains of the Basilica	
5	Complete high-quality grouting of the remaining part of the walls with mortar recommended by experts in conservation work	
6	In the inner part of the Basilica, pave the entire space or only the middle part with broken stone slabs that would be pressed into the ground, to get the effect of a floor with greenery again, without damaging its original condition or using white gravel that will not keep rainwater	
7	In the case of constructing a floor made of broken stone slabs, a larger amount of water would be created and retained inside the Basilica itself, and it is necessary to provide drainage to lead water outside the walls of the Basilica	
8	Around the Basilica itself, make a walking path for visitors, which will be slightly higher than the terrain around the Basilica, and have a slope towards the terrain, and which will also be made of the same materialization as inside the Basilica	

(continued)

Table 3 (continued)

#	Steps in Progress	Benefits
9	Dismantling the existing fence, whose height and shape do not emphasize the importance of what the fence physically protects, but makes this place more accessible to acts of vandalism	Steps will make the area of the Basilica more accessible to visitors and give importance to the building itself
10	Installation of a new fence that will be transformable or movable in case of need for rehabilitation events	
	To ensure that the location of the new fence is following the project of the wider arrangement of the Basilica and the possible future construction of accompanying buildings but also following the projects for the rehabilitation of the archaeological site	
11	Installation of an entrance gate with automatic ticket automat for visitors with a symbolic ticket price	Safer environment for archeological site Collecting assets for conservation fund
	Collected money should be transferred to the fund for Basilica's conservation and rehabilitation managed by Municipality	
12	Installation of video surveillance, covering all parts of the Basilica	
13	Installation of lighting on the entire archeological site with highlighting Basilica. Lightning should be done under the project by professionals	
14	Installation of the static or interactive table with information about the site and brief history of this area in the domestic and English languages	Emphasizing the importance of the Basilica Arousing the interest and attention of visitors Introducing visitors to the Basilica to the cultural heritage and history of the area
15	Installation of static or interactive panels with drawings to showcase the Basilica's aesthetic formation during its best time. Interactive panels or holograms can promote the Basilica and present a 3D model, as well as serve as an announcement platform for upcoming events and notifications	
	Rehabilitation Measures	
1	Create/build a shared space for indoor museums, exhibits, and events	
	The building could also house a local radio station that has been existed for 30 years	
2	Create connecting paths to the swimming pool complex and cinema building, which is a center of cultural and social activities and has an outdoor children's playground	
	At the moment the city's swimming pool complex is not in use, but according to the announcements of the Municipality in future years, it will be revived again *In mentioned cinema building that is held at this moment by the BCC "Preporod", are located other cultural and social societies, such as the local radio station "Radio Breza", Library "Muhamed Kantardžić", chess club, CCS "Napredak" and newly opened social house as a part of Municipality of Breza*	
	Creating a working plan for seasons such as workshop month, Breza's summer week, etc. That will be a permanent list of activities that will collect funds and pay for salaries to those that will be working here as curator or similar	

significance to the inhabitants of Breza extends beyond its historical value. Despite its central location within the municipality, its potential has yet to be fully realized.

As a result of investigating the second research question, it was found that the state of the Late Antique Basilica in Breza has partiall deteriorated compared to its state when it was declared a national monument. The exterior walls have collapsed or been eroded due to the passage of time and lack of care, which is evident in the low-quality binding material used to preserve the Basilica. Although the shape of the walls has remained the same mostly, the top of the walls has been altered in almost all places.

As the answer to the third question about possible interventions and measures that would bring this monument into an evitable state, it is necessary to follow the steps in Table 3. Those steps are about conservation works mostly, even though there is a proposal for a few rehabilitation measures that could be implemented. All of this is to bring the destroyed or deteriorated parts of the Basilica's remains into a decent state. Furthermore, it should be noted that reconstruction is a conservation method that is not recommended in cases where authentic materials are lacking, as is the case with the Late Antique Basilica in Breza.

The documented level of decay and deterioration of the stone remains serves as a warning for cultural heritage institutions, the Municipality, and residents of the settlement. The damages are largely due to human negligence, either through direct destruction or indirect disregard for this important heritage and identity of Breza.

Before the intervention and the construction of the glass cube above the archaeological remains of the Basilica, how it is planned by the latest project, recommendations are to do a scientific study should first, which would present how the future structure will affect the Basilica in the course of its further life. This study should make a special reference to greenhouse gases due to closing the structure in a potential greenhouse.

Recommendations for future researchers include a deeper consideration of environmental and social sustainability principles when approaching the protection of cultural heritage. Additionally, this paper should serve as a starting point for further research and increased recognition of the significance of this cultural heritage site.

References

1. Efendić E (2022) Breza—Geoekološki problemi (diplomski rad), https://dokumen.tips/documents/elvir-efendic-breza-geoekoloski-problemi-2006-diplomski-rad.html, last accessed: 2022/05/31
2. Ćorović V (1913) Izvještaj o iskopavanjima u Brezi. Glasnik Zemaljskog muzeja u Sarajevu XXV, 409–420
3. Commission to preserve national monuments in bosnia and herzegovina, Decision of designation of properties as national monuments: Late antique basilica at Crkvina, the archaeological site, http://old.kons.gov.ba/main.php?id_struct=50&lang=4&action=view&id=2481, last accessed 2022/06/01
4. Studio URBING d.o.o. Sarajevo: Glavni projekat vanjskog uređenja prve i druge zaštitne zone kasnoantičke bazilike u Brezi, Sarajevo (2020)

5. Rypkema DD (2006) Economics, sustainability and historic preservation. Forum J 20(2):1–27
6. Gražulevičiūtė I (2006) Cultural heritage in the context of sustainable development. Environ Res Eng Manag 37(3):74–79
7. Imamović E (1997) Antički kultni i votivni spomenici na području Bosne i Hercegovina. Veselin Masleša, Sarajevo
8. Patsch K (1906) Arheološko-epigrafska istraživanja povijesti rimske provincije Dalmacije, VII dio. Glasnik Zemaljskog muzeja u Sarajevu XVIII, pp 151–181
9. Paškvalin V (1975) Kamenjača, Ul. 6.aprila, Breza kod Sarajeva. Arheološki pregled 17, pp 57–62
10. Čremošnik G, Sergejevski D (1930) Gotisches und römisches aus Breza bei Sarajevo. No-vitates Muzei Sarajevoensis 9, Sarajevo
11. Proleksis Enciklopedija: Spolij—Struka likovne umjetnosti i arhitektura, https://proleksis.lzmk.hr/46574/, last accessed 2022/06/22
12. Bojkanovski I, Čelić D (1969) Kasnoantička bazilika u Brezi. Naše starine XII, pp 7–25
13. Looijenga T (2003) Texts and contexts of the oldest runic inscriptions. Brill, Leiden
14. Komisija/Povjerenstvo za očuvanje nacionalnih spomenika BiH: Odluka o proglašenju graditeljske cjeline—Star grad Ostrožac, općina Cazin, nacionalnim spomenikom BiH, http://aplikacija.kons.gov.ba/kons/public/uploads/odluke_bos/Cazin_Stari%20grad%20Ostrozac%20BOS.pdf, last accessed: 2023/02/13
15. Conjunto Arqueológico de Baelo Claudia (2022) The first great restoration of baelo claudia, google arts and culture, https://artsandculture.google.com/story/the-first-great-restoration-of-baelo-claudia/yAXBQeN-KQy2LQ, Last Accessed: 2022/09/05
16. Conjunto Arqueológico de Baelo Claudia (2022) Presentacion conjunto, propuestas de recorrido, consejería de cultura y patrimonio histórico, https://www.museosdeandalucia.es/web/conjuntoarqueologicobaeloclaudia/propuestas-de-recorrido, last accessed: 2022/09/06
17. Ragin CC (1998) The logic of qualitative comparative analysis. Int Rev Soc Hist 43:105–124
18. Simister N, Scholz V (2017) Qualitative Comparative Analysis (QCA), Intrac
19. World Meteorological Organization: Greenhouse gases, https://public.wmo.int/en/our-mandate/focus-areas/environment/greenhouse-gases, Last Accessed 2022/10/03
20. Environmental protection ageny united states: sources of greenhouse gas emissions, https://www.epa.gov/ghgemissions/sources-greenhouse-gas-emissions, Last Accessed 2022/10/03

Perspectives and Challenges of Sustainable Preservation of the Hotels from the Socialist Period in Bosnia and Herzegovina: Focus on Hotel Visoko

Lejla Džumhur, Lejla Kreševljaković, and Nermina Zagora

Abstract Preservation of the socialist-built heritage in the countries of the former Yugoslavia has become a trending topic in the recent architectural discourse, surrounded with controversies due to a bias towards the legacy of the past socialist regime from 1943–1992. The architectural heritage of Bosnia and Herzegovina from this period is abundant, yet it is considered endangered. Consequently, this research strives for a scientifically based approach to the assessment and sustainable preservation of the socialist architectural heritage in Bosnia and Herzegovina. Architectural typologies designated for commercial purposes are particularly vulnerable. Affected by the ideals and aesthetics of the post-socialist reality, tourism facilities have been sacrificed in the process of transition, as in the case of the remarkable hotel architecture designed by the renowned Bosnian architect Zlatko Ugljen. The only exception is the hotel Visoko because it still remains almost intact. Based on the premise that the *status quo* of the hotel Visoko can be regarded as an opportunity, this paper seeks the most appropriate approach for its sustainable preservation. The assessment of the designated hotel as a "socialist architectural heritage" encompasses the impacts of the socialist ideology and the architectural aesthetics of modernism/postmodernism. Recognition of the values of the Hotel Visoko as an outstanding example of town-hotel architecture from the socialist period is the basis for sustainable preservation of its physical structure and reinvention of its social identity in a dialogue with the community. The new vitality of the hotel Visoko can be generated from the inside out. The introduction of participative, iterative methodologies in interior design can simultaneously strengthen the existing and authentic qualities and create a new social value inherent to the socialist architectural heritage.

Keywords Modern heritage · Socialism · Renovation · Adaptive reuse · Hotel architecture · Interior design

---

L. Džumhur (✉) · L. Kreševljaković · N. Zagora
University of Sarajevo – Faculty of Architecture, Patriotske lige 30, 71000 Sarajevo, Bosnia and Herzegovina
e-mail: lejlas@af.unsa.ba

M. Arslanagić-Kalajdžić et al. (eds.), *Interdisciplinary Advances in Sustainable Development II*, Lecture Notes in Networks and Systems 804,
https://doi.org/10.1007/978-3-031-46692-2_15

1 Introduction: Assessment of Value and Sustainable Preservation of the Town-Hotels from the Socialist Period in Bosnia and Herzegovina

The leading drive to conduct a scientific assessment of value and analytically point at the correct and sustainable path of restoration and protection of the Hotel Visoko lies in the fact that the attitude towards socialist heritage in Bosnia and Herzegovina has not yet been articulated.[1] This is particularly evident in the case of architectural heritage with a high share of commercial functions, such as tourism architecture. In the process of transition, it has suffered extreme transformation and adjustment to the aesthetic and functional visions of post-socialist realities. The majority of the exceptional opus of hotel architecture by architect Zlatko Ugljen is irretrievably lost.[2] Only by chance, due to auspicious circumstances, the Hotel Visoko has remained almost untouched in its exterior appearance.

By now, the most comprehensive overview of the regional and historical development of tourism architecture on the territory of former Yugoslavia was presented in Nebojša Antešević's doctoral dissertation entitled "Arhitektura modernih turističkih objekata Jugoslavije (1930–1985)" (Architecture of modern tourism buildings of Yugoslavia (1930–1985) [1]. He points to the inability to draw clear territorial or stylistic borders in terms of typology development.[3] The town hotels, according to Antešević, represent a distinctive type whose primary feature is determined by the capacities of the respective urban environment. They reflect the context of the city, regional, social, and economic development of the urban environment [1]. The Hotel

[1] The legacy of tourism architecture from the socialist period has been attracting more and more attention from the scientific community and cultural heritage agencies and organizations in recent decades, particularly in those segments that are dealing with the phenomenon of tourism architecture on the Yugoslav Adriatic coast. Besides the architectural perspective, the highlight of the discourse is on the role of tourism capacities in constructing the ideology of integral Yugoslav identity. In other regions in the former Yugoslav federation and typological variants outside the primary tourism infrastructure along the seaside, the study of values and characteristics, and consequently their protection, has been almost completely left out. For a deeper insight into the issue see [2, 3].

[2] Architect Zlatko Ugljen designed the following hotels in Bosnia and Herzegovina: *Orient* in Travnik (1973), *Visoko* in Visoko (1974), Motel *Vukosavci* (1976), *Ruža* in Mostar (1977), *Bregava* in Stolac (1977), *Kalin* in Bugojno (1983), *Vučko* on the Jahorina Mountain (1983). So far renovated hotels of the architect Zlatko Ugljen, due to an inadequate approach to renovation, testify of the partial or complete degradation of their original values. The renovation of the Ruža Hotel was very much in the public eye, but despite UNESCO and ICOMOS's involvement, the hotel was demolished. A new hotel was built in its place, ignorant of the scale, material, design, and spirit of the original building [4]. Hotels Kalin and Vučko experienced a similar fate. The last renovated hotel designed by Ugljen was the Hotel Orient in Travnik, in 2022. The redesigned hotel has a completely new character that does not communicate with the previous one. The designer Bernard Bostjančić introduced new formal elements on the facade and in the interior, related to oriental arabesque motifs, which were not present in the original building. These and other design elements have given a completely new character to the object.

[3] In his doctoral thesis, Nebojša Antešević reviews the development of the tourism architecture typology according to regional and cultural peculiarities through five main chronological units: 1930–1941; 1947–1955; 1956–1966; 1967–1975 and 1976–1985 [1].

Visoko, designed by architect Zlatko Ugljen, which represents the case study in this research, is an example of a town-hotel typology [1].

To conduct the valuation study of such typology of town hotels, it is necessary to introduce some additional assessment parameters, building upon Antešević's theoretical framework mentioned. Namely, in the process of assessment of socialist built heritage, two notions contained in the term "socialist architectural heritage", are to be considered: the notion of "socialist particularities" as an aspect of value related to practices and logics of the socialist system, and the notion of "architectural and functional design layout", which is independent of the previous one, and describes functional particularities of the type, aesthetic of modernism/postmodernism, and author's implications on the value of his creation.

Consequently, an inevitable segment of the evaluation of socialist architectural heritage is the determination of its "socialist" attributes. By all means, it is not to be found in the formal expression. Namely, as Zarecor [5] states, "architects across ideological systems embraced the postwar international style and its urbanism as appropriate to their era and circumstances." Also, the simplified statement of formal or visual essence of "socialist modernism" in the expressions of brutalism, aesthetics of monuments, or attributions outside stylistic characteristics, such as "dilapidation" or "ugliness", "has as much sense as the attempt to identify inherent aesthetic characteristics of capitalist modernism" [6]. When considering this problem, Zarecor offers a surprisingly simple summary of the issue: "In the simplest terms, all socialist cities were modernist cities, but not all modernist cities were socialist" [5]. Such a view, to an extent, translates to individual buildings and typology subsets, and, while keeping a small question mark, we can say that all socialist buildings were modernist buildings, but not all modernist buildings were socialist. What is considered "socialist", still remains unarticulated.

Zarecor [5] has embarked on the path of qualitative analysis when determining socialist attributes of urban systems. Her theory may also be adopted as a methodological template for questioning the place of typological categories in the system of socialist cities.

2 Research Methodology

In the first instance, this study seeks to establish a general framework for the value assessment of town-hotel architecture from the socialist period in Bosnia and Herzegovina (from 1943–1992), as a basis for its sustainable protection. Following the proposition of embodiment of specific socialist features in socialist architecture, we conduct the extraction of such characteristics, employing the methodological/theoretical framework determined by Zarecor [5]. She unfolds and questions the logic and practices of spatial, economic, and urban development innate to socialist societies and its spatial consequences—interaction designated by Zarecor as a *socialist scaffold*. According to this theory, we track, bottom-up, the socialist features of infrastructural thinking and the logic of the socialist scaffold in the building practice of town hotels in Bosnia and Herzegovina.

The problem is specifically addressed in the case study of the Hotel Visoko in Visoko. Its value assessment is based on a separate analysis of socialist features and characteristics of the author's modernist creation.

Such elements of the established value, according to recommendations from UNESCO Operational Guidelines [7], will determine the decision concerning the level of intervention in the protection and preservation program and its estimated role in sustainable development.

From this point of view, a particularly challenging task is the search for the right approach to interior restoration. Namely, the original interior is seen as an important component of the value, but also as the component that has suffered the greatest degree of destruction, also lacking original project and photo documentation.

Questioning to what extent the original form and design of the interior support the established value-interrelation proposed by UNESCO doctrine [7]—and acknowledging the potential of the future renaissance of the hotel, will allow us to formulate a scientific-based attitude towards its role in sustainable preservation.

3 Assessment of Value

As recommended in Approaches to the Conservation of Twentieth-Century Cultural Heritage [8] to use accepted heritage identification and assessment criteria, this research starts with the study of tangible attributes, including the site and location, the views, architectural forms and spatial relationships, color schemes, construction systems, material finishes, technical equipment, as well as research of the significance that lies in "use, historic, social, scientific or spiritual associations, or evidence of creative genius and/or in its intangible values" [8].

'Socialist' Attributes of Value

According to Zarecor the infrastructural thinking and logic of the socialist scaffold are the analytical frameworks used for the observation of socialist cities outside of the dominant paradigm of an urban structure as a direct manifestation of a political top-down system.

In the context of the revolutionary demographic, social and production transformation in the first half of the twentieth century and the decades after World War II, reflections on urban structure occupied many great thinkers of urbanism and architecture. What makes the difference when comparing capitalistic and socialistic infrastructural thinking is not the question of scale which such thinking has reached, having in mind the proportions of Le Corbusier's syndicalist Ville Radieuse [9], but also the scale and continuity of its spatial and temporal implementation, which was in socialist societies integrative, comprehensive, and long-lasting. Exactly such level and scale were enabled by the ideological narrative of the socialist system, where the city was seen as a critical element in the project of profound socialist transformation. An economic model of development was inseparable from the ideological discourse.

Zarecor refers to the implementation of socialist infrastructural thinking as a socialist scaffold. Such ideologically based mechanism of implementation has included and intertwined the political, economic, social, and environmental systems and instruments at various levels, specifying how and in what way the society would continue to develop. It operated with spatial and ideological tools. "The socialist scaffold became material and perceptible through urban systems, but it also had an overarching ideological thrust that allowed a set of decisions to cohere into a recognizable strategy. […] Its purposefulness did not derive from an aesthetic expression of power that could lose or change meaning over time, but rather from relationships manifest between objects, images, and people in space" [5].

We can distinguish the spatial and temporal implications of this mechanism as specific socialist particularities. Spatial implications, as opposed to the scaffold of another political system, were comprehensive and "outside the field of real estate development and its motives to earn profit". The city was one of the nodes of Kotkin's "single entity" [10]. Namely, it represented an economic and social ideology of "distributing responsibility of producing specialized equipment across the region to avoid redundancies in the network. By extending this concept to the design of socialist cities, each one can be seen as a node in its country's single entity, reinforcing the master narrative of the national economy by positioning itself as a generator of specific material goods and general prosperity, while also reproducing the idea of the single entity at multiple scales—the urban region, the city, the district, and the neighborhood" [5].

The temporal implication of such ideological discourse required gradual implementation of the mechanism. "Designers at state-run design institutes were tasked with setting in motion an urban system that would come into being over decades, rather than in the few years of a typical capitalist real estate development cycle. Such step-by-step growth was desirable given that the five-year planning increment required development to be envisioned as a linear, mechanized process that proceeded according to a predetermined plan, which came from the logic inherent to the scaffold. Projections for large new housing estates, for example, with their equitably distributed shops, schools, and public services, as well as extensions of public transportation routes to reach them, appeared in plans years before there were material, financial, or labor resources to support their construction. This was necessary to propel the logic of the scaffold into the future." [5].

Bosnia and Herzegovina and its urban centers were regarded as the mentioned nodes in a "single entity". Distribution of labor among republics was a part of the general strategy of economic development. Within the wider Yugoslav framework after World War II, and after the split from the USSR in 1948, Bosnia and Herzegovina was given the economic role that determined its fate in terms of development. Bosnia and Herzegovina was designated for the production of raw materials, resources, and semi/products for the needs of the whole federal economy, and it was appointed for the investments into capacities of military and resource/basic industry. This distribution of roles required supporting infrastructure that would keep up and catalyze the development of the economic and social instances in spatial dimensions—characteristic of the socialist scaffold.

Additionally, we also found that the Yugoslav scaffold acquired unique characteristics when observed in a broader socialist context.

Namely, the gradual implementation of self-management and social planning in the late 1950s considered the bottom-up approaches, trying to diminish the policy of state intervention. Self-management economic and social units, at least in theory, had to provide a base for the overall development of communes.[4] This implied ongoing negotiation among levels of government regarding the control over the production assets and labor surpluses. All these new patterns in development planning certainly had an impact on infrastructural thinking. In the case of Yugoslavia, it was implemented through a scaffold structured by economic (social) plans at various levels. The federal plan was based on the plans of the self-management administrative and economic units at the level of community and republic. In addition, the structure and scale of social plans implied an in-depth reorganization of urban planning too that, after 1961, started looking at a broader spatial framework—region. Based on Hilberseimer's theory of regional planning, the main goal of the spatial planning methodology was to dismiss disparities between the rural and urban areas and link all of them into a coherent regional space, a model that is in line with the theories of dis urbanization [11]. Here, the developmental achievements of self-management communes were supposed to be aligned and harmonized while at the same time alleviating the pressure of developmental trends on regional economic and administrative centers. The spatial regional plan thus puts in its center the commune and its social and infrastructural relationship with other communes [11]. The commune, with its functions, played a key role in spatial strategies of the transformative development of society.

We can conclude that the pattern and density of spatial distribution of communes and their intended social role required appropriate form, density, and repetition of infrastructure at the level of the region and of each commune/municipality. Contrary to other socialist scaffolds, produced by the greater level of centralization, the Yugoslav scaffold was very peculiar in its spatial expression, but not on a temporal and spatial scale, which we see as a general socialist characteristic.

As an initiative and investment of the self-management enterprise Velepromet Visoko, the construction of the hotel Visoko in 1976 followed and catalyzed the socio-economic plan at the level of the Visoko commune, but also of higher regional

[4] According to the 1963 Constitution of former Yugoslavia, a commune is defined as "a basic socio-political community in which material and other conditions are ensured for people's labor and the development of productive forces; directs and coordinates the development of the economy and social services; determines and distributes funds for the needs of the municipality; creates conditions for meeting the material, social, cultural and other common needs of citizens; harmonizes individual and joint interests with general interests; realizes social self-management as directly as possible; authorities, social self-management and social services of common interest are organized" [12].

and federal instances. In particular, the five-year social plan of the municipality of Visoko, which preceded the construction of Hotel Visoko (1971–1975), based on the Laws on the Social Development Plan of Yugoslavia and Bosnia and Herzegovina [13, 14], the adopted social plan of Yugoslavia and the social development plan of SRBiH for the period from 1971–1975 [15, 16], as well as on programs and adopted medium-term plans for the development of economic organizations and non-economic activities in the area of the municipality of Visoko [17], was intended to control the flows of labor, resources, and investments, within the essential Yugoslav socialist framework.

The critical points of the presented analysis determine the Hotel Visoko as an integral part of the spatial expression of the commune's development strategy. It can be regarded as an architectural reflection of the structure of the self-governing social system. Such a feature embodied in the urban system designates this building as a distinctive socialist heritage.

Assessment of the Architectural Attributes of the Hotel Visoko

Architectural program and typology. Gracefully rising from a heterogeneous, natural, and built environment and overlooking the river Bosnia, Hotel Visoko stands out as one of the key landmarks in the panoramic views of the city. Its remarkable architectural silhouette is composed of two solids, whereupon the vertical segment is juxtaposed with a platform-like horizontal block. The tower, representing the vertical segment, contains the accommodation units, while the horizontal segment, which is rooted to the ground level, acts as a platform for all the public and social facilities (see Fig. 1).

In the same way that the hotel was originally intended to induce the economic progress of the former socialist commune in the second half of the twentieth century, its elegant upright silhouette epitomizes the modern urban identity of Visoko to the present day. The horizontal block, on the other hand, contained all the public spaces that were accessible to the hotel guests and external visitors alike. The main hotel common spaces were situated on the ground floor, such as the hotel reception and lobby, lounge area with a panoramic view, reception area, modern café, traditional Bosnian café, and international and national restaurants, arranged in a radial layout revolving around the centrally located kitchen. The underground floor is comprised of a nightclub, a pizzeria, and technical facilities. The hospitality areas altogether provided 450 seats, serviced by the kitchen that was capacitated for 500 meals. One of the most significant architectural features of the hotel was the roof terrace located on top of the horizontal block of the hotel, which acted as a Piazzetta, the central gathering space with spectacular vistas of the river Bosnia and the surroundings, as well as offering the close-up views of the hotel tower.

The hotel tower housed 125 beds in total, distributed in 52 rooms, 40 of which were double and 12 single rooms, 15 apartments, and one exclusive roof-top 300 m^2 apartment accompanied by a 200 m^2 terrace. Hotel Visoko was labeled as a higher

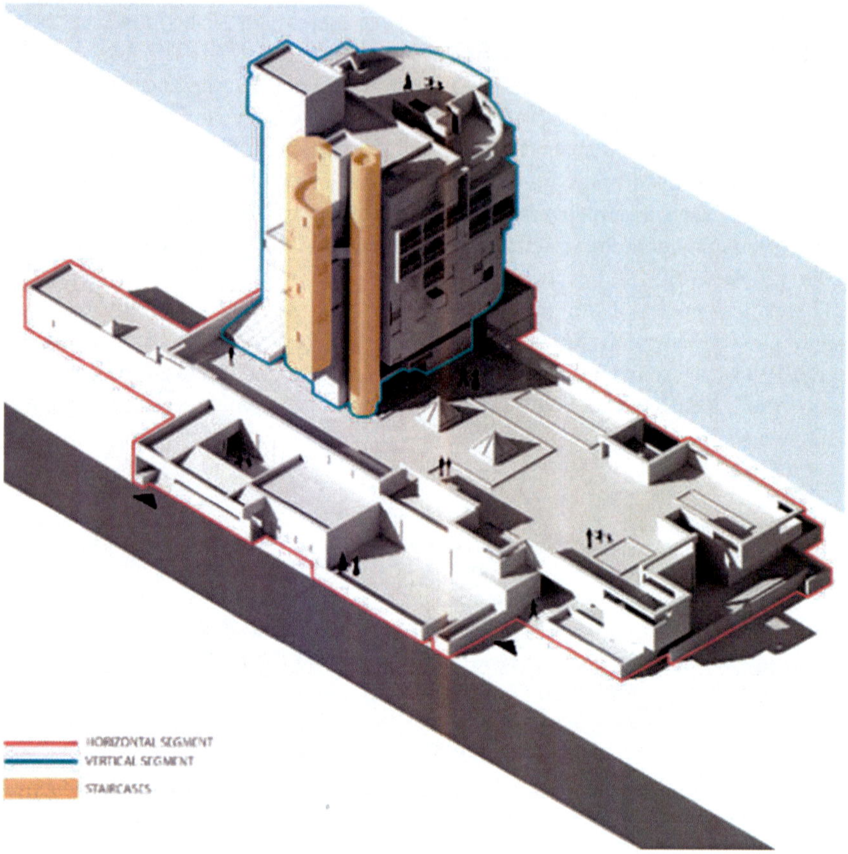

Fig. 1 Spatial diagram of Hotel Visoko. *Source* Illustrated by Zejneb Basarik

B-category hotel. The layout of the residential block of the Hotel Visoko, important for the typological division of hotel facilities,[5] especially of hotel buildings created in that period, is atypical compared to the architectural production of the time. Its standard residential floor layout aligns with both linear and atrial typologies, and, in addition, the form of the residential block associates it with the tower hotel typologies. The horizontal and the vertical blocks are connected via a recessed technical floor, which acts as an indirect link, resulting in the optical effect of floating of the tower block. Unlike the typical floorplans with centralized vertical communication blocks, in the case of Hotel Visoko vertical communications are extruded from the building volume creating a distinctive sculptural effect. Yugoslav architects lag in following

[5] Depending on their standard floorplan layout, hotels can be categorized in linear, atrial and tower hotel typologies (Author's comment).

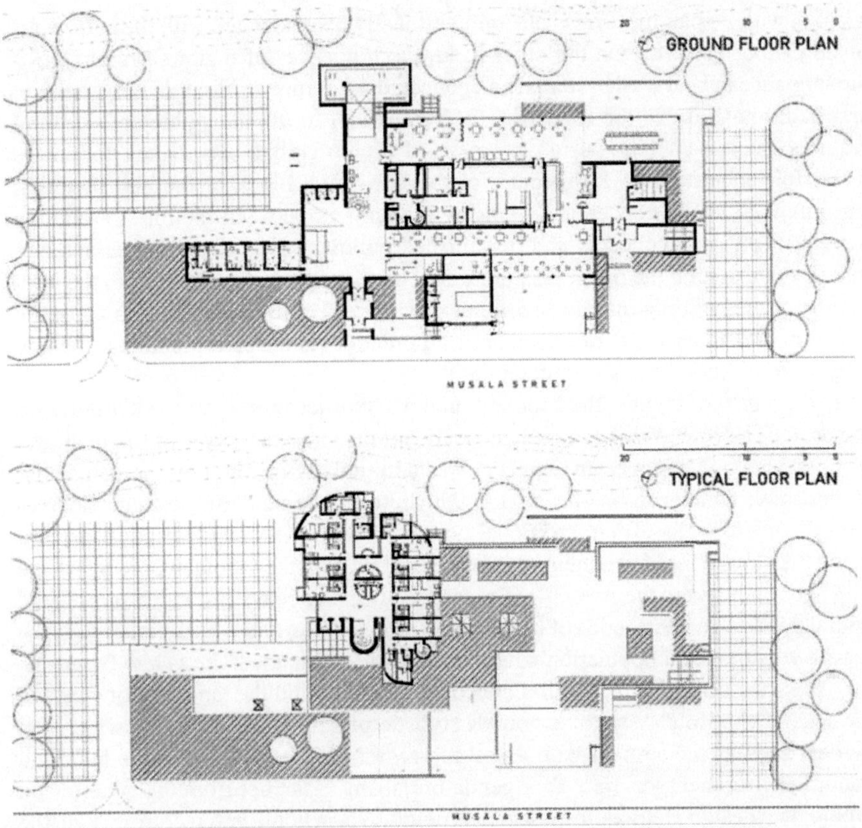

Fig. 2 Ground floor and typical floor plan layout of the hotel Visoko. *Source* Illustrated by Zejneb Basarik

the Brutalist manner of decomposing the function through the form and one of the earlier examples of consistent execution can be observed in the case of Hotel Visoko.[6]

When it comes to the characteristics of its layout, the typological identification of the hotel Visoko is ambiguous. The hotel rooms are arranged bilaterally along the linear corridor, with a small atrium in the center of the floorplan. Due to its uncommon floorplan layout, Hotel Visoko defies a clear categorization in either of the three hotel typologies, which is considered a unique architectural value and an expression of a site-specific design approach (see Fig. 2).

Context and concept. Examination of the architectural language and style of the Hotel Visoko reveals ambiguity and multivalence. Modernist identity is distinctly expressed in pure, Euclid geometry and white façade rendering: "Town-hotel in

[6] Brutalism tendencies were somewhat delayed in the architectural production of the former Yugoslavia. One of the most notable examples of the Brutalist principle related to functional grouping of masses can be identified in the mixed-use tower Genex in New Belgrade neighborhood, built in 1980 and designed by Mihajlo Mitrović.

Visoko, with its playful forms, can unintentionally be associated with the early opus of Le Corbusier" [18]. On the other hand, the influences of regionalism and post-modernism can be traced in spatial and conceptual interpretations of history, culture, and tradition. Besides the reflections of genius loci in the intangible architectural features, the site-specific design approach of Zlatko Ugljen was manifested in the masterful integration of architecture of the horizontal block of the hotel with its surroundings, while conceptually erasing the borders between the interior and the exterior: "Ugljen's concepts and his finished buildings [...] speak of the author's mastery in reaching the integral saturation of all visual manifestation, from the integration of the building into its environment with respect for the context in the sense of a harmonic in growing, of its semantic articulation, to the careful integral shaping of the interior and the smallest detail in it" [19].

In the words of Ugljen, the geometry of the horseshoe represents the leitmotiv for the design of Hotel Visoko, simultaneously, reminiscing the tower of the medieval fortress Visoki. Moreover, the concept was additionally inspired by the vernacular architecture, which can be observed in the white façade rendering which resonates with the white walls of the traditional houses from the Ottoman period in Bosnia, while the playful façade elements can be associated with the Ottoman porches on the first floor giving onto the street [4]. The facade finish in béton Brute can be correlated with the ethics and aesthetics of brutalism in architecture, and white rendering is the reason why the local population gave the hotel the nickname "The White Beauty".

Assessment of the architectural concept, materials, and the language of space in the case of the Hotel Visoko, expounds stylistic pluralism and layering, which can be described as modernism with a peculiar regionalist sensibility for the built and natural environment, or even, the "gentle brutalism". Distinct from the generic and typological design methodology, which is ubiquitous in hotel architecture design, the site-specific approach of Zlatko Ugljen can be read both in the exterior and interior, giving the building unique, idiosyncratic architectural values.

Gesamtkunstwerk. As in other projects authored by Ugljen, the hotel Visoko is an example in which the three design levels and scales, from architecture and interiors to furniture, are all united by principles of total design and represent an inseparable, organic whole, a total work of art: "Interiors, of course, often present the original conception and basic measure of the architectural unit, which thus manifests itself outwards through spatial dynamism, its existential content, and in its articulation through a programmatic, richly conceived usability" [19].

Bespoke furniture and fixtures, general and decorative lighting of the common spaces and the hotel rooms and the apartments were all specifically created and custom-made for Hotel Visoko, with a distinguishable Ugljen's signature, expressed in the use of the local material—solid wood: "Thus it is that Ugljen is a very sophisticated designer as well, using that principle to define the interior along with partitions and built-in fitting, furniture and finished products so that the constructiveness of material, perfect usability, and the semantic impression are recognizably integrated in the formulated sense [...]. In this view, we should maybe define Ugljen's approach rather as a unique process of design, since, after functionally defining the plasticity of inner walls as an architect, he carefully turns his attention to original furnishings, to the furniture that he designed for each building separately, in accordance with the overall formal composition that has been established" [19].

Historical and Social Value of the Hotel Visoko

Strategic planning of small industrial towns in Bosnia and Herzegovina typically comprised the development of industrial facilities in the first place and was followed by the construction of a railway station, a municipal building, a post office, a health center, a cultural center, and a hotel.[7] In addition to public facilities, the urban setting of a small Bosnian town included a hotel that had a commercial character in a capitalist socio-economic context. However, this type of urban hotel in small industrial socialist towns was not a typical commercial hotel, in relation to the tourist offer. It was primarily associated with large enterprises and local industries. These enterprises were frequently visited by experts and politicians from all over Yugoslavia, and the hotels were mainly built for their accommodation. On the other hand, public hotel facilities were occasional or often the only ones of this kind in small towns in Bosnia and Herzegovina. For that reason, together with the local cultural center,[8] they were the central gathering places for citizens. Following the liberalization and internationalization of the Yugoslav market in the 1970s, Bosnian companies started to attract foreign trading partners. Even then, these small-town hotels were not strictly economically viable and market-oriented, but they had their logical place in the wider socialist infrastructural thinking.

The hotel Visoko was officially inaugurated on 18 December 1976. The investor was the social business enterprise "Velepromet" from Visoko. The contractor was the construction company "Hercegovina" from Mostar, together with several subcontractors "Kovina" from Visoko, construction company "6. April" from Ilijaš, "Monter" from Split, the companies "Lik" and "Univerzal" from Sarajevo and "Visočica" from Visoko [20]. The construction of the hotel itself primarily brought together local entrepreneurs and businesses, but also the construction companies from wider areas of Bosnia and Herzegovina and Yugoslavia.

The socialist era of the Hotel Visoko lasted for sixteen years (1976–1992). During this period, we can observe the peak of the economic development of socialist Yugoslavia in the late 1970s, followed by the economic crisis of the 1980s and the definitive collapse in the early 1990s. At the same time, Yugoslavia was increasingly opening up to the Western market and culture.

The socio-cultural environment of the epoch was vividly portrayed in an article that appeared in Visoko's local paper in 1981. It tells about the declining popularity

---

[7] Many small towns in the central part of Bosnia and Herzegovina have been urbanised under this principle of planned construction, among them Tešanj, Maglaj, Zavidovići, Žepče, Gračanica, Srebrenik, Travnik, Pucarevo (todays Novi Travnik), Donji Vakuf, Zenica, Visoko, Kakanj etc.

[8] For more information on the role of cultural centres in local communities during the socialist era, see [21].

of traditional meeting places for socialist youth, such as the House of Culture, the JNA (Yugoslavian National Army) House, and the student clubs. Instead, the local youngsters were more drawn by the three most popular local cafes of the time. The participants in the citizens' survey underwent in 1981 even requested from the authorities for culture in Visoko and the management of the hotel Visoko to transform the hotel bar into a nightclub to meet the contemporary social needs of the youth. The hotel management replied that the bar operates well and that the hotel does not have sufficient funds to renovate the space and purchase the necessary equipment, and for that, they asked for help from others. The Self-Government Interest Community for Culture and Information also confirmed that due to the lack of resources, the request by the youth could not be granted [22]. This example clearly shows the existence of direct communication between the citizens, the Self-governing Community, and the Hotel Visoko, which was made possible via the local media. The survey shows that the local population considered that the hotel facilities belonged to them, and they had the need and the right to express and propose their ideas for remodeling the interior space of the hotel.

Hotel Visoko was built and managed by the socialist enterprise "Velepromet" from Visoko. As the enterprise itself was socially owned, the hotel also had a specific role within the local community. It was, in a way, a building that was managed by the local community using the unique Yugoslav self-governing system. The idea of self-governing socialism is part of the historical and social value of socialist buildings, and it was directly reflected in the Hotel Visoko.

"The Hotel Visoko served as the venue for all important meetings, business meetings, and weddings. The new year was welcomed, Women's Day was celebrated, and graduations and graduation anniversaries were celebrated. The students had their concerts in the bar of the hotel" [23]. From its inauguration in 1976 until its closure in 1992 the Hotel Visoko, in addition to its hospitality objectives, played a particular social and cultural role and represented the central gathering hub for the local community.

The essential socio-cultural values of the town-hotel Visoko result from the close relationship between the local community and its premises, which is inherent in the socialist architectural heritage. Therefore, the recognition of the historical and social significance of the Hotel Visoko originates from its historical and ideological affiliation to the socialist' context.

4 Assessment of Authenticity

The UNESCO Operational Guidelines from 2021 define authenticity as the degree to which the attributes supporting the value of the property can be considered truthful and credible i.e., "properties may be understood to meet the conditions of authenticity if their cultural values are truthfully and credibly expressed through a variety of attributes including: form and design; materials and substance; use and function; traditions, techniques and management systems; location and setting; language, and other forms of intangible heritage; spirit and feeling; and other internal and external factors" [7].

The essence of cultural property significance, therefore, is not contained in the cultural property itself but in its ability to convey value. Furthermore, authenticity is not a value itself but is linked to attributes that describe the value. Confirmed authenticity is an established appropriate measure or degree of truthfulness of such attributes, that plausibly succeeds in conveying a defined value. The assessment of authenticity and, consequently, the direction for its restitution, if it has been damaged, requires: (1) formulation of value, (2) extraction of attributes that support it (form, design, setting, purpose/intention, etc.) and finally (3) assessment of the degree of their truthfulness/authenticity, which has direct implications on the decision on the degree of intervention.

The clarification of the conditions for the authenticity assessment is most exhaustively elaborated by Jokilehto [24–26] and Stovel [27]. Jokilehto [25] groups different aspects of authenticity into three units depending on the artistic, historical, social, and scientific dimensions of the cultural property:

1. Authenticity of the creation—this aspect refers to form and design, materialization, location, and setting. Given that the architectural creation implies a specific program basis, both use and function are treated. In this case, Jokilehto suggests relying on Paul Philippot's definition: "the authenticity of a work of art is a measure of the truthfulness of the internal unity of the creative process and the physical realization of the work" [25].
2. Historical-material authenticity—is based on recognizing the historicity of heritage. It refers to materialization, substance, location, and setting, but also tradition and techniques, use, and function, taking into account the perspective of historicity. The category of originality is not a precondition of authenticity.
3. Socio-cultural authenticity is relevant especially when it comes to a continuous cultural tradition. It refers to the intangible aspects of heritage, such as use and function, traditions and techniques, and the spirit and feeling of what constitutes a continuing tradition. Also, "Attributes such as spirit and feeling do not lend themselves easily to practical applications of the conditions of authenticity but are important indicators of character and sense of place, for example, in communities maintaining tradition and cultural continuity" [7 par. 83].

Determining the appropriate degree or threshold of the authenticity of individual attributes varies from case to case; it depends on the heritage category, but primarily on the share of attributes in the structure of the value.

Authenticity of the Hotel Visoko

Following the assessment of value carried out in this study we can state the basic significance of the Hotel Visoko as follows:

> The Hotel Visoko, in Visoko, by the architect Zlatko Ugljen, is an exceptional example of the type of town-hotel architecture from the socialist period. It is a testimony of the Yugoslav socialist peculiarity in the development of urban systems - the representation of the particular social and spatial planning discourses where the commune is seen as the axis of the transformative development of society.
>
> The role of the Hotel Visoko in the local communities' social life and the mediatory position between the commercial and the collective domains within socialist system represent a specific socio-historical value.
>
> Design, architectural, and functional concept is a supreme example of the authorship of Zlatko Ugljen and his distinctive approach to the principles of Gesamtkunstwerk. Interior and exterior complement each other formally and aesthetically in a sensitized and site-specific modernist approach, expressing the spirit of the local context and regional characteristics of the place.

The principal attributes describing the socialist peculiarities are the setting and location at the urban system level. Social value is primarily associated with the attributes of use and function, as well as with the sense and spirit of the place, which has evolved from the relationship between the local community and the hotel. The design unity of exterior and interior is considered the main attribute informing on the artistic and creative value and the author's original contribution.

The main challenge is to identify the threshold of truthfulness that the restoration process has to gain for the renewed building to successfully convey its whole same, integral value. Although the socialist and social components of the value are not supported by the formal or material attributes, these components still directly impact the decision concerning the restoration of interior form and design which would establish a sufficient level of truthfulness. Namely, even if Ugljen's design concept of the hotel directs towards a more consistent restoration i.e., a more complete reconstructive procedure following the description of creative artistic authenticity as an "internal unity of the creative process and the physical realization of the work", reconstruction, in this case, can be discarded. The original mediatory role between commercial and social aspects (a social component of value), and the place in the socialist scaffold of the Hotel Visoko (socialist component) allow and encourage greater interpretative freedom and a creative approach to the reconstruction. In other words, since the Hotel Visoko originally was intended to support overall socio-economic development, a contemporary design gesture as a reflection of the noticeable progress in the past almost half a century, concessions to market demands through reflection on the design, and finally, the involvement of the local community in the decision-making process on restoration and protection actions would eventually support the integral value.

On the other hand, consistent reconstruction of interior elements would truthfully convey only the creation as a component of value. How social significance can be an important and even a crucial factor in decisions about restoration methods is best illustrated by the example of the rehabilitation of the UNHQ (United Nations Headquarters) building in New York, the largest modern mid-century complex, whose designers were, among others, Le Corbusier and Oscar Niemeyer. The symbolism of the building is contained in the idea of civilizational progress; the emphasis is on its dynamics, the process of constant change. The rehabilitation guidelines promoted the requirement that, despite the historicity of the material, which must be carefully preserved, the restoration should "reflect progress made in areas like life safety, universal access, and especially sustainability in the fifty-five years since the complex opened" [28]. The essence of the authenticity of the UNHQ building was impossible to separate from the symbolism of the social significance. The language used in the implementation of these changes, the language of contemporary modernism, itself evolved in the meantime and was used to further underline the aesthetics of modernism of mid-century architecture [28].

This approach is in line with the position of Cesare Brandi [29] on the actual historical act of restoration: "[…] it must not be secret and out of time, but rather allow itself to be punctualised as the historic event it is, due to its being the product of human artifice and its insertion into the process of the work of art's transmission to posterity."

As has been demonstrated, the most appropriate approach in the rehabilitation and sustainable preservation of the Hotel Visoko is consistent reconstruction and restoration of the exterior and external appearance, including the urban setup, location, form, design, and materialization. The recommended direction in the rehabilitation of the interior and equipment is an interpretative process and consistent reconstruction of fragments, components, and elements of design quotations paying homage to the original author's work, with the aim to integratively convey the socio-historic, formal-functional characteristics and socialist peculiarities of the hotel's significance (Table. 1).

Continuity and New Vitality of Old Hotels

This section of the article presents a critical analysis of the contemporary practice in the domain of sustainable preservation of modernist hotel facilities while focusing on interior design. The selected projects share common stylistic and chronological and other comparative references, such as the importance of authorship in design and authentic value, indicating the prospective approach and tactic in interior remodeling of the Hotel Visoko.

One of the most prominent examples of modernist hotel renovations is the SAS Royal Hotel of Arne Jacobsen in Copenhagen,[9] which was designed by the architectural studio Space Copenhagen. During the renovation, the architects were guided by the original concept, especially in the hotel's public spaces, preserving the highlighting modernist impression and ambiance of the interior. The most important aspect of the new project was the preservation of the furniture designed by Arne Jacobsen. This was completely feasible since the production of Jacobsen's chairs and armchairs designed for this hotel has never ceased. Several interior details have been adapted and modernized, but the reference to the original ambiance has been

[9] SAS Royal Hotel was built in 1960 and is one of the first tower hotels in the world, recognizable by its suspended facade and total design by architect and designer Arne Jacobsen.

Table 1 The relationship between the components of Hotel Visoko value and their defining attributes seen as a regulatory factor in the decision on the level of authenticity in the restoration process

Components of value		Socialist particularities	Social value	Artistic creation	
				Exterior	Interior
The principal attributes conveying the value		Setting; location of urban structure	Use; function	Form; design	Form; design
Destruction of original features		No	Yes	No	Yes
The ability of attributes to transfer value	Through consistent reconstruction	No	Yes	Yes	-
	Through creative interpretation	Yes Continuity of the original socio-economic role, achieved with designing tools	Yes Continuity of the original collective-private mediatory role and concessions to market requirements, achieved with designing tools and involvement of the local community in the decision-making process	-	Yes Quotations of the original design

preserved subtly, with up-to-date materials and interior elements. The interior of the rooms was modified, except for room 606, which is the only one to have been restored according to the original project. Room 506 by the Spanish designer Jaime Hayón represents an innovative, creative resonance of the iconic Jacobsen-inspired interior and furniture design.

The presented case of the preservation of the old hotel designed by world-class architects such as Arne Jacobsen demonstrates that the author's name became an important part of the marketing and branding strategies. This example also shows how the original design can be used to create new values, when it is interpreted creatively and innovatively, thus incorporating the hotel into the framework of the sustainable symbolic economy.

Hotel Marcel[10] is an example where the name of a famous architect, the author of the building, became the essential element in the branding of the hotel. While the

[10] The current Hotel Marcel, Connecticut is a building designed by architect Marcel Breuer in 1970 as an administrative building for the Armstrong Rubber Company. In 1988, the building was transformed into the North American headquarters of the tire company Pirelli, until 2003, when it was bought by the Swedish furniture giant IKEA. The Brutalist administrative building was designated a historic monument and protected thanks to the efforts of the New Haven Arts Council Architectural Alliance in 2000. In 2022, the Connecticut architectural firm Becker & Becker completed the renovation and transformation of the building to a hotel.

building was originally an administrative building, it was converted into a hotel with the latest reconstruction completed in 2022. The architectural firm Becker&Becker has entirely preserved the brutalist facade of the building. One of the most important features of the project is reflected in the application of the principles of sustainable renovation with the Passive Home certificate and the application for the LEED Platinum certificate. The interior design is authored by the Dutch studio East Design, in which the warm and humane atmospheres of the hotel rooms and public facilities are intertwined. The brutalist spirit was also kept inside the building, although the function of the building was completely altered [30].

In some recent projects originating from the countries of the former Eastern bloc, the socialist heritage is used as a source of inspiration to create completely new environments that exude the retro atmosphere of the communist era, as in the Pure Hotel in Krakow.[11] It is a newly built hotel, designed by the Polish architecture office Paradowski Studio. Although it comprises internationally branded contemporary pieces of furniture, the interior was inspired by the pure functionality of modernist mid-century hotels in Kraków. Unlike the previous cases which illustrated the process of renovation and interior remodeling, this hotel shows how modernist heritage can be used as a source of inspiration in the branding strategy of the new hotels.

Another interesting case of interior renovation is the Maestral Hotel in Brela, Croatia. The hotel was built in 1965, according to the design of architects Ante Rožić, Julius De Luca, and Matija Salaj, while the interior was designed by Bernardo Bernardi. The restoration was supervised by the Croatian Preservation Office During the reconstruction, the existing structure and interior elements were preserved and the Croatian design studio Franić Šekoranja was appointed to perform creative interior remodeling, instead of a verbatim restoration of interior design and equipment, while preserving the modernist atmosphere of the interior.

The preceding analysis has shown that the architectural heritage of the Modern Movement is used in the contemporary branding of the hotel and interior design. In some cases, the new interior concepts pay tribute to the modernist architects and designers, in a creative, interpretative, and innovative manner. Such branding strategies of modernist spawn from the symbolic economy, and contribute to better visibility of the hotel, its commercial offer diversity, and ultimately, its economic viability. In heritage hotel rehabilitation cases, exterior appearance elements are consistently reconstructed and/or preserved. On the other hand, the interiors are designed as creative and contemporary interpretations of the modernist spirit, paying tribute to the authenticity of the host building and its author, and ultimately, preserving and even highlighting the memory of the place. Renovation projects of the hotels designated as modern heritage in the capitalist socio-economic environment, highly prioritize economic sustainability over substantial social value. Nevertheless, it is important to keep in mind that simply exploiting authorship, or the aesthetic value of modern architectural heritage to reinforce the economic pillar of sustainability, does not necessarily mean contributing to the sustainability of the overall renovation, if educational, social, and other recognized values are overlooked in this process.

---

[11] See: [31].

5 Discussion

The Hotel Visoko case study has disclosed the dual character of the hotel architecture typologies designated as the socialist modernist heritage. On one hand, its function implies commercialization for the symbolic economy, and on the other hand, it addresses the local community of the post-socialist society.

In response to the demands of the market, the socialists' attributes of the hotel may be mobilized for certain city marketing strategies in the sense that Zukin [32] speaks of culture as an instrument of urban governance. "The spaces of the symbolic economy", in this case, socialistic architecture and design, are a means of commercialization, "images or symbols used to sell the city to outsiders, to encourage them to visit, spend their money, invest, or relocate their residence or business" [32].[12] A legacy of socialism, town-hotel complexes have the potential to become a part of the overall urban experience. Nevertheless, there is a need to avoid the pitfalls of defining otherness [33] and refrain from trivializations through the creation of representations.

Restoring the physical structure by conservation principles and fully reactivating the building will preserve its material values. In this research, we have demonstrated that the Hotel Visoko represents one of the most remarkable landmarks and values of the city, as a result of the close relationship between the community and the architecture and, the space in general, which is intrinsic in the socialist legacy. The privatization of the Hotel Visoko, as well as the overall transition process, has considerably weakened the position and past relationships between the community and the hotel premises. Thus, rehabilitating and reinventing the social value of the hotel, derived from the socialist past, represents the major challenge in the contemporary, post-socialist society. However, society and the local communities have recently been drawn to the focus of experts on the matter of the protection of cultural heritage.

The role of the community in the cultural heritage preservation process, which includes the concepts of participative and collaborative protection, has been a part of the conservation doctrine since 1964 [34] and has continuously been questioned in a series of international documents. Numerous documents, declarations, conventions, and charters emphasize civic participation in the process of protecting architectural heritage. One of the outcomes of these efforts is the Convention on the Value of Cultural Heritage for Society of the Council of Europe, signed in Faro in 2005 [35]. The entire process of the European Cultural Heritage Strategy for the twenty-first century [36] is based on the Faro Convention. All clearly underline the social dimension of heritage restoration and may be used to help restore the social value of the architectural heritage of the socialist era.[13]

---

[12] Editor's introduction to Zukin's article.

[13] The new EU Agenda for Culture, introduced in 2018, promotes a wider understanding of heritage. Puts people and communities at the center and engages them in heritage decision-making i.e., empowers them to make decisions. The agenda calls for the inclusion and empowerment of many actors who care about heritage through research, experimentation, and practice. It will test new models of participatory management and cultural heritage management, as well as encourage social innovation and linkages with other sectors [39].

In the case of the integration of socialist heritage into post-socialist communities, where the public's attitude toward socialism is still ambivalent and largely generationally conditioned [37], the design solutions can play a role in the progressive creation of what Tweed & Sutherland [38] refers to as'Heritage by appropriation'. This is about democratizing culture because the space is left open to the public i.e., the local community to "choose" its own heritage. "Heritage by appropriation generally emerges from public behavior rather than through organized lobbying" [38].

If we recall Arnstein's scale of citizen participation,[14]not all forms of participation would guarantee the sustainability of architectural heritage. For this reason, it is extremely important to encourage and involve the local community in all stages of the preservation process. Three actors were recognized in the restoration process of the Hotel Visoko: the entrepreneur, heritage institutions and agencies, and the local community. The level of participation of the former will impact the social value that can be strengthened by the restoration process itself, not simply the use of a renewed hotel.

The local people represent an excellent source of information about the hotel, which might have been undiscovered, if there were not for the social tools designed to gather the testimonies of people, documents, photos, etc. Thanks to its outstanding artistic values, Hotel Visoko can become the venue for numerous professional scientific events, such as art and architecture workshops, exhibits, group discussions, lectures, etc. In all these listed events, it is possible and necessary to include citizens, volunteers, and citizens' associations operating in this area.

In Visoko, which is a progressive economic and business town in Central Bosnia, the private sector plays an important role in society. Appropriate means must be found to involve the civic, public, and private sectors of society in the joint analysis and elaboration of action plans. All stakeholders should be aware of their role and contribution to the broader interests of the community or society. And in this process, again, we recognize interior design as a tool that can be used to re-establish authenticity in a contemporary and creative manner, maintain historical continuity, and promote social values. The possibility to generate a new social value is an open discussion about the interior design of the Hotel Visoko and its functional and formative meaning. The discussion should be initiated by the community of architectural-design-preservation experts, whose creative methods of preservation would encompass the civil and private sectors of society.

---

[14] Sherry Arnstein, author of the highly influential 1969 article "The Ladder of Citizen Participation", pointed out that participation without a redistribution of power is an empty and frustrating process for the powerless. Therefore, the redistribution of power allows citizens who are not in a position of power and who are currently excluded from political and economic processes to be included in the future. In short, it is how they can initiate significant social reforms to share in the inexhaustible benefits of society. Therefore, she highlighted 8 levels of participation with different shares in the distribution of power: Nonparticipation (1. Manipulation, 2. Therapy); Tokenism (3. Informing, 4. Consulting, 5. Placation); Citizen power (6. Partnership, 7. Delegated power, 8. Citizen control). If there is a partnership and/or representatives of the citizens in the decision-making process and/or citizen control, there is also participation in the distribution of power, that is, concrete results can be expected in achieving the goals for which citizens took part [40].

Having acknowledged the guidelines and the principles for sustainable preservation of the physical structure of the Hotel Visoko, the restoration of its functionality and reintegration into contemporary society rely on its successful reactivation from the inside out. Due to its proximity to all the involved stakeholders, the discipline of interior design in the case of Hotel Visoko can be regarded as the most tangible and the most operative collaboration tool which can bring all the stakeholders in the rehabilitation process. The interior design should provide the missing link and re-establish the much-needed historical continuity, respectfully cherishing the memory and the importance of socio-cultural legacy, the architectural and design integrity of the total work of art, while at the same time, infusing and reinventing the new vitality of interior spaces, forms, and materials in a contemporary idiom. This mediation between the past, the present, and the future by means of interior design, may be performed in an iterative and collaborative process, in the series of creative workshops and exhibitions involving the universities, the designer professionals, the industry professionals from interior and furniture design sectors, together with the economy specialists and entrepreneurs, the public sector and the community. In this case, the absence of the original pieces of interior layers and components and the sincere and brutalist-like bareness of the interior shell may be regarded as an opportunity. The interior may be recreated in a series of in situ art installations paying homage to the original project, which may emerge from the original research and conversations with the author and can gradually be presented to the public in a participative process, to ensure that the relevant economic and social feedback is included in the final design. The proposed methodology takes on the versatility and flexibility of the interior design discipline in a more thoughtful and socially responsible manner. In the case of socialist modernist heritage hotel preservation, interior design can be used as a tool to communicate socio-cultural values with the wider public, as well as to establish a realistic and responsible balance between the dynamics of the market with the demands of the local community.

6 Conclusion

This research addressed the subject of sustainable preservation of hotels from the socialist period in B&H, focusing on the Hotel Visoko designed by Zlatko Ugljen. Based on the analysis of this remarkable, representative, and yet endangered piece of architecture dating back from 1976, the objective of this research was to set the general standards for the assessment of the value of the socialist urban hotel architecture in Bosnia and Herzegovina and to yield the guiding principles for its sustainable rehabilitation.

As the starting point of sustainable preservation is the assessment of value, in the first part of the paper we conducted a comprehensive evaluation of the Hotel Visoko based on UNESCO doctrine. Assessment of the value begins with the exploration of socialist peculiarities, that surpass the prejudiced perception of socialist architecture as a product of the totalitarian paradigm.

In the light of the expounded historical and socio-cultural context, the assessment of the value has outlined the following idiosyncratic attributes of Hotel Visoko: The social value of the Hotel Visoko emerged as a result of mediation between the communal and commercial development strategies of the time. Thanks to the close relationship that the local community established with the hotel for over two decades since its construction, it gained certain distinctive "socialist"—social attributes and a sense of belonging, a phenomenon which cannot be encountered in the commercial and tourism architecture of the capitalist provenience.

Assessment of architectural attributes of the Hotel Visoko accentuated its stylistic pluralism and layering, which can be described as "gentle brutalism", or modernism with a peculiar site-specific sensibility, simultaneously echoing nature and the historical context. The authorship of the architect Zlatko Ugljen can be identified in his total design approach, and the spatial unity of exterior and interior, the space, and the objects, resulting in unique, idiosyncratic architectural values.

In the quest for a prospective approach and methodology in the interior remodeling of the Hotel Visoko, the research included a case study of sustainable preservation projects focusing on the contemporary interior design of modernist hotel facilities. The case study has shown that the architectural heritage of the Modern Movement can be used in the contemporary branding of hotels. Interiors are designed as creative and contemporary interpretations of the modernist spirit, paying tribute to the authenticity of the host building and its author, and ultimately, preserving and even highlighting the memory of the place.

Reestablishing the authenticity of tangible and intangible attributes was determined as the essential criterion in the process of sustainable preservation of the Hotel Visoko. The discussion on authenticity has delineated the imperative of consistent restoration of the building exterior and its adjoining surroundings. Interior design, on the other hand, as the most variable and zeitgeist-related design domain, allows greater interpretative freedom and a more flexible creative approach. A creative and interpretative approach in interior design, rather than consistent reconstruction, may truthfully support and convey the authenticity of defined integral significance. Acknowledging the sensitivity of the proposed methodology, the precondition is to establish a suitable balance between the verbatim restoration of authentic spaces, fragments, and pieces, and the creative translation of authentic interior ambiance and objects into a contemporary and innovative idiom.

The proposed guiding principles for future action in the case of Hotel Visoko are structured at three levels: (1) consistent restoration and conservation of its exterior structure, (2) rehabilitation of its social meaning, and (3) remodeling of the interior. Restoring the exterior structure in accordance with heritage protection doctrine, allows for the form and design of exterior artistic creation, as an important segment of the integral value, to be truthfully and authentically preserved and conveyed. Rehabilitating and reinventing the social value of the hotel, originating from the socialist past, implies an active involvement of the local community in all stages of the preservation process. This involvement is crucial for the local community to be able to sense the heritage as their own, a process described as "heritage appropriation". It is closely related to and interwoven with the problem of authenticity preservation of the interior, which has to be functionally adapted and redesigned/restored to serve not only commercial but also public interests associated with its socialist identity.

The interior design should provide the missing link and re-establish the much-needed historical continuity, respectfully protecting the memory and the socio-cultural legacy, the architectural and design integrity of the hotel as the *total work of art*, while at the same time, infusing a new vitality in interior spaces, forms, and materials using a contemporary idiom and sustainable approach.

References

1. Antešević N (2021) Arhitektura modernih turističkih objekata Jugoslavije (1930–1985). Univerzitet u Beogradu, Filozofski fakultet
2. Mutnjaković A (ed) (2014) Identitet jadranske turističke arhitekture: međunarodni znanstveni skup: zbornik radova. Hrvatska akademija znanosti i umjetnosti, Znanstveno vijeće za turizam, Zagreb
3. Grandits H, Taylor K (eds) (2013) Sunčana strana strana Jugoslavije. CEU PRESS, Zagreb
4. Niebyl D The Lost Legacy of Zlatko Ugljen's exquisite hotel architecture. Retrieved from Spomenikdatabase. https://www.spomenikdatabase.org/post/the-lost-legacy-of-zlatko-ugljen-s-exquiste-hotel-architecture
5. Zarecor KE (2017) What was so socialist about the socialist city? Second world urbanity in Europe. J Urban Hist 44(1):95–117; 98–103. https://doi.org/10.1177/0096144217710229
6. Thaler W, Mrduljaš M, Kulić V (2012) Modernism in-between. The mediatory architectures of socialist Yugoslavia. Jovis Verlag GmbH, Berlin, pp 17
7. UNESCO: The operational guidelines for the implementation of the world heritage convention. Annex 5. World Heritage Centre, Paris, p 107. https://whc.unesco.org/en/guidelines/
8. ICOMOS-ISC20C (2017) Approaches to the conservation of the twentieth-century modern heritage: Madrid—New Delhi document. ICOMOS, Paris. https://isc20c.icomos.org/wp-content/uploads/2022/03/MNDD_ENGLISH.pdf
9. Fishman R (1982) From the radiant city to Vichy: Le Corbusier's plans and politics, 1928–1942. In: Walden R (eds) The open hand: essays on Le Corbusier. The MIT Press, pp 244–283. https://doi.org/10.1162/a8667414.add39a6f
10. Kotkin S (1997) Magnetic mountain: stalinism as a civilization. University of California Press, Berkeley
11. Bojić N (2018) Social and physical planning: two approaches to territorial production in socialist Yugoslavia between 1955 and 1963. Architect Hist 6(1). https://doi.org/10.5334/ah.309
12. Mileta V (1967) Komuna u novim uvjetima. Politička misao 4(4):647–654; 651. https://hrcak.srce.hr/115464
13. The Official Gazette of Former Yugoslavia no. 13/69 (1969)
14. The Official Gazette of Former Bosnia and Herzegovina no. 20/69 (1969)
15. The Official Gazette of Former Yugoslavia no. 35/72 (1972)
16. The Official Gazette of Former Bosnia and Herzegovina no. 23/72 (1972)
17. Društveni plan razvoja opštine Visoko za period od 1971. do 1975. godine. Skupština opštine Visoko, Visoko (1973)
18. Štraus I (2010) 99 arhitekata sarajevskog kruga 1930–1990. Šahinpašić, Sarajevo
19. Bernik S (20202) Arhitekt/Architect Zlatko Ugljen. Međunarodna galerija portreta, Tuzla, pp 41–42
20. Istorija Visokog: Godišnjica otvorenja Hotela "Visoko". https://visoko.ba/video-istorija-visokog-godisnjica-otvorenja-hotela-visoko/

21. Kreševljaković L (2021) Community Centres in Bosnia and Herzegovina; places of personal emancipation through social and cultural activities. In: Galjer J, Lončar S (eds) The caring state and architecture: sites of education and culture in socialist countries, pp 199–218, Hrvatsko etnološko društvo, Zagreb (2021). https://hrvatskoetnoloskodrustvo.hr/wp-content/uploads/The-Caring-State-and-Architecture.pdf
22. Beatović Z (1981) Gdje su izlazili Visočani prije 40 godina? Naš život, pp 10. https://visoko.ba/nas-zivot-gdje-su-izlazili-visocani-prije-40-godina/
23. Dedić S (2020) Tragom starih tekstova iz "Našeg života": Hotel Visoko. https://www.vipromo.ba/tragom-starih-tekstova-iz-naseg-zivota-hotel-visoko/?fbclid=IwAR0098JDNmSud2y7KxSn9euKmuht5w1R6UT-ZNfVgc4YTIp6QxMqQJ2yjFQ
24. Jokilehto J (2006) Considerations on authenticity and integrity in world heritage context. City & Time 2(2). http://ceci-br.org/novo/revista/docs2006/CT-2006-44.pdf.
25. Jokilehto J (2006) World heritage: defining the outstanding universal value. City & Time 2(2):2
26. Jokilehto J (2009) The Complexity of authenticity. Kunstiteaduslikke Uurimusi 18(3–4), 125–135 (2009). https://ktu.artun.ee/articles/2009_3_4/ktu_2009_18_3_125-135_jokilehto.pdf
27. Stovel H (2007) Effective use of authenticity and integrity as world heritage qualifying conditions. City & Time 2(3):21–36. http://www.ceci-br.org/novo/revista/docs2007/CT-2007-71.pdf
28. Fixler D (2008) Is it real and does it matter? rethinking authenticity and preservation. J Soc Archit Hist 67(11–13):13. https://doi.org/10.1525/jsah.2008.67.1.11
29. Brandi C (1996) Theory of restoration. In: Stanley-Price N, Kirby Talley M, Melucco Vaccaro A (eds) Historical and philosophical issues in the conservation of cultural heritage. The Getty Conservation Institute, Los Angeles, pp 232–233
30. Hotel Marcel. https://www.dutcheastdesign.com/interiors/hotel-marcel
31. Puro hotel. https://www.architonic.com/en/project/piotr-paradowski-studio-puro-hotel/20162533
32. Zukin S (2008) Whose culture? whose city? In: Oakes T, Price P (eds) The cultural geography reader. Routledge, London & New York, pp 431–438
33. Kulić V (2018) Orientalizing socialism: architecture, media, and the representations of eastern Europe. Architect Hist **6**(1). https://doi.org/10.5334/ah.273
34. ICOMOS (1964) International Charter for the Conservation and Restoration of Monuments. ICOMOS, Venice
35. Council of Europe (2005) Convention on the Value of Cultural Heritage for Society. Council of Europe, Faro
36. Council of Europe (2018) European Cultural Heritage Strategy for the 21st century. Council of Europe, Bruxelles
37. Hann C, Humphrey C, Verdery KV (2002) Introduction: Postsocialism as a topic of anthropological investigation. In: Hann C (eds) Postsocialism. Ideals, ideologies and practices in Eurasia. Routledge, London and New York, pp 1–28
38. Tweed A, Sutherland M (2007) Built cultural heritage and sustainable urban development. Landsc Urban Plan 83(1):39–49. https://doi.org/10.1016/j.landurbplan.2007.05.008
39. European Commission: New European Agenda for Culture 2018. European Commission, Brussels (2018). https://culture.ec.europa.eu/document/a-new-european-agenda-for-culture-swd2018-267-final
40. Arnstein S (1969) A ladder of citizen participation. J Am Inst Plann 35(4):216–224. https://doi.org/10.1080/01944366908977225

Sustainable Spaces and Designs

Persistence of Socialist Apartment Buildings: Functionalist Design Approach Over Time and Usage

Lejla Kreševljaković and Mladen Burazor

Abstract The most significant part of the housing stock that Bosnia and Herzegovina has today is residential buildings from the socialist period. The buildings have been in service for 50–70 years, but they are not recognized as architectural heritage of value. Over that period, many aspects of life have changed, and the rising question is how relevant and accommodating those buildings are now. It is important to reflect on the inherited values but also the limitations of the socialist apartment buildings. There are lessons to be learned especially in the design phase when economic, environmental, and political aspects affect architectural solutions. The quality of architectural design can be perceived through its' ability to endure changes over time and adapt to new requirements and in that sense, grounds itself in the sustainability paradigm. Research performed on 4 residential high-rise buildings built in 1964 in Sarajevo demonstrated that the functionalist layout of apartments has been maintained and has directly contributed to the persistence of socialist apartment buildings to date. As well as that there is an urgent need to establish design guidelines for adapting apartment buildings to the contemporary needs of tenants and, in due process, to preserve the values of socialist apartment buildings.

Keywords Apartment buildings · Socialism · Modern heritage · Functionalist design

---

L. Kreševljaković (✉) · M. Burazor
University of Sarajevo – Faculty of Architecture, Patriotske lige 30, 71000 Sarajevo, Bosnia and Herzegovina
e-mail: lejla.kresevljakovic@af.unsa.ba

M. Arslanagić-Kalajdžić et al. (eds.), *Interdisciplinary Advances in Sustainable Development II*, Lecture Notes in Networks and Systems 804,
https://doi.org/10.1007/978-3-031-46692-2_16

1 Introduction

The architectural heritage from the socialist period, which is considered part of the modern heritage of Bosnia and Herzegovina, is the most numerous compared to previous periods.[1] The cause of its great number is mass construction. It was made possible by the development of industry, and the growing industrialization of the construction process: prefabricated, semi-finished, and finished building components, such as prestressed reinforced concrete slabs, walls, staircases, etc. For the first time in the history of Bosnia and Herzegovina, thanks to new construction technologies, entire residential areas have been built.

While our obligation to preserve the modern heritage of the twentieth century is equal to the important cultural heritage of earlier periods,[2] the architectural heritage of this period was not critically considered to the extent that values would be generally recognized and protected.[3] As a result, the modern heritage in Bosnia and Herzegovina is most often endangered. This applies to residential establishments and their facilities too. Most of the socialist apartment buildings are still in active use today and over time, they have been adapted to the needs of their users in several ways. A great number of residential buildings suffered damage during wartime between 1992 and 1995. After the war, property relations changed from social property to private property, and many repairs and reconstructions were done individually. Finally, adaptation to modern energy efficiency standards and regulations, etc. is currently underway.

Depending on when the apartment buildings from the socialist period were built, they bear witness not only to the principles by which they were built but also to different ways of using residential space over the past 50–70 years. The research questions are how apartment buildings from the socialist period have survived to this day, how does each of the aspects of sustainability (economic, ecological, and social) reflect on the housing construction from the socialist period, furthermore, what has contributed most to their persistence to this day.

There is a need, on the one hand, to define the values of socialist apartment buildings as part of the modern heritage, and on the other hand, to acknowledge and explain users' needs during the socialist time and now. The task of the research is to recognize how contemporary users' economic opportunities, energy efficiency, and individual and social needs are realized in the housing space. The aim is to give

[1] The historical periods of Bosnia-Herzegovina which has left a significant amount of architectural heritage are Ottoman (1463–1878), Austro-Hungarian (1878–1918), Kingdom of Serbs, Croats, and Slovenes (1918–1929), and Socialist Yugoslavia (1943–1992).

[2] See [14].

[3] The books of academician Ivan Štraus [7, 8] give an insight into the Yugoslav and Bosnian architectural production of this period. Monographs were published on the architectural opus of Ivan Štraus [9], Zlatko Ugljen [12], Hamdija Salihović [11] and Hasan Ćemalović [10]. Martino Stierli and Vladimir Kulić curated the exhibition "Toward a Concrete Utopia: Architecture in Yugoslavia, 1948–1980" [13]. Authors of monographs and exhibition presented, analyzed, and reviewed the work of these architects, and significantly contributed to the evaluation of the architecture of this period.

recommendations that would preserve and protect the values of socialist apartment buildings in this process.

2 Methods

The research was performed on four identical residential high-rise buildings (skyscrapers), which form a set of residential buildings (see Fig. 1). The apartment building type of skyscraper was chosen for the research, as the number of repetitions of the same elements on the facades is higher. This makes the indicators of change in the exterior appearance of buildings clearer.

The selected buildings were built in 1964 in Sarajevo's settlement Grbavica II in Bosnia and Herzegovina, designed by architect Hamdija Salihović [1]. These facilities have been in existence for over sixty years. They were selected because the research focuses on changes that tenants have made within living spaces over time. These objects were destroyed in the war of 1992–95. They were on the front line and were devastated. After the war, the buildings were shielded from the outside. New façade cladding and new windows were provided by the humanitarian organizations. However, postwar apartment owners had to completely renovate the interior living spaces. There were no doors, and the partitions were partially destroyed, as were the plaster on the walls and ceilings, the plumbing, the floors, etc. Only structural elements were left after everything else was stripped down. As a result of these circumstances, tenants were provided with the opportunity to better adapt the apartments to their needs. It was assumed that the results of the research, in terms of adapting the facilities to the modern needs of tenants, would be more accurate.

The research focuses on the preservation of architectural heritage, so the structure of the work combines the methodology of the approach of the assessment of the value of the architectural heritage and IMRAD. Therefore, the results will first elaborate on the original condition of the buildings, and their social, economic, urban, and architectural context, thus defining the original values of the architectural heritage. Next, the results of the research on the found condition will be presented. The discussion develops the outcomes toward the guidelines for the future management of this type of architectural heritage.

The methodology framework is based on multiple methods. Physical or non-physical context plays an important role in architectural definition. Thus, the historic and descriptive method was used to observe the original condition of a given housing typology. The comparative method allows objective processing of the data of the condition found in the given typology, and from the results, conclusions are drawn on the changes that have occurred. Deductive reasoning, cause, and social effects in transitions were observable in the given architectural discourse.

Finding the original condition was done by examining the literature, while the research for the found condition was carried out according to the availability of certain segments of the search. The exterior aspect is entirely accessible, and the photographic method and observation were used. The complete results for all objects

Fig. 1 Four residential skyscrapers along the Miljacka River

are collected for this segment of the research. Research on changes in residential spaces was carried out on a sample of 52 apartments of one skyscraper, with an insight into the found condition of the residential spaces and surveying the tenants. The survey focused on tenant property relationships, spatial transformations (demolition of walls, modification of space usage), and changes to increase energy efficiency (change of windows, closure of balconies). It was considered whether the actions were consistent with the original condition of the apartment buildings.

3 Results: Original and Found Condition of Socialist Apartment Buildings

Original Condition: Values of Socialist Apartment Buildings

The socioeconomic context of housing construction during the socialist period. The socialist concept of society was greatly based on idealism where the needs of many were ranked higher than the individual needs. In the former Yugoslavia, the first step after the Second World War was to nationalize private assets and give them to the "working class" of people. The essence of the socialist idea was that the workers worked for society, so they had all the right to decide on how to utilize benefits accordingly. The right to housing was seen as one of the basic human rights of socialism for all working people. Therefore, one of the most common ways to spend accumulated capital was to create housing for employees. Development priorities and solidarity funds have been established. That required everyone to set aside a certain amount of their salary to fund infrastructure projects such as schools, sports, culture facilities, or housing. Many socialist enterprises have been successful in improving their housing plans and addressing the housing needs of the working class. This increasing demand for housing in cities has been caused by the large-scale migration of people from rural to urban parts of the country posed by a process of industrialization after the Second World War.[4]

The capital of Bosnia-Hercegovina, Sarajevo went through the process of socialist urbanization and grew rapidly. In the municipality of Novo Sarajevo, where residential buildings, the subject of research are located, the plan was to build 4760 apartments over a period of 4 years (1967–1971). A total of 2175 was built which is equivalent to 543 apartments per year [2]. Yet so many homes produced every year could not meet the needs of all workers and their families.

Mass housing was made possible through industrial development, the growing industrialization of the construction process, prefabricated, semi-finished, and finished building elements, such as prestressed reinforced concrete slabs, walls, staircases, etc. Using new construction techniques, entire residential areas were constructed.[5]

The urban layout of socialist apartment buildings. Urbanization processes in socialist Yugoslavia have been part of the modernist movement. There were two primary reasons for that. The first was to cut the links with the previous regimes and to build on new ideas, which fit perfectly in the vision of the new society. The second reason was that most school-educated architects studied in other European countries and were influenced by their peers. In the case of Bosnia and Herzegovina, the Faculty of Architecture was established in 1949 and the academic staff consisted of people who were educated abroad, mainly in Vienna and Prague. "Several talented

[4] For more about socialism in Yugoslavia and Bosnia and Herzegovina, see Perović [15] and Kamberović [16].

[5] For more about industry of mass housing development in socialist Yugoslavia see Jarić et al. [17].

individuals, led by Prague students, with their avant-garde modernist activities in the interwar period, pulled Bosnia and Herzegovina out of professional lethargy, introducing it in great steps on the European architectural scene. Dušan Smiljanić, Jahiel Finci, Muhamed Kadić, and Emanuel Šamanek are the key protagonists of the 'Sarajevo School of Architecture." [3]. Furthermore, some prominent architects had worked alongside leading Modernist icons. For example, Juraj Neidhardt moved from Peter Behrens' office to Le Corbusier's office and came to Bosnia-Hercegovina in the years before the Second World War. All these architects who came to Bosnia and Herzegovina contributed their skills and expertise and used them to shape the new architectural and urban visions in the post-war country. However, global architectural trends were delayed in Bosnia and Herzegovina. Apartment blocks started to be mass-built in the 1960s and skyscrapers were introduced into the city fabric. But soon, due to criticism of this type of housing, this construction concept was almost entirely abandoned by the end of the 1970s [4].

The urban setting of apartment buildings like Grbavica I and II in Sarajevo clearly shows modernist principles. A range of linear configurations, many green areas between them, sunlight exposure, and unprotected views, are the main principles behind the spatial positioning and the height of the buildings. High-rise buildings have negative effects in urban settings, they cast long shadows but, more importantly, can be obstacles to natural ventilation in cities. Their form, which is usually rectangular with more than 15 stories high, often prevents air circulation if it is not well positioned. "In terms of the layout and shape of the buildings, Sarajevo has several examples of very good construction practices. Regarding the positioning of the buildings in the Hrasno settlement, several skyscrapers represent an example of a good arrangement of buildings in terms of urban ventilation … Four skyscrapers are arranged in a grid of two columns and two rows, leaving a large space between each of them." [5].

The basic features of the urban setting of the observed apartment buildings are that they are located at an adequate distance, not interfering with the sunlight of the other. The urban setting of the structures follows the riverbed. They are surrounded by greenery. The outdoor design of residential buildings also exhibited uniformity and equality. Equal windows, fences not balconies, their repetition in the vertical plane, and symmetry on the façade panels are the main characteristic of the external appearance of residential buildings. The above is part of urban architectural and educational value. The urban architectural form of the socialist apartment buildings is a testimony to the construction method characteristic of socialist and functionalist urban planning in Bosnia and Herzegovina.

The functional layout of the apartment organization. The author of the buildings is the architect Hamdija Salihović, a professor at the Faculty of Architecture in Sarajevo. In 1960, he won an internal competition for 4 apartment buildings along the Miljacka River (see Fig. 1) and they were built in 1964. Three types of flats have been designed in the residential blocks that have a ground floor and 12 floors. Based on the layout similarities, they are designated as A1 and A2 (54 m^2), B (58 m^2), and C (72 m^2). Differences in the area are minimal, and all are based on the same functional scheme. Even Apartment C, which has room more with a balcony than the others (see Fig. 2), has the same layout as the other apartments. All apartments are with two-sided orientations, which significantly contributes to their sun exposure and the possibility of natural ventilation. Apartments B and C have south and east or west orientations, which is why they are extremely exposed to sunlight. Apartments A1 and A2, apart from the eastern or western orientation, have a northern orientation (see Fig. 2). However, they have a view of the river, which also makes a significant contribution to the quality of life (see Fig. 1).

The apartments were designed for unknown users. An equal organizational pattern, as well as surfaces, contributed to the sense of equality among tenants through usage. The idea was not to separate citizens in terms of financial or educational status since all "workers" were supposed to have equal rights. While different "codes" or

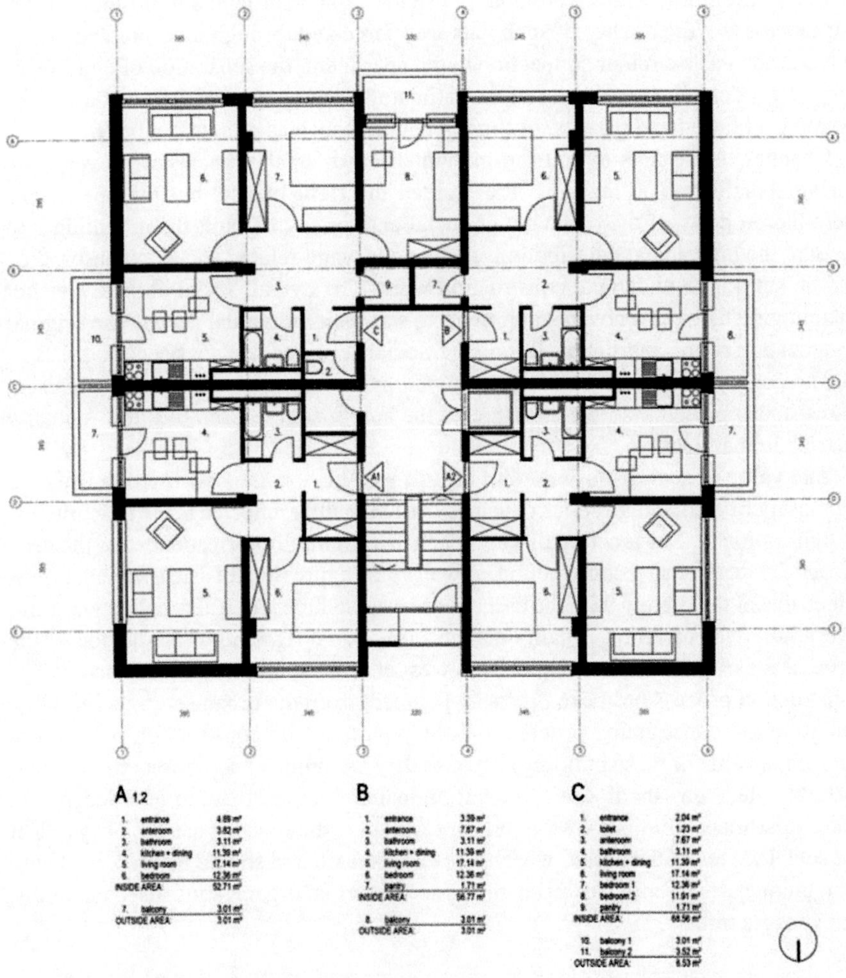

Fig. 2 Typical layout of a residential floor of a skyscraper

names have been used to describe these units, we may also see objections to such labeling: "[…] any "categorization" such as: "workers' flats", "dedicated flats", and "solidarity flats" is not accepted. An apartment is an apartment, and solidarity, for example, is seen in the society providing an appropriate apartment to those whose income does not allow it." [6].

The apartments were assigned to families who were on the priority list, according to the number of family members, the number of employees, and the number of children. The kitchen and dining area have been designed separately from the living area. In the case of numerous families, the living room was used for sleeping. To respect intimate life, the kitchen with the dining room often served as a living room and was the main gathering place for the family. The users were unknown, and the priority was the great functionality of the apartments, which was reflected in the fact that different families were able to use the same apartments in different ways. For this reason, the quality of such flats was perceived through incorporated layout features such as a circular connection in the apartment, the separation of children of various genders, the possibility of the intimate life of tenants of different ages (for example, children and parents, or third generation such as grandparents) [6].

Changes in the flats to suit the individual needs of the tenants were very rare during socialism. The tenants were granted the right of use, but the apartments were the property of the company or social enterprise that built them. Taking into account the high social contributions for various wage-related social expenses, they had no surplus to undertake construction work. The overall social climate was not encouraging users to renovate apartments to suit their individual needs. The original organization of the apartments during the Socialist period has not been modified.

The concept of equality widely promoted by the socialist state was also present in the housing sector, so the structure of the apartments reassembles the "equality matrix" in many ways.

The value of socialist apartment buildings. The housing construction from the socialist period contains a series of values that are mainly reflected in the functionalist design approach. The functionalist approach in design fully corresponded to the need for accelerated housing construction, as well as the expression of a new socialist spirit which meant the rupture with the past and the establishment of a new collective spirit. Everything was based on equality and ensuring equal rights and satisfaction of the basic needs of all working people. That was reflected in housing development. The construction process has been precisely planned, from the urban setting to building design to the construction process. All three parts of the construction process are part of the value of these buildings because they are unique and nonrepetitive. Their integral value consists of a material and immaterial nature that are interdependent. These apartment buildings are a significant artifact of the socialist period. They reflect the social values of the period when they were created and should be preserved with their unique urban setting, exterior appearance, internal organization, used materials, and construction.

Found Condition of the Apartment Buildings

Tenants. After the 1992–95 war, in the transition period, through the process of the privatization process, previous social ownership became privately owned. The former right to use apartments in the socialist period changed to the right to buy apartments. This meant that all apartments first became the private property of former tenants which they could then offer them in the housing market. Nearly 30 years have passed since the war, we do not know on a general scale the proportion of apartments used by their owners and those apartments that are rented out. For this research, it was important to investigate such a ratio because it may affect spatial changes. The search revealed that out of 52 apartments, their owners live in 30 apartments (57.69%) (See Table 1). The rest of the apartments are either not used at all or have been rented.

The exterior appearance of apartment buildings. The research on the exterior appearance was carried out by immediate observation and facades were photographed for additional checks. Changes in the exterior appearance of buildings resulting from various tenant interventions are reflected in the: glazing of balconies, installing outdoor blinds, installing air conditioners, and replacing windows (compared with the originals, this can be a change of color and geometry of the window). The data was collected based on all facades of the 4 apartment buildings. The results are shown in Table 2.

Glazed balconies are found in 40.63% of the apartments and two approaches can be observed. In the first case, the glazing of the balcony meant transforming the balcony area into an indoor living area. In the second, glass balconies are still unheated areas. The tenants have made individual glazing interventions, so in principle, the balcony glazing is not uniform. The details of the existing balcony fence and the materials, colors, and geometry of the glass supports differ from one case to another (see Fig. 3). The way the balconies were glazed caused the greatest negative impact on the exterior appearance of buildings.

The results show that 43.27% of the apartments have installed air conditioners, therefore the external air conditioners are visible on the facades. When installing the air conditioning units, some tenants were careful to place the outdoor air conditioning unit where other tenants had previously installed it. However, that was not always the case. The appliances themselves are of similar size and material. Thus, for the

Table 1 The number of owner-occupied apartments

The apartment is used by the owners															Solitaire	A
Floor		HGF	I	II	III	IV	V	VI	VII	VIII	IX	X	XI	X11	No.	%
Type of apartment	A1	●					●			●		●	●	●	6	46.15
	A2		●	●		●	●			●		●	●	●	8	61.54
	B		●	●	●	●	●							●	6	46.15
	C		●	●	●	●	●	●	●		●	●		●	10	76.92
Total:															30	57.69

Table 2 Changes in the exterior appearance of buildings caused by various interventions by tenants

Solitaire	A		B		C		D		Total		
	No. of apartm	%	No. of apartm	%	No. of apartm	%	No. of apartm	%	No. of apartm	Total No. of apartm	%
Glazed balconies	11	22.92	25	52.08	25	52.08	17	32.69	78	192	40.63
External blinds	5	9.62	6	11.54	10	19.23	17	32.69	38	208	18.27
Air conditioners	26	50.00	20	38.46	22	42.31	22	42.31	90	208	43.27
Window color	8	15.38	2	3.85	1	1.92	14	26.92	25	208	12.02
Changes geometry	5	9.62	1	1.92	2	3.85	15	28.85	23	208	11.06

Fig. 3 Different approach and realization of glazed balconies on the façade of a skyscraper

reasons mentioned above, despite the high percentage of air conditioners introduced, the exterior units did not significantly affect the exterior appearance of the buildings (see Fig. 4).

After the 1992–1995 war new windows were installed on all the skyscrapers with funds from donations. Due to the lack of original industrial production of window frames, the installed window profiles differed from the original profiles, but the original color and geometry of the window frames have been retained. Because of the poor quality, the donated windows were soon changed by the tenants. The results show that in 12.02% of cases they changed the color and in 11.06% the geometry of the windows (see Figs. 3 and 4).

The internal organization of the apartments. The change in the internal organization of the apartment, which was registered in 15 out of 52 apartments, or 28.85% refers to the combined kitchen, dining room, and living room into a single space. We see that this shift is almost equally reflected in all types of apartments (see Table 3, Figs. 2, 5, 6, 7 and 8).

Minor corrections led to the expansion of the bathrooms, which according to the initial design were only 3.11 m^2, so the need for their expansion was widely expressed (see Figs. 6 and 8). An interesting case is the transformation of the Apartment B in which, through the fusion of the kitchen, dining room, and living room and the closure of the balcony, the living room was divided and so the number of rooms in the apartment increased (see Fig. 7).

Fig. 4 Exterior air conditioners on the skyscraper façade

Table 3 Apartments with combined kitchen, dining, and living area

Connected kitchen, dining and living room															Solitaire	A
Floor		HGF	I	II	III	IV	V	VI	VII	VIII	IX	X	XI	X11	No.	%
Type of apartment	A1				●	●		●		●					4	30.77
	A2				●				●	●					3	23.08
	B	●							●	●					3	23.08
	C					●		●	●		●			●	5	38.46
Total:															15	28.85

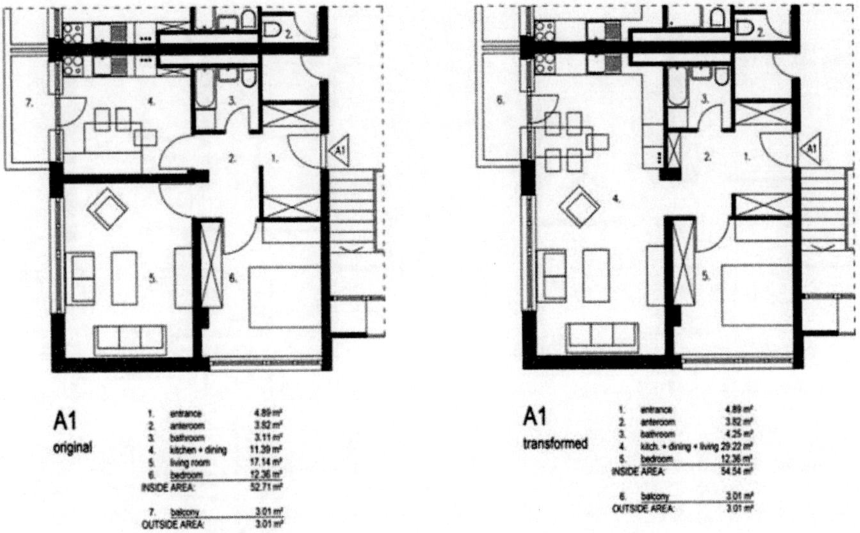

Fig. 5 Original and generally altered condition found in apartment A1 with combined kitchen, dining room, and living room

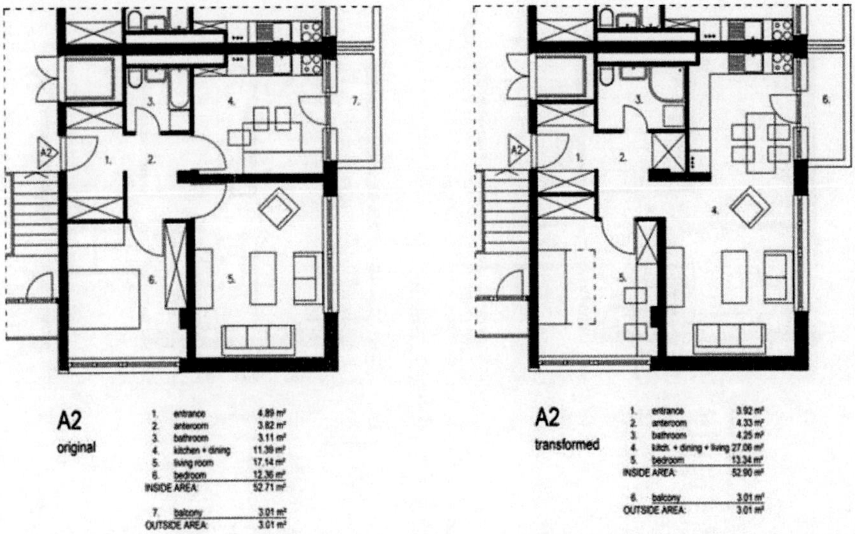

Fig. 6 Original and altered condition found in apartment A2 with combined kitchen, dining room, living room, and extended bathroom

Fig. 7 Original and altered condition found in apartment B with an increased number of bedrooms in the structure of the apartment

Fig. 8 Original and generally altered condition found in apartment C with combined kitchen, dining room, living room, and extended bathroom

4 Discussion

The results of the survey on original condition indicated that the apartment buildings that were observed, with their original urban setting, external appearance, internal organization, used materials, and construction, represent an artifact of the period in which they were created. The investigation of the found condition showed that several factors influenced the changes in the initial condition. They led to the degradation of the values previously defined. The discussion goes in the direction of recognizing how the modern economic possibilities of the users, the needs of the users for energy efficiency, modern individual and social needs are realized within the apartment's space, and the recommendations that would help the preservation and protection of the values of the socialist apartment buildings in this process.

Tenants

Contrary to the socialist period where the tenants were the users of the apartments, and the company was the owner (i.e., a social enterprise), research has shown that now 100% of apartments are private, with their owners living in 57.69% of apartments. This raises many questions about the intensity of individual space interventions that apartment owners have done or will do in the future. The intensity of space changes in the socialist period was characterized as low. While private ownership of apartments gave property owners much more rights to make spatial changes based on their individual needs. Tenants who rent flats are less prone to space interventions. The high proportion of homeowners (57.69%) living in their apartments speaks of a more direct relationship between space and users that will continue.

Exterior Appearance of Apartment Buildings

Research findings clearly show that the original appearance of the facade is mainly damaged by the glazing of the balconies, and partly by the addition of external air conditioning units and changes to windows that differ in color and geometry.

The high glazing percentage of the balcony indicates the actual need of tenants to increase the indoor living area and to close the balcony to make more use of the balcony space in the winter period. This is mainly because of the large number of cold days in the year and high air pollution. The need of the tenant for balcony glazing is obvious but the glazing that has been done so far has been done with no design guidelines. Individual interventions on the façades significantly impaired the original exterior appearance of the building (repetitive openings and balconies, symmetry of façades, etc.).

In addition, a high percentage of air conditioner units indicates that tenants need to cool the space too. The buildings are well exposed to the sun, and in summer there is a need for shade, on the windows openings facing east and west.

The small percentage of changes in the color and geometry of the windows during the changes carried out by the tenants themselves can be explained by the fact that the donated windows were uniform for all buildings and that they were a kind of design guideline. Tenants surveyed said (100%) they would follow instructions during interventions if they existed. The problem of damage to the exterior appearance of the building can be resolved by providing design guidance.

Therefore, the adoption of design guidelines would be effective for the preservation of original design values. Guidance for users on how to proceed with future renovations, particularly concerning:

- Balcony glazing,
- Installation of air conditioners,
- Changing the windows,
- Installation of external blinds.

Internal Organization of the Apartments

The changes in the internal organization reflect the different ways of living in socialist times and today. Originally there was a need for a larger number of separate rooms, so the living room was separated from the kitchen and dining room. The results show a high percentage (28.85%) of changes where the partition wall was demolished and the kitchen, dining, and living areas were joined. The need of today's tenants for a larger living space is made possible by advances in furniture fabrication technology. The kitchens become an integral part of the residential interior. Before, in socialist times kitchens and apartment equipment elements were unified. As a result, not only the layout of the apartments but also the interiors of the apartments have been unified as well. Today, it is nearly impossible to purchase ready-made kitchens that fit the space; therefore, they are now custom-built. There is a wide range of different furniture elements as well. The difference within the apartments is an expression of this time, which can represent the new value of the apartments.

The original functionalist approach to apartment design has allowed space transformations. In substance, it was retained and as such turned out to be very usable and comfortable for life even today. In the surveys, tenants have often mentioned the right sunlight, the possibility of ventilation, and the view as benefits that the apartments have, which are part of the original design values.

During the renovation, tenants are guided by other tenants' experiences and changes they have previously made to the apartments. But, if design guidelines are to be developed, they should also include possible transformations of the internal organization of the space. This would give tenants a better insight into the potential transformations that would meet their needs.

The results showed that changes in the interior space also go in the direction of adding rooms, which can also be reflected in the exterior appearance of buildings. Based on experience with socialist residential buildings, when designing new buildings alternative solutions should be considered in advance and the opening positions should be planned on the façade so that the process does not affect the building's exterior appearance.

5 Conclusion

The values of socialist apartment buildings, which were reflected in equality, are threatened in the changed social and economic environment. Therefore, the issue of the sustainability of socialist apartment buildings is one of sustaining their values. The apartment buildings from the socialist period in Bosnia and Herzegovina are not under institutional protection, and their values are at risk of tenant interventions. Interventions do not involve unified design solutions, so it is urgent to establish design guidelines. By providing clear design guidelines for socialist apartment buildings, the contemporary needs of tenants can be met, while protecting the original values. These guidelines should refer to internal home conversion opportunities, as well as to interventions on the facades of buildings such as glazing of balconies, the introduction of air conditioning units, replacement of windows, and installation of external blinds. However, research has emphasized the persistence of the functionalist design approach over time and usage. The functionalist arrangement of the apartments proved to be good for the contemporary use of the apartments. Such a design approach has made it possible to make the necessary corrections in the layout of the apartments, without disrupting the stability of the buildings, and at the same time respond to the modern needs of tenants. The functionalist layout of the apartments has been maintained to date and has directly contributed to the persistence of socialist apartment buildings.

References

1. Štraus, I (1964) Interni konkurs za stambenu kulu Grbavica II u Sarajevu—1960. ARH 5(2):20. Društvo arhitekata Sarajevo, Sarajevo
2. Žabčić B (1982) Izgradnja stambenog naselja Hrasno I u Sarajevu. In: Jarić M, Bajlon M, Marković S (eds) Stan i Stanovanje: Časopis "Izgradnja" posebno izdanje. Savez društava arhitekata Srbije i Savet za građevinarstvo Republičke privredne komore SR Srbije, Novi Sad, Savez građevinskih inženjera i tehničara SR Srbije, pp 111–124
3. Salihović E, Burazor M, Islambegović V (2022) Intercultural impact of Czech architects on architectural discourse and education in Bosnia and Herzegovina. In: WMCAUS 2022 7th world multidisciplinary civil engineering-architecture-urban planning symposium, world MESS Ltd., pp 10, Prague
4. Arnautović-Aksić D et al (2016) Typology of residential buildings in Bosnia and Herzegovina. Faculty Architecture University of Sarajevo, Sarajevo

5. Hadžiabdić M et al (2019) Zeleni akcioni plan Kantona Sarajevo: Studija o urbanim ventilacionim koridorima i uticaju visokih zgrada, pp 34, Evropska banka za obnovu i razvoj, Sarajevo. https://mpz.ks.gov.ba/sites/mpz.ks.gov.ba/files/studija_o_urbanim_ventilacionim_koridorima_i_uticaju_visokih_zgrada.pdf
6. Bajlon M (1982) Neka pitanja u vezi sa upotrebnom vrijednosti stana. In: Jarić M, Bajlon M, Marković S (eds) Stan i Stanovanje: Časopis "Izgradnja" posebno izdanje, 27–38, pp 28. Savez građevinskih inženjera i tehničara SR Srbije, Savez društava arhitekata Srbije i Savet za građevinarstvo Republičke privredne komore SR Srbije, Novi Sad
7. Štraus I (1998) Arhitektura Bosne i Hercegovine, 1945.-1995. OKO, Sarajevo
8. Arhitektura Jugoslavije 1945–1990 (2013) Fondacija za razvoj održivog dizajna, Sarajevo
9. Ivan Štraus arhitekt (2002) '52-'02. Akademija nauka i umjetnosti Bosne i Hercegovine, Sarajevo
10. Husetić A (2019) Arhitekt Ćemalović. IKD University Press, Sarajevo
11. Salihović H (2002) Arhitekt i umjetnost graditeljstva. Arhitektonski fakultet Univerziteta u Sarajevu, Sarajevo
12. Bernik S (2002) Arhitekt Zlatko Ugljen. Međunarodna galerija portreta, Tuzla
13. Stierli M, Kulić V (eds) (2018) Toward a concrete Utopia: architecture in Yugoslavia, 1948–1980. MoMa Publications, New York
14. ICOMOS (2017) Approaches to the conservation of the twentieth-century modern heritage: Madrid—New Delhi document. ICOMOS, Paris
15. Perović L et al (eds) (2017) Jugoslavija u istorijskoj perspektivi. Helsinški odbor za ljudska prava u Srbiji, Beograd
16. Kamberović H (2000) Prema modernom društvu: Bosna i Hercegovina od 1945. do 1953. godine. Centar za kulturu i obrazovanje, Tešanj
17. Jarić M, Bajlon M, Marković A (eds) (1982) Stan i Stanovanje: Časopis "Izgradnja" posebno izdanje. Savez građevinskih inženjera i tehničara SR Srbije, Savez društava arhitekata Srbije i Savet za građevinarstvo Republičke privredne komore SR Srbije, Novi Sad

Atrium as the Element of Spatial Configuration in HPS

Senaida Halilović-Terzić, Alma Hudović Kljuno, and Zulejha Šabić-Zatrić

Abstract Taking care of our health has become a priority nowadays. Atrium spaces can be considered as a valuable asset in architecture design because of their dual role. They can contribute to the enhancement of the physical conditions within the building, but they also serve as places of socialization. The focus of this paper is on atrium spaces in educational facilities and on their impact on the physical and psychological health of children. The paper gives a brief overview of atrium development through history and its use in different building typologies focusing on the educational buildings. The second part of this research introduces three schools with atrium spaces in Sarajevo, dating from the socialist period. The current use of those atrium spaces and their potential are discussed. Finally, this paper presents a case study from Sarajevo, a school project with two open atria designed in the way to serve as a multifunctional space with the aim to promote healthy school environment and new, informal ways of learning. The design solution of the introduced school project in Sarajevo follows the strategy of health promoting school (HPS).

Keywords Health promoting school (HPS) · Atrium · Informal learning · Sarajevo

S. Halilović-Terzić (✉) · Z. Šabić-Zatrić
Faculty of Architecture, University of Sarajevo, Patriotske lige 30, Sarajevo, Bosnia and Herzegovina
e-mail: senaidah@af.unsa.ba

Z. Šabić-Zatrić
e-mail: zulejha.sabic.zatric@af.unsa.ba

A. H. Kljuno
Faculty of Engineering and Natural Sciences, International University of Sarajevo, Hrasnička Cesta 15, Ilidža, Bosnia and Herzegovina
e-mail: ahudovic@ius.edu.ba

M. Arslanagić-Kalajdžić et al. (eds.), *Interdisciplinary Advances in Sustainable Development II*, Lecture Notes in Networks and Systems 804,
https://doi.org/10.1007/978-3-031-46692-2_17

1 Introduction

The definition of health, according to WHO, is "Health is a state of complete physical, mental and social well-being and not merely the absence of disease or infirmity" [1].

This definition is dated in 1948, but it is still valid, maybe today more than ever. In 1986 the definition is updated, with further clarification as "A resource for everyday life, not the objective of living. Health is a positive concept emphasizing social and personal resources, as well as physical capacities" [2].

Our health is not only relative to us, individually. If anything, Covid-19 pandemic show us that we live in codependent relation with the totality of living world. Some research on this topic gives more comprehensive approach, as one from Georges Canguilhem. He offers us a definition of health which "includes the animate and inanimate environment, as well as the physical, mental, and social dimensions of human life" [3].

All explanations could be considered questionable, from many points of view [4], but what is evident is that the conjunction of physical, mental and social wellbeing remains relevant to this day. Also, we are tending to understand how environment that surrounding us is important for our health, both on individual and community level.

As we know, the quality of our urban environment has a strong impact on our daily life. Since 1960s the quality of life is connected to health through so many dimensions of our environment [5].

Health is a concept that we need to understand and appreciate since our birth. Every age has its own needs which must be fulfilled in order to remain healthy individual. School experiences are among the first notions we have about ourselves as individuals, members of our community, and social beings. The educational environment is directly connected to our health, and vice versa, healthier children achieve better educational outcomes. Therefore, the link between health and education is crucial for development and growth, both, on an individual and social level.

Health promoting school (HPS) is a strategy, initiated by the WHO, together with UNESCO, launched in 1995. The goal is to develop global standards and indicators for health promoting school which are "environments that constantly strengthen their capacities as healthy settings for living, learning and working" [6]. In those schools we are creating a healthy environment which plays a vital role in the health of their occupants, in total of physical, mental and social wellbeing.

Strategy of HPS creates a system of global standards which apply on: government policies and resources, school policies and resources, school governance and leadership, school community and partnerships, school curriculum, school social and emotional environment, school physical environment and school health services. School physical environment rely on a variety of disciplines, from architecture to pedagogy. In implementation of this standard, we need to provide safe and clean areas, with access to clean water, food, green outdoor spaces, learning resources, sport outdoor spaces etc.

School infrastructure needs to provide, primary, places for learning. However, nowadays, learning is undergone through numerous changes, and influences how we design physical environment in schools. Individualized and self-paced learning in combination with recognized importance of collaborative learning demand diverse physical learning environments where formal but also informal learning happens. One of the defining characteristics of informal learning is that it tends to defy organization and location [7]. Traditional classroom can't be the only available space for learning since it doesn't take into account learner's individual characteristics. Contemporary schools are using a space dedicated to social activities to enhance learning experience. They made shift from formal way of learning in classrooms to informal way of learning in social and other spaces. In that term, in architectural typology of school, we may consider all spaces that are outside of the classroom, where children don't meet the teacher, as a possible place of informal learning [8].

In a general sense, the atrium is a space that meets all above stated requirements. The atrium design type has been considered, from ancient times, as a privileged space of meeting, circulation and interior lighting, and with regard to school building typology dates from early beginnings of school design [9]. Atrium is an element which could bring many positive health impacts to the building users and at the same time be a place for social gathering, simultaneously offering the spatial possibility for informal learning.

2 The Development of Atrium and Its Significance in Architectural Design

The formation of atrium space as we know it today is a process that lasted for over three millennia. The roots of the atrium space can be traced back to the domestic architecture of ancient civilizations (Mesopotamia, Egypt, Persia, Crete, Greece and Roman Empire). Dictionaries such as Encyclopedia Britannica and Merriam-Webster define atrium as an open central space in the Roman house, even though this spatial configuration was present in earlier cultures and also in other building typologies such as temples, *gymnasion* (school) or later on in the Christian Basilica. Nowadays, atrium space in architectural design is common for both, residential and public buildings and can be found as an open or closed space in different forms and positions within the building volume.

Atrium spaces have dual role in the building design. By providing natural light and fresh air to the interior, they regulate the microclimate, but they also serve as places of meeting and circulation. This social function of atrium spaces was present from the very beginning. In the Roman house, atrium was serving as a domestic place with hearth and kitchen including a marble water basin *impluvium* in which the rainwater was collected that on the other hand was used for the household. With the further development of the Roman house *domus* domestic spaces like kitchen and storages were moved to the rear part of the house and atrium began to function

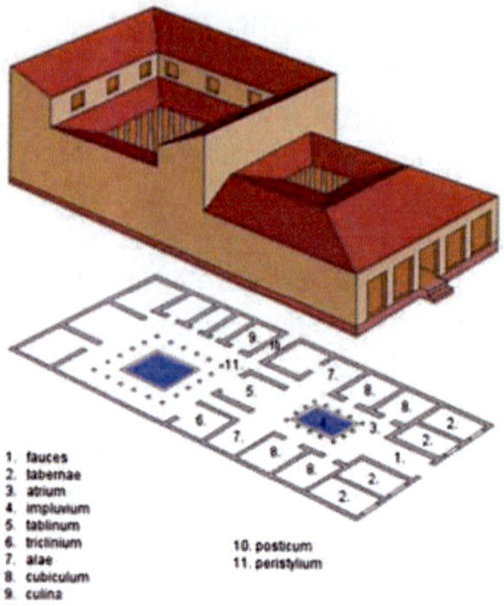

Fig. 1 Plan of a typical Roman house (*domus*). *Source* https://commons.wikimedia.org/wiki/File:Domus_latina.png

as a formal reception room [10]. With such a strong centralized position and semi-public character (space was not anymore used by the family members only, but also by the guests and business clients of the house owner), the atrium changed its appearance. The columns were added together with the altar of the family gods so that the whole space became more representative (*peristil*); a sort of a semi-open showroom displaying a family history and wealth. Due to its central placement within the house, the atrium (or atria in bigger houses belonging to wealthier families) had a function of a buffer zone between the house indoor space and the public exterior space/street (see Fig. 1).

Through the history of architecture and creation of new building typologies, the concept of atrium space changed and along with that its function as well. Looking back at the newer history of western architecture, atria became covered courts, arcades, galleries, and winter garden. One of the pioneering examples of conversion of an interior courtyard into a covered atrium is John Nash's project of the picture gallery at Attigham Park in 1805. The space should serve as a gallery hall illuminated with a zenithal light only, so Nash proposed a curved cast iron frame covered with glass panels.

With technical development of industrial era in the late nineteenth century, many technical possibilities in use of iron and glass contributed to creation of covered private and public spaces of representative character (exhibition and entrance halls, railway stations, retail passages, winter gardens, etc.). One such representative public project was the Crystal Palace in London by Joseph Paxton in 1851.

At the end of the nineteenth century, atrium became part of office buildings design as well. A symbolic example is the Rookery building in Chicago by Burnham and Root built in 1888. This atrium is a two-story lobby space surrounded by shops in the ground and mezzanine floor, with a large representative staircase in the center [11].

This trend continued in the early twentieth century through some of the Wright's office building projects, such as the Larkin Building in Buffalo built in 1903 or Johnson Wax headquarters in Racine from 1936. Frank Lloyd Wright applied this concept in many of his buildings, including the V. C. Morris store in San Francisco from 1949 or the Guggenheim Museum in New York from 1959. The innovation of the latter two was the circular ramp connecting the gallery levels, which besides the achieved aesthetics underlined the visual connection and interaction between the users.

In the twentieth century, the concept of atrium space was changed. These were not any-more only spaces positioned in the center of the building and covered by a glass roof, starting from the ground floor level, but could be positioned also on upper levels of high-rise buildings (with or without skylight) or on the edge of the building volume rather than in the center of it [12].

Regarding the design of school buildings in Europe, their construction relates to the establishment of public schools and the secularization of education in the nineteenth century. The first built town schools were grand, with multiple floors and classrooms attached to both sides of long dark corridors. As the methodology of education has changed and the physical and mental wellbeing of children became of the upmost importance, the school design offered numerous varieties of spatial organization. Pavilion type schools, one to two floors high, became popular. Atrium was also an element widely used in the design of school buildings. The classrooms became brighter and better ventilated, and children had access to atrium spaces to rest and socialize or to use them for informal learning. Some of the remarkable examples representing school buildings with multiple atrium spaces are Munkegaard School at Gentoft, Denmark designed by Arne Jakobsen in 1957 and Hans Scharoun's village-like schools in Lünen from 1962 and Marl from 1970, both in Germany [13].

From the second half of the twentieth century, the use of the open or closed atrium space became very popular in the design of educational facilities-kindergartens, schools and universities with the aim to create a healthier learning environment and to remove the border between the natural-exterior and manmade-interior space.

Use of Atrium Space in Bosnia and Herzegovina

Generally, atrium houses are typical for the areas with warm and moderate climate, such as the Middle East and Northern Africa. However, they are also present in the regions with the harsher climate such as in Europe, China or Latin America.

Bosnia has a continental climate, whereas Herzegovina (the southern part of the country) belongs to the moderate Mediterranean climate. Yet in both parts of the country, buildings with atrium spaces are present.

Even though at the very beginning of the atrium creation its role was to protect the house inhabitants and to provide a more intimate and safer place within the urban pattern, the significance of use of atrium space in residential and later in public buildings rose.

Atrium is serving not only as a buffer zone between exterior and interior in terms of a thermal comfort (air circulation, ventilation, natural lighting) but also in terms of spatial organization and the level of privacy inside the building.

In Bosnia and Herzegovina, houses with internal courtyards that were built during the Ottoman period are good examples to show multiple roles of atrium space.

Those were houses with two internal courtyards, each serving for male and female part of the house. The courtyard within the male part of the house was more public and accessible for visitors and clients, whereas the courtyard within the female part of the house was used only by family members, female servants, and female visitors. It had more domestic character, with greenery and gardens used for relaxing and family gatherings. An existing representative example for this kind of architecture is Svrzo's house in Sarajevo [14]. Apart from the use of internal courtyards in the residential architecture of the Ottoman period, they were common in lodging facilities (*han/karavansaray*) and Islamic religious schools (*madrasa*) as well. Madrasas were built in Bosnia and Herzegovina in the early sixteenth century. Kuršumlija or Gazi Husrev bey's madrasa in Sarajevo was built in 1537 and provides a typical spatial organization of such facility [15]. This madrasa has a symmetric layout with one large common classroom and 12 study rooms located around the common internal courtyard, a semi-open atrium space with a water fountain in the middle (see Fig. 2). Similar to the residential architecture of the Ottoman period, also madrasas had introverted building design with spaces oriented towards the center of the building—the atrium space that would provide privacy for the users. The centralized atrium space of Kuršumlija madrasa was serving as a communication and socialization space among students. The water fountain is a meeting point within the atrium because it was used not only for drinking water but also for the obligatory ritual washing before the prayer.

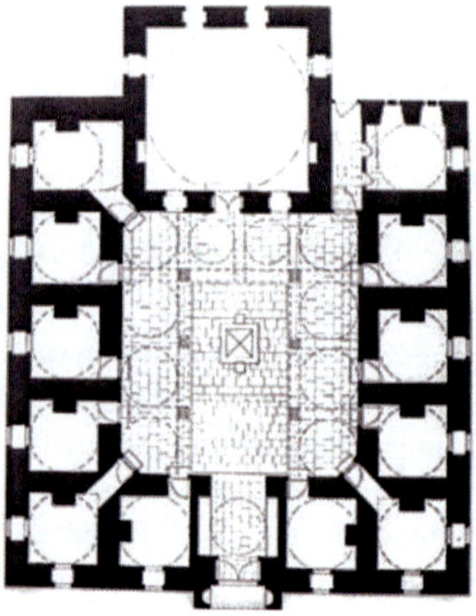

Fig. 2 Plan of Kuršumlija madrasa in Sarajevo, 1537. *Source* https://commons.wikimedia.org/wiki/File:Gazi_Husrev-begova_medresa_2015_tlocrt.jpg

3 Case Studies—Sarajevo, Bosnia and Herzegovina

Atrium, as an element of spatial configuration, appears in few schools in Sarajevo but rarely as an element that promotes any aspect of health (physical, mental, or social). For the purpose of proving this claim, authors visited three elementary schools located in Sarajevo metropolitan area analyzing the current and past use of their atria spaces. In addition to the above-mentioned examples, this paper also presents one school project which is currently under construction where atria were incorporated into design with the aim to contribute to the healthier environment in schools following the HPS strategy.

Current State of Elementary Schools and Use of Atriums in Bosnia and Herzegovina

Most elementary schools that are used today in Bosnia and Herzegovina are in architectural sense typical twentieth century schools built between 1920 and 1992. A significant number of schools from this period were built as a part of the construction effort 'Program for the construction of elementary school buildings', which was popularly called '1000 schools' and lasted from 1974 to 1981 [16]. The neglect of public infrastructure and the dominance of private investments in Bosnia and Herzegovina led to the fact that the number of elementary schools built after the war, since 1995, is very small.

These postwar schools' spatial configurations are most often organized around a central hall with blocks of corridors with classrooms on one or both sides of the corridor. To date, most of them have undergone minor or major changes in architectural terms. These changes are, more often than not, initiated, designed and executed without the help of an architect. They are usually part of the effort to convert auxiliary and other spaces into classrooms. The reason for this is the lack of teaching spaces due to the increase in the overall number of students, and the decrease in the number of students per class (the average number of students in the class decreased from 23.2 in the school year 04/05 to 21.2 in the school year 08/09) [17].

The analysis of the original architectural projects shows a clear spatial differentiation between classes and other activities. The way the space is distributed indicates that the schools' spatial design belongs to an approach that promotes a strict separation of socialization and social activities of students from teaching activities, which is contrary to contemporary pedagogical tendencies and the HPS trend.

Elementary School "Fatima Gunić"

Built and founded in 1977 "Fatima Gunić" elementary school is located in the residential settlement Alipašino Polje. The school was designed by Nikola Nešković, an architect with significant experience in designing educational buildings. The school is a two-story building with rectangular layout. Two corridors are positioned along and on the opposite sides of the entrance hall, multifunctional space, daycare center, atrium and sports hall at the end of the corridors.

In this case, long corridors, which appear in almost every elementary school in Sarajevo and have problems with the lack of natural light, are absorbing enough light through the atrium strategically located between them. Apart from usual benefits of the atrium, cross ventilation and added natural light on the south and north side to the corridors, this location allows proper natural lighting for four more classrooms on the east side of the atrium. On the west side, two large spaces for daycare are located.

Up until a few years ago, the atrium was used as a rose garden without students having access to it. This was an aesthetically pleasing space but without added functionality. In 2019 the atrium was transformed into a children playground with standard equipment and natural greenery was replaced with artificial grass (see Fig. 3).

Since its transformation, the atrium is a place of socialization for children, with access provided from each daycare room and corridors. It is a controlled space which allows students from daycare to spend time outside under the watchful eyes of their teachers.

Fig. 3 Atrium in "Fatima Gunić" elementary school, after transformation into children's playground. *Source* Authors

Even though the size of the atrium (dim. ca. 20×7 m) is big enough to accommodate the playground and a small garden it seems that these two functions are mutually exclusive, thus the full potential of this space is not realized.

Elementary School "Meša Selimović"

The elementary school "Meša Selimović" was built in 1981 and is located in the residential settlement Alipašino Polje. The school was designed by Nikola Nešković, the same architect who designed "Fatima Gunić" elementary school. Similar to "Fatima Gunić", this school as well has a large compact layout but it is a one-story building. The large layout is punctured by four atriums, strategically placed so that the indoor spaces get enough natural light. Teaching spaces are organized around the central multifunctional space, which is naturally illuminated and directly connected, on its north side, to one of the larger atriums (dim. ca. 16×10 m). This atrium (see Fig. 4) is paved, with two large planters, and it is closed for students. Apart from lighting and ventilation, the atrium is used as a temporary storage for construction equipment. The central position and its connection to the multifunctional space opens up opportunities for different kinds of uses that are not being explored in this particular case.

The second atrium is positioned on the south side of the multifunctional space (see Fig. 5). The size and layout of these two atria are the same, except for the second atrium not having direct access to the multifunctional space. Around the second atrium, teaching spaces and hallways are organized. With its proximity to the school's daycare center this space has the same potential of usage as the atrium in "Fatima Gunić" school.

Fig. 4 Atrium in "Meša Selimović" elementary school, with direct access to multifunctional space. *Source* Author

Fig. 5 Atrium in "Meša Selimović" elementary school, without direct access to multifunctional space. *Source* Author

The remaining two atria are smaller in size (dim. ca. 11 × 6 m and ca. 7 × 4 m) and are positioned on the west side of the layout, with library and school's administration organized around them. Like the other two atria in this school, also these spaces don't have a specific use and are not accessible to students. Since atriums are in direct proximity to functions (administration and library) that require controlled conditions in order to function properly, the outdoor spaces are organized as green oases (see Figs. 6 and 7).

Even though the four atrium spaces in "Meša Selimović" elementary school are currently not being used as social and gathering spaces, their potential lies in their strategic positioning within the building volume, close to the functions that could benefit from having direct access to outdoor spaces using them for socialization and possible informal learning spaces in the future.

Fig. 6 Atria in the south-east part of "Meša Selimović" elementary school currently in use as a green oasis. *Source* Authors

Fig. 7 Atria in the north-east part of "Meša Selimović" elementary school currently in use as a green oasis. *Source* Authors

Elementary School "Malta"

The elementary school "Malta" was built in 1980 in residential settlement Dolac Malta as part of the construction effort 'Program for the construction of elementary school buildings'. Just like the elementary school "Fatima Gunić" also this school's two-story layout consists of two blocks of corridors positioned parallel to each other with central hall, atrium and daycare center in between, and sports hall at the west end of the corridor blocks. The only atrium (dim. ca. 12×4 m) is located on the west side of the layout. With classrooms in which the windows are positioned above the line of vision (there is no visual connection with the atrium) and auxiliary spaces organized around, the atrium is a neglected space to which only service staff has access to see Fig. 8. It is used as a staff break area and temporary storage space. The position of the atrium, far from other social activity spaces and its direct connection with the kitchen, makes it hard for the atrium to be used for other purposes involving students of the school.

Case Study of Eleventh Elementary School, Ilidža, Sarajevo

Newly designed elementary school is located in an urban area, mostly occupied by individual residential buildings. The surroundings are characterized by small urban structures, with an average height of up to two-story. The plot has an irregular shape, with a slight semicircular extension to the south, and an extension in the western direction that can be seen as separate from the whole, and used for other purposes. It is bounded on the north side by Dobrinjska street, on the south side by the river Dobrinja, and on the sides, east and west, by neighboring plots. The approximate

Fig. 8 Atrium in "Malta" elementary school as a neglected space. *Source* Author

surface dimensions of the plot are 93.0 m by 110.0 m. The area of the plot is minimal for educational building purpose, which led to specific design solution.

The concept behind the design is creating diversified spaces that range from public to semiprivate and private spaces, which only students can use. This logic is used for the outdoor and indoor spaces. Positioning of the school on the outside edges of the plot allowed the architects to maximize the use of allotted space and to use the central part of the plot for two atria (see Fig. 9).

The first atrium (dim ca. 30 × 20 m) on the north side of the plot is a semipublic space that can be accessed through the passage on the north side. It is a square for gatherings of students in front of the main entrance on the east side of the atrium. In addition, this space can be used when the central circular multifunction space is rented by the local community as an extension of the social spaces. The position of

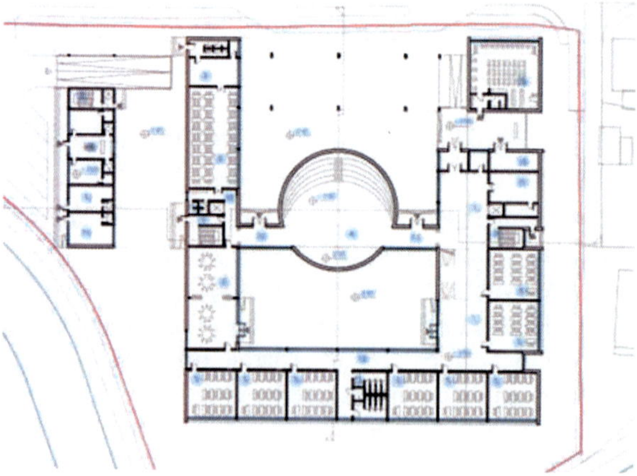

Fig. 9 Ground floor of 11th Elementary school, Ilidža, Sarajevo. *Source* Authors

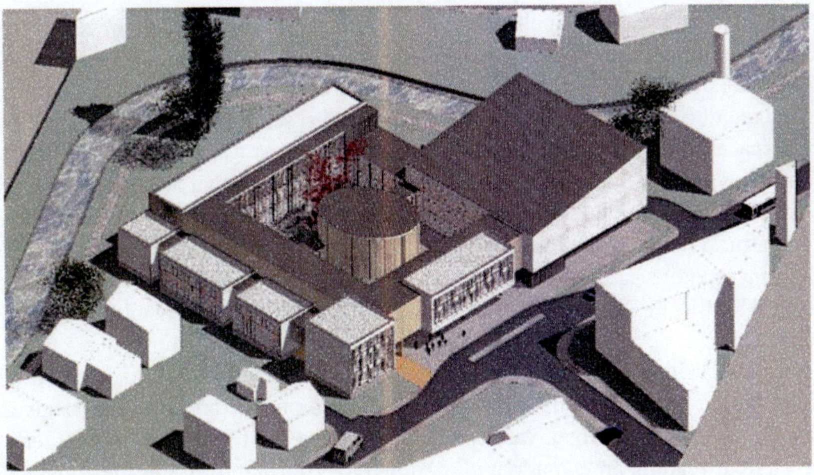

Fig. 10 3D visualization of 11th elementary school, Ilidža, Sarajevo. *Source* Authors

the school's dining hall on the west side that can easily be reached through the atrium and library on the east makes the atrium an active space throughout the day.

The second atrium (dim ca. 30 × 15 m) is located on the south side of the building, and it is planned to be used by younger students. On the ground floor the atrium is surrounded by a daycare center on the west, multifunctional space on the north, and hallways of the classrooms for students attending 1st, 2nd, 3rd and 4th year of education. Because there are no classrooms on the ground floor that have direct connection with the atrium, this space can be used during the whole day without disrupting ongoing lectures. Hallways between classrooms and atrium act as a sound buffer, isolating the noise from the atrium.

The sizes of the atriums (see Fig. 10) were carefully considered, since one of the main reasons why they were not used by the students in previously analyzed schools are their inadequate dimensions. They were too small for the potential number of students that could access them. Since the schools do not have staff available to control the use of atrium spaces, they find it easier to restrict access to them.

Both atria are imagined and designed as multipurpose spaces where students can play, learn, relax and spend time during the day.

Discussion

This paper addresses the importance of the use of atrium spaces in educational buildings as spaces of socialization and informal learning that can contribute to the physical and psychological health of children.

The paper reflects on the development of atrium spaces and their use in different building typologies focusing on educational buildings. One part of the research is the historical overview on application of atrium spaces in the design of residential and educational buildings of Ottoman period present in Bosnia and Herzegovina.

The main part of the research is conducted in three elementary schools in Sarajevo constructed during the socialist period. All presented schools have one ore multiple atrium spaces within their building volume.

The analyses of the case studies show that the atria spaces, that have great potential for diversifying physical learning environments, are almost not being used at all.

The atrium in "Fatima Gunić" elementary school functions as a playground but fails to incorporate natural elements that offer respite from the urban environment the school is set in. Most likely this is due to lack of staff that would tend to the garden.

The three atriums of the "Meša Selimović" elementary school, which are strategically located near functions where direct access to outside areas would be advantageous, might be used for a variety of purposes e.g., outdoor play and gathering area, outdoor reading area, outdoor classroom etc. This demonstrates the architect's foresight at the time the building was designed and built. Unfortunately, this is not the case most likely for two reasons: lack of staff and different pedagogical approach to learning and teaching.

Atrium in "Malta" elementary school is a neglected space located far from other social activity spaces and in direct contact with the kitchen. Its positioning makes it hard to be used for other purposes involving students of the school.

The last presented example is the elementary school which is still under construction. This school is designed following the HPS guidelines and the two atria have the potential to be used as multifunctional spaces. This school, once finished, can serve as a case study for testing the impact of its design in terms of the spatial organization on the informal learning and general well-being/health of its users.

4 Conclusion

Today, more than ever, the architects have a challenge to design schools as sustainable and comfortable spaces of (informal) learning that will provide children with security and enjoyable use of space. Taking into consideration the dynamic of educational process in schools and the fact that we all are very much connected with the nature that was our primarily habitat, atrium spaces in school design can provide a quality space for informal learning, relaxation or socialization among children. This can directly influence the physical and mental wellbeing of school children. Depending on the educational program, in many countries children spend more than 5 h daily at school. Therefore, the design of school buildings in terms of spatial organization, the size of spaces, and in particular the size of classrooms, light and lighting, airflow, visual connection between users, interior design (materials, colors), but also in terms of connection to the external, natural environment, became very complex and challenging. Thus, school design solutions that integrate atrium spaces can contribute to creation of pleasant and healthy educational spaces. At the same time, such design concepts have potentials to fulfil the requirements in creating HPS.

Development of new pedagogical concepts is changing the way the schools are designed today. Bosnia and Herzegovina's primary education still heavily relies on behavioral learning theory and ex-cathedra approach in teaching and with that leaves very little unstructured time for students. With this in mind, one could argue that currently there is no need for places for informal learning in primary schools. Even though pedagogical approach in primary schools in Bosnia and Herzegovina is still very much a traditional one, there will inevitably come a time in near future when this needs to change. Since history of school buildings design shows us that the development of school's spatial conditions is wholly bound by the dictates of the world of education [18] new schools have to be built with this thought in mind and the existing ones have to adapt. In order to adapt to these changes, existing school have to use the entirety the space available to them.

In this paper, authors presented three examples of elementary schools in Sarajevo built in ca. 1980s which do have one or multiple atrium spaces integrated into the design concept. However, their position within the building, their size or blocked/ restricted access, do not permit to use them as spaces that could enhance physical and mental health of children. There are many ways to use atrium spaces (space for playing or relaxation, as an open-air classroom, space for informal learning) but so far, the presented schools did not use the full potential of their atrium spaces. The last example presented in this paper shows the involvement of HPS strategy in the design. The two atrium spaces are foreseen for the active use and not just as decorative elements within the building volume. Hopefully, when the school construction is finished, the atria will be used to their full potential promoting a healthy environment and new ways of learning.

References

1. WHO homepage. https://www.who.int/about/governance/constitution. Accessed 15 Feb 2023
2. Ottawa Charter for Health Promotion. https://www.euro.who.int/__data/assets/pdf_file/0004/129532/Ottawa_Charter.pdf. Accessed 15 Feb 2023
3. Editorial, What is health? The ability to adapt. Lancet 373(9666):781
4. Sartorius N (2006) The meanings of health and its promotion. Croat Med J 47(4):662–664
5. Making every school a health-promoting school: implementation guidance. World Health Organization and the United Nations Educational, Scientific and Cultural Organization, Geneva (2021)
6. Wu X, Oldfield P, Heath T (2020) Spatial openness and student activities in an atrium: a parametric evaluation of a social informal learning environment, building and environment, vol 182
7. Johnson M, Majewska D (2022) Formal, non-formal, and informal learning: what are they, what are they, and how can we research them? Cambridge University Press & Assessment Research Report
8. Tarek A, Fouad Z, Sailer K (2019) The design of school building potentiality of informal learning spaces for self-directed learning. In: Conference: 12th space syntax symposium, Beijing

9. Heitor JV (2013) Modular courtyard school: one concept, two projects. https://fenix.tecnico.ulisboa.pt/downloadFile/395145839580/extendend%20abstract.pdf. Accessed 15 Feb 2023
10. Atrium architecture. https://www.britannica.com/technology/atrium-architecture. Accessed 1 Feb 2023
11. The Rookery. https://www.architecture.org/learn/resources/buildings-of-chicago/building/the-rookery/. Accessed 1 Feb 2023
12. Hung WY, Chow WK (2011) A review on architectural aspects of atrium buildings. Archit Sci Rev 44(3):285–295
13. School Buildings and Architecture—Europe. https://www.encyclopedia.com/history/historians-and-chronicles/historians-miscellaneous-biographies/school-buildings. Accessed 22 Feb 2023
14. Salihbegović A (2012) Atrium in architecture—a transparent structure integrated into the volume of the building. PhD dissertation, Sarajevo
15. Baylon M (1972) Škole-školske zgrade, Građevinska knjiga, Beograd
16. Pašić A, Sivčević N, Zečević M (2014) Design strategies for future rural schools in Bosnia and Herzegovina. U K. Vitale (Ur.), Typology for healthy learning environments. University of Zagreb, Zagreb
17. Centar za politike i upravljanje, Osnovno obrazovanje u Bosni i Hercegovini - kvalitet, kreativnost i inovativnost? Centar za politike i upravljanje, Sarajevo (2010)
18. Hertzberger H, de Swaan A (2009) The schools of Herman Hertzberger. 010 Publishers, Rotterdam

How to Improve Spatial Adaptability in Small Inflexible Apartments with Minimum Investment?—A Case Study from Sarajevo

Amela Šljivić

Abstract In this paper, we have analyzed newly constructed apartments in multi-storey residential buildings in Sarajevo through the aspects that influence the adaptability and flexibility of living spaces. Conducted analyses have shown that analyzed apartments present entirely rigid, inflexible units and emphasize the crucial parameters essential for housing flexibility. Nevertheless, those apartments have excellent sales due to the perfectly polished conceptual 3D visualizations of the interior spaces. This paper investigates how household members arrange life in those apartments and how real life and interior design differ from the projected 3D visualizations. Through a case study that follows changes in adaptability in one newly constructed rigid apartment in Sarajevo for five years, we have concluded that household members often become interior designers for their apartments due to the difficult economic situation in the country and low incomes. Therefore, the final results of spatial adaptability depend only on their education, desire for change, and capability to adjust the space to their needs with a limited budget. This paper demonstrates that it is possible to adapt space to living needs with minimal interventions. Simple methods such as remodeling and reusing existing furniture using various self-adhesive foils and other cheap methods will improve the adaptability of the space with minimum investment. Therefore, free education about these methods for citizens is needed. However, additional instruction for the investors is urgently required because the previously mentioned and analyzed problems will be solved when they recognize housing flexibility as one of the basic requirements for newly constructed apartments.

Keywords Flexibility · Spatial adaptability · Reusing · Affordable apartments · Limited budget

A. Šljivić (✉)
Faculty of Architecture, University of Sarajevo, 71000 Sarajevo, Bosnia and Herzegovina
e-mail: amela.sljivic@af.unsa.ba

M. Arslanagić-Kalajdžić et al. (eds.), *Interdisciplinary Advances in Sustainable Development II*, Lecture Notes in Networks and Systems 804,
https://doi.org/10.1007/978-3-031-46692-2_18

1 Introduction

Sarajevo, the capital city of Bosnia and Herzegovina, faces rapid urbanization and intensive population migrations from rural areas to the city center [1]. The previously mentioned has caused the demand to construct many residential settlements in the city. However, new residential settlements in Sarajevo are often built without facilities for primary education, preschool childcare, primary health care, or social care facilities. Also, these settlements are constructed without spaces for recreation or socialization, and potential public space is mainly reduced just to regulated communications and access to the facilities, without any additional spaces for the socialization of residents [2]. Nevertheless, it is important to mention that the number of registered sales of apartments in Sarajevo is almost equal to the registered sales of apartments in all other Cantons in the Federation of Bosnia and Herzegovina (Fig. 1).

Despite all the previously mentioned facts, these apartments have excellent sales due to the fantastic 3D visualizations of interior spaces. In addition, young people and married couples are often forced to buy smaller apartments with the funds they

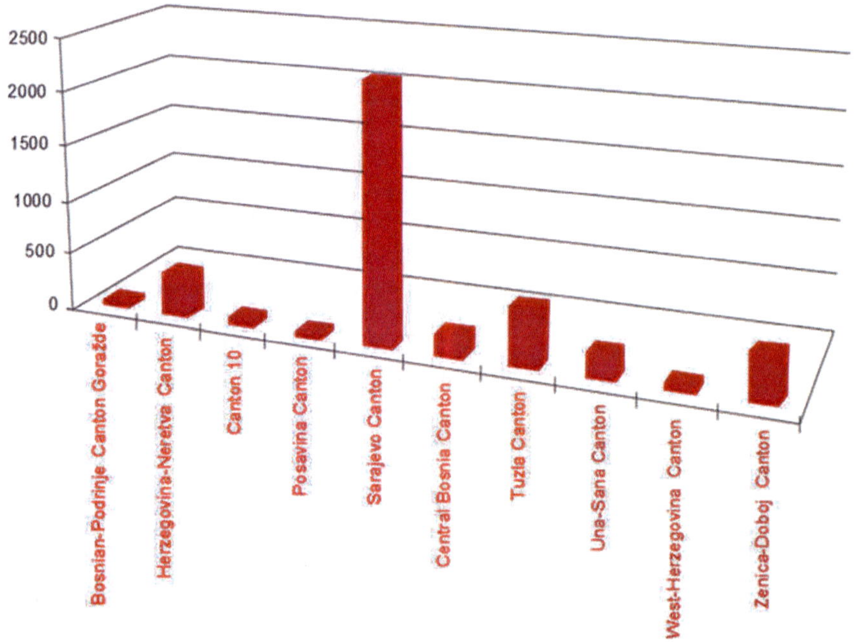

Fig. 1 Number of registered sales contracts of apartments classified per cantons, the data for this chart was used from [3]

can get from a bank due to the country's high prices of apartments[1] compared to average net wages.[2]

It is a well-known fact that the house is considered one of the essential spaces contributing to individual and group's quality of life [4]. However, the living space and living conditions in the apartments and houses changed over time. Previously mentioned changed the traditional understanding of the interior, which is now presented as a living organism that should "grow" according to the needs inside a provided context [5].

When discussing the flexibility and adaptability of architectural space, we often associate these terms with different partitioned walls formed of elements such as shelves, libraries, panels, or pieces of furniture that can be transformed and used in various scenarios. However, concepts of flexibility and adaptability are considerably more complex. Therefore, many studies have defined essential characteristics of these terms. For example, some researchers analyzed changes in the spatial organization of housing units [6], and spatial flexibility is gaining importance engaged with mobility and sustainability, which are the primary critical issues in contemporary architecture [7]. Flexibility was widely discussed in the 1950s [8], and since the Modern Movement, this term has been one of the most attractive words in architecture [7]. According to [9], flexibility is an innovative approach that sets variability as a relevant parameter in designing architectural space, especially in the housing field with complex relationships between users and their immediate environment.

The term "flexible architecture" could generally be referred to the opportunity to modify the layout according to different patterns and most often is connected with the extension of spatial functionality [5]. Today, flexibility in the housing area has become vital, and flexible architecture arose from an authentic need for movement, lifestyle changes, and constantly growing needs [10]. In the paper [11], "flexible housing" refers to housing designed for change over its lifetime. According to [4], three different types of flexibility (structural, functional, and cultural) significantly influence the flexibility of the residential building.

It is important to emphasize that there is no considerable difference between the terms "flexible space" and "adaptable space" in many languages (for example, the Bosnian language is one of them). According to [12], architects and researchers use the term "adaptable" for non-physical changes, while "flexible" is usually used for physical changes. Adaptability is defined as a building's ability to adapt to the changing needs of its context to maximize its value over a lifetime [13]. In this paper, we have chosen the definition from [14] that flexibility is referred to the utilized changes during the design process, and adaptability refers to the changes in an apartment during its exploitation process. Therefore, flexibility allows buildings to be useful for an extended period using adaptations that guarantee continual

[1] According to the Agency for Statistics of Bosnia and Hercegovina, the average price of a new apartment has risen by almost 450 BAM in a year in Sarajevo. In 2022 the average price of a residential apartment in Bosnia and Hercegovina reached 2.587 BAM/m^2.

[2] According to the Agency for Statistics of Bosnia and Herzegovina, average net wages was 1.209 BAM in April of this year.

utilization [8]. However, many problems relate to this subject because many buildings designed for flexible use are practically inflexible due to irresponsible planning and inadequately chosen building systems [7].

Most newly designed buildings in Sarajevo have characteristic floor plans repeated through almost all floors in the building. Unfortunately, these floor plans affected the choice of rigid structural load-bearing systems that do not meet the basic requirements for lifelong use of space, significantly affecting the space's adaptability and flexibility. However, changes such as family growth and the need for new and different rooms open various approaches that end-users (household members) use to transform utterly rigid architectural units into flexible and adaptable ones.

Figure 2 compares characteristic floor plans of the newly constructed apartments in Sarajevo (a) with an example of a flexible floor plan (b) of Greenwich Millennium Village by architects Proctor and Matthews presented in the paper [11]. This example shows a flexible apartment where identical plan forms can accommodate different functions using sliding walls as subdivision elements. The same space can be adapted for a family (1), a couple that uses this space also as a workspace (2), and three independent people that share this space (3). Comparing these examples (a) and (b) shows that the analyzed apartments from Sarajevo are mostly built with transverse load-bearing walls that prevent additional space flexibility (Fig. 2a). Analyses have shown that analyzed flexible apartment (b) does not have the limitations of space flexibility because the interior space is entirely free of load-bearing walls, and in this case, giving special attention to defining the positions of service spaces such as bathrooms and kitchens is crucial.

This paper presents a case study that follows changes in adaptability and flexibility for five years in one newly constructed rigid apartment in Sarajevo. During these years, there were significant changes in the lifestyles and habits of family members, the family's growth, and some unexpected circumstances, such as the Covid-19 pandemic. Guided by the fact that design theorists have recognized the importance of users through the active and creative roles they play in design processes and the fact that there is a persistent gap between design visions and lived realities of buildings [15], we have examined coupled questions in this paper. These questions are: "How in the rigid apartment can household members create space that will satisfy their everyday needs that come with time with a minimum investment?" and as it was discussed in [16] "Whose position is more important in the adaptability of architectural spaces: architects or end-users?" These analyses have been done to find guidelines to improve the quality of life in analyzed buildings.

Fig. 2 Comparison between newly constructed apartments in Sarajevo and flexible floor plans Greenwich Millennium Village, adapted according to [11]

2 Analysis of Newly Constructed Apartments in Sarajevo from the Aspect of Flexibility

The interior design of residential buildings, in particular, and the architectural design, in general, play an essential role in balancing the determinants of the engineering production of residential apartments [4]. In this paper, we have randomly selected four typical apartments from different newly constructed settlements in Sarajevo for further analysis. Selected apartments were analyzed through parameters established by [9] that significantly affect space flexibility. Finally, analysis has been conducted to establish whether these apartments provide space flexibility and which of the analyzed parameters has proven to be the most critical. We have concluded that the typology of newly designed and constructed apartments in Sarajevo does not exist because they are most often created without compliance with the basic regulations needed for these structures.

The results of analyses (Table 1) have shown how the wrong steps in the conceptual stage of the design process will negatively affect and determine end-users' lives. For example, the presented analyses in Table 1 discovered fundamental problems in analyzed apartments, such as:

- the design of deep housing units that exceed the maximum depth of 6.0 m,
- the one-side orientation of the apartments,
- massive load-bearing walls that are placed between all rooms,
- access to bedrooms is usually from the living room (which prevents multifunctional use of the living room).

Also, architects usually draw the furniture on the floor plans without any more profound analysis, and these spaces' lifelong living and use were not considered at any design stage. The small budget for construction usually conditions the structural system selection in analyzed buildings because the investor wants to return the invested funds as soon as possible. Therefore, dedicating a unique analysis to the choice of load-bearing structural systems in the earliest design stages is necessary.

We have concluded that a massive structural system is applied to all analyzed apartments. Moreover, according to [9] massive structural system with longitudinal and transverse walls minimizes space flexibility. Analyzing the presented apartments through the set criteria and recommendations of the mentioned authors, we have established that the new dwellings in Sarajevo are rigid and do not allow almost any space flexibility. Incorrectly selected structural load-bearing systems are one of the critical problems that prevent space adaptability and flexibility. In addition, the positions of bathrooms and kitchens as service spaces further blockade the possibilities of adaptability, and one-sided orientation precludes the division of rooms due to the impossibility of natural lighting and ventilation.

Table 2 shows the general principles that should be applied when designing units with the capacity for adaptability and flexibility. Unfortunately, the presented principles have not been respected in the newly constructed apartments in Sarajevo.

3 Case Study—The Necessity for Space Flexibility and Adaptability in One Small Inflexible Apartment

In this case study, we have analyzed the changes that happened over time in one of the previously analyzed apartments for five years. When they bought the apartment, the family members were a mother (27), a father (29), and a 2-year-old girl. This family has some unique requirements, such as that their new home should have a children's room with many toys and a workspace for parents who occasionally finish their work at home. Although the subject apartment did not fulfill their requirements, this young couple bought the property because of the credit terms and perfect 3D interior design visualizations (Table 3). This case study is presented because one family member (mother) is an architect with three years of experience, and the father, an economist,

Table 1 Analyses of aspects that influence apartments flexibility in residential buildings according to [9]—examples of apartments from newly constructed buildings in Sarajevo

Selected examples are representatives of recently constructed apartments in Sarajevo	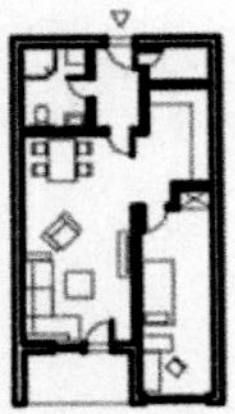	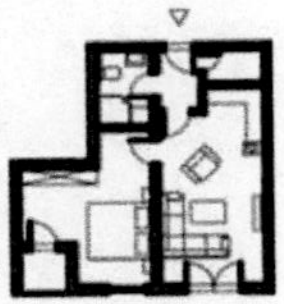	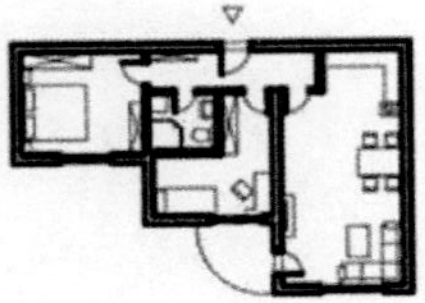	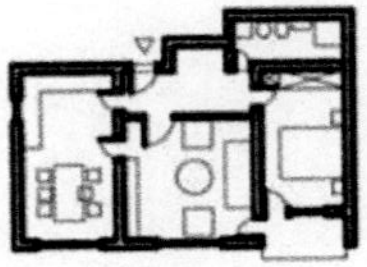
Parameters	Apartment 1 Area = 56 m^2	Apartment 2 Area = 45 m^2	Apartment 3 Area = 60 m^2	Apartment 4 Area = 57 m^2
Orientation of housing unit	One-side	One-side	One-side	Two-sided
Geometry of plan	Compact form	Dispersed form	Dispersed form	Compact form
Structures and size of the flat	Two-rooms	Two-rooms	Three-rooms	Three-rooms
Number and disposition of the entrance	One central entrance	One central entrance	One central entrance	One peripheral entrance
Position of technical services	Individual placed	Individual placed	Individual placed	Individual placed
Building structures	Massive load-bearing walls	Massive load-bearing walls	Massive load-bearing walls	Massive load-bearing walls

(continued)

Table 1 (continued)

Selected examples are representatives of recently constructed apartments in Sarajevo	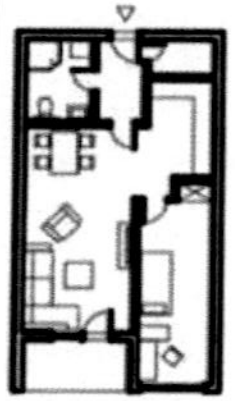		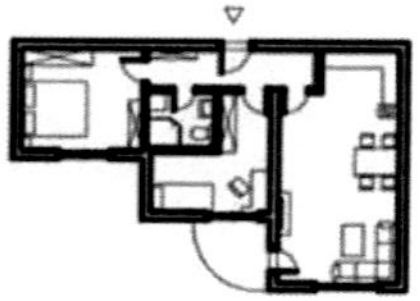	
Parameters	Apartment 1 Area = 56 m^2	Apartment 2 Area = 45 m^2	Apartment 3 Area = 60 m^2	Apartment 4 Area = 57 m^2
Archived degree of freedom of interior space	Zero degrees of freedom of interior space	Zero degrees of freedom of interior space	Zero degrees of freedom of interior space	Zero degrees of freedom of interior space
Potential for multifunctional use of space	None potential	None potential	None potential	None potential
Changes in the number of the rooms	There is no possibility of changing the number of rooms	There is no possibility of changing the number of rooms	There is no possibility of changing the number of rooms	There is a possibility of changing the number of rooms

Table 2 Generic flexibility principles applied to newly constructed apartments in Sarajevo, adopted to [8]

Generic flexibility principles Note: The principles are proposed by Schneider and Till [11]		Principles applied to analyzed apartments 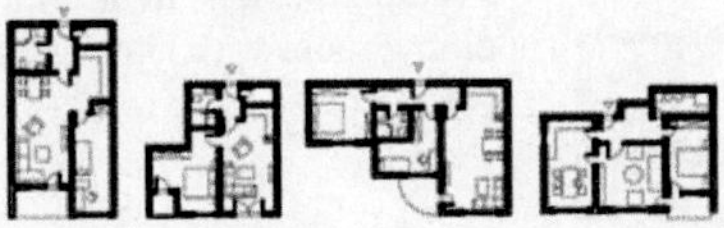
Space	Increased capacity and free use of space as less specified	Accurately defined rooms with furniture that is placed without respecting the basic principles of space organization
Construction	Structures allowing easy access for intervention and maintenance	Massive load-bearing systems do not allow any additional adaptability of space and completely block the possibilities of transformability and adaptability of space
Design for adaptation	Capability to predict future scenarios and room options	Based on the floor plans, it is possible to conclude that the architects were not thinking about future room scenarios and options
Layers	Structure, skin, services, internal partitions, and finishes	These factors were not considered while developing the floor plans and other drawings
Typical plan	Generic space without specification	Accurately defined and divided rooms prevent any additional adaptability
Services	Location planning for future changes	The wrong positions of service rooms completely blocked all possibilities for future apartment transformation and changes

has specific knowledge about various crafts. They started their life together and decided to buy the apartment based on the loan they got from the bank, but they needed more funds to equip it with furniture. In addition, they had to develop a specific approach to interior design, which was why we chose this apartment for a case study.

We have concluded that the only difference compared to other analyzed apartments through the parameters proposed in Tables 1 and 2 is that this apartment has a two-sided orientation of one room due to its position in a building. However, it was essential in this case because it allowed the division of the dining room into two rooms, as shown in Table 2. Therefore, it is necessary to have at least a two-sided orientation of the apartments.

Furnishing the Apartment with Furniture

The living space should live and adapt to its users, and they should enjoy living in their apartments. However, some problems happen at the beginning of life in new apartments in Sarajevo. One of these problems is the lack of funds for hiring

Table 3 Floor plan of the apartment proposed by architects and changes done by end-users

Floor plan	Description
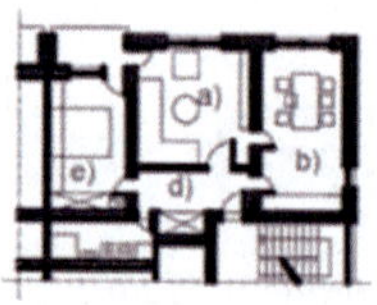	The apartment is divided into the following rooms: (a) living room, (b) dining room with kitchen, (d) corridor, (e) bedroom, (f) bathroom, and (g) balcony
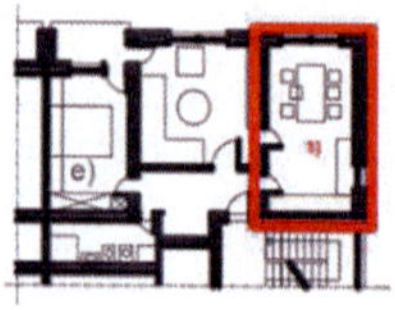	There was a limited possibility for space flexibility in this apartment. The only possibility is to partition and reduce one of the rooms because the apartment is in the corner of the building, so the last room (dining room) has a two-side orientation. In this case, there were two different scenarios for reducing the interior space of the dining room and creating a children's room from the existing one
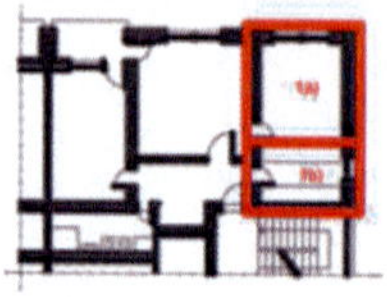	The first variant involved "losing" the dining room and creating a children's room that would be reasonably regular in size. However, the entrance from the living room to the children's room was not an ideal solution
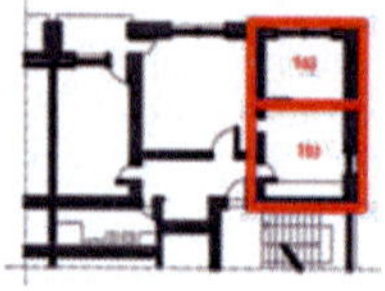	The owners divided the kitchen and dining room into small children's room and a small dining room with a kitchen. Since the child was a baby and would share a bed with her parents for some time, they created a home office in the children's room. This way, they could observe the child in play while working

professional interior designers due to the poor economic situation of citizens who buy apartments using bank credits. Therefore, the owners of this apartment decided to recycle certain pieces of furniture and thus give the old furniture another chance to be used due to their limited budget for interior design. They bought furniture from old stocks and second-hand stores. One of the most significant investments was the kitchen made by the owners with little help from artisans and significant help from Pinterest (Table 1). Chairs in the dining room were restored from the 1950s, and the dining table was made of metal profiles with chipboard because the prices of similar products in stores were extremely high. The interior space satisfies a young family's essential everyday life and work needs. Table 4 shows the difference between the 3D visualization of the apartments and the actual interior design after the household members move into it.

Table 4 A 3D visualization versus actual interior design of the apartment

<table>
<tr><th colspan="5">Floor plan of the apartment</th><th colspan="2">Description of adaptability</th></tr>
<tr><td colspan="5">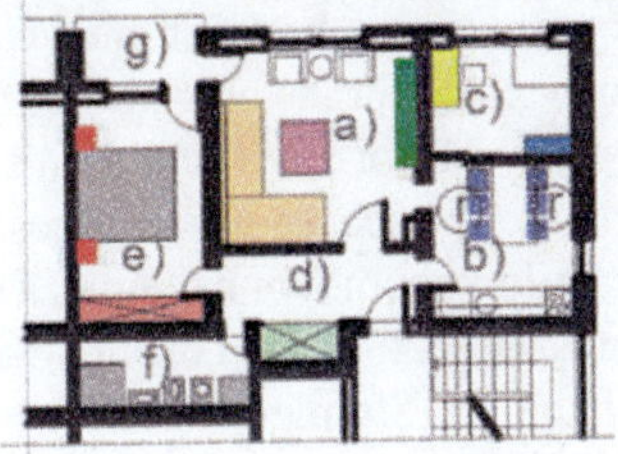
Legend label
new furniture
used furniture
furniture that will be reused in another apartment
(r) repaired furniture</td><td colspan="2">The apartment is divided into the following rooms:
a) living room; b) dining room with kitchen, c) home office and children's playroom, d) corridor e) bedroom, f) bathroom, g) balcony

Note: The furniture that was used before is marked with specific colors. In this way, its movement and application can be traced in different rooms in the apartment</td></tr>
<tr><th colspan="7">3D visualizations of the apartment</th></tr>
<tr><td>a)</td><td>b)</td><td colspan="2">d)</td><td colspan="3">e)</td></tr>
<tr><td></td><td></td><td colspan="2">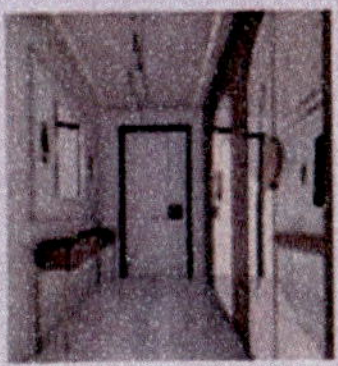</td><td colspan="3">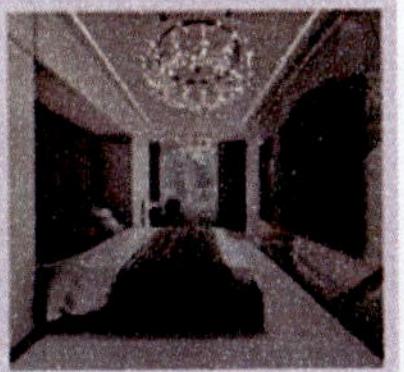</td></tr>
<tr><th colspan="7">The actual interior design of the apartment</th></tr>
<tr><td>a)</td><td>b)</td><td>c)</td><td>d)</td><td colspan="3">e)</td></tr>
<tr><td></td><td></td><td></td><td></td><td colspan="3"></td></tr>
<tr><td></td><td></td><td></td><td></td><td colspan="3"></td></tr>
</table>

The Adaptability of Space Through the Years

There was a considerable need for a children's room with a bed just one year after moving into the new apartment, which meant a new place for the home office needed to be found. Well-preserved recycled furniture was used for a new children's room. The children's room allows for certain transformations because it was necessary to find a new place for some old furniture from the child's room. For example, a simple element with five shelves now becomes an integral part of the corridor with the mirror previously located at the doors of one old closet. Simple frames of strip roll moldings made the mirror fastened directly to the wall with screws look great. The home office has become an integral part of the bedroom, and for the next certain period, the bedroom was divided into two functional units (Table 5).

Over time, the family expanded and got a new member. Therefore, the new member required new space for the baby cot and clothes. The shelves from the corridor were reused for baby clothes.

The Covid-19 pandemic caused a situation where everyone began to work from home. However, work from the bedroom was impossible because the baby cot replaced the home office desk. Working from home with two children was complicated, so providing a new workspace in an apartment of 57 m^2 was necessary. Also, unlike the previous home office, the new space required special silence, allowing a mother with two children to work from home without disturbing each other. Therefore, a new space completely isolated and separated from other parts of the apartment has been found on the balcony (Table 6). After glazing the balcony using PVC windows, a completely new, physically separated workspace was obtained.

Due to the small space in the children's room, it was necessary to provide a new space where children could engage in creative work. For this reason, the dining room is undergoing an entirely new transformation, and with certain corner sofa elements, it provides an entirely new place for family socialization. The problem arose over time when the older child did not want to sleep alone in the children's room because of a new baby, which caused a complete bedroom transformation where another extra bed was added. During that period, all family members slept in the small bedroom, while the children's room was transformed into a playroom again (Table 7).

One year after the last changes in the apartment were made, household members concluded that two completely non-functional rooms were in the apartment. Therefore, they have decided to switch the children's room and bedroom and to use only their existing furniture without buying new ones in new rooms. As a result, the bedroom was equipped with a bed modified by changing its side elements using ordinary wooden plywood and previously used shelves. Furthermore, according to the desires and requirements of the younger family members, the furniture for the children's room was restored by using cheap self-adhesive foils and other simple methods (Table 8).

After creating a functional children's room, children no longer needed to use the living room as a playroom, which enabled its transformation. Although corner sofas are trendy, the owner bought two more oversized sofas to accommodate more people

Table 5 The first adaptability of space (one-year after moving in)

Floor plan of the apartment	Description of adaptability
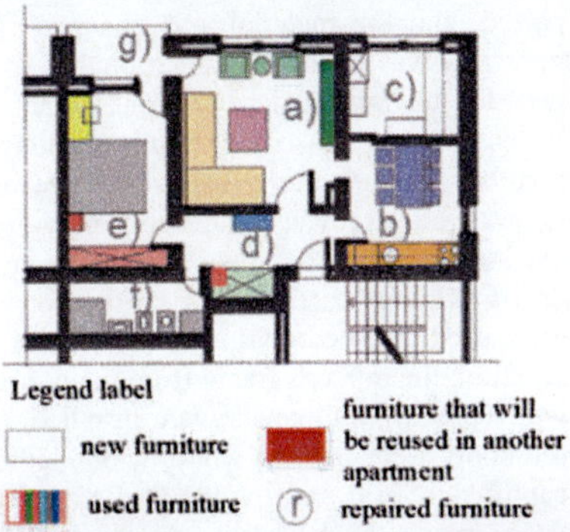	The living room gets new details on the wall and furniture elements. The children's room ultimately becomes children's without space for the parents' home office. So instead, a home office was created inside the bedroom. Recycled furniture previously used in another room is now set in the corridor

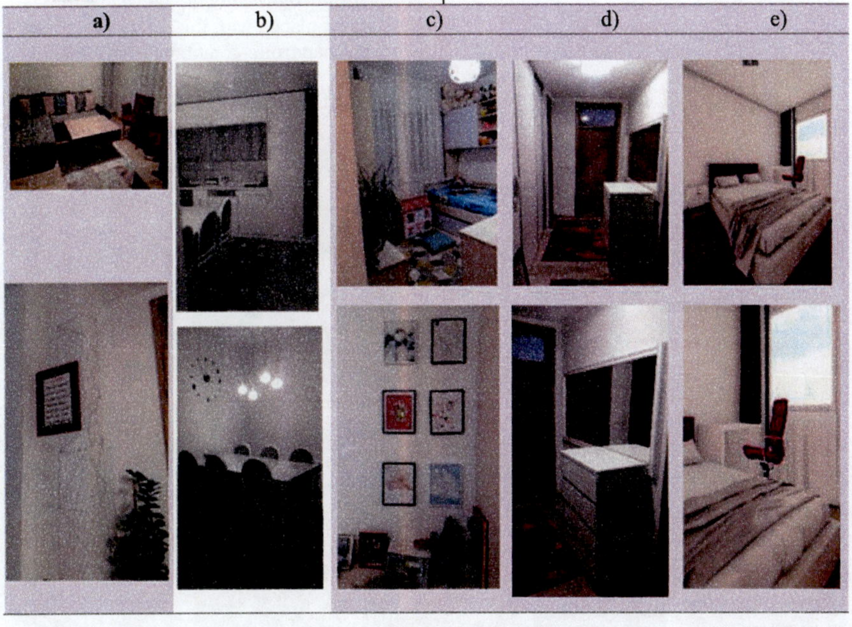

in their apartment. In addition, some mirrors bought cheaply in nearby shops are combined on the wall in the living room, and some old shelves were painted in a new color. Also, it was necessary to provide space to store groceries because it is constantly questioning whether it will be possible to procure basic groceries due to the war in Ukraine. Therefore, the glazed balcony used exclusively for sitting, then as a home office, is now divided into a space for rest and storage space for groceries.

The necessity for space flexibility and adaptability in this apartment will continue because two little girls will require their separate rooms very soon. In that scenario, the new bedroom could become a children's room again, and parents will have to

Table 6 The second adaptability of the space

<table>
<tr><th colspan="3">Floor plan of the apartment</th><th colspan="2">Description of adaptability</th></tr>
<tr><td colspan="3">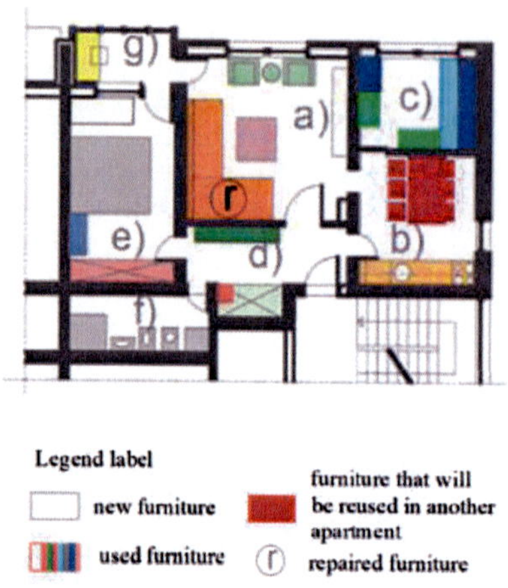
</td><td colspan="2">The corner sofa in the living room was repaired, and the material was changed. The new table was set in the living room due to the need for the lightest piece of furniture that can be easily moved, so the family's youngest member can use the living room area as a playground. Recycled furniture previously used in the living room was used in the corridor. Furthermore, shelves from the corridor were used in a bedroom as a baby closet. In addition, the mirror's frame from the corridor was replaced, wallpapers were installed, and the doors were painted white, which significantly refreshed and modernized the space. Also, inside the bedroom, the home office desk was replaced with the baby's cot. However, the pandemic-conditioned caused full-time work from home, so the balcony was glazed, and a new home office was created</td></tr>
<tr><td>a)</td><td>b)</td><td>c)</td><td>d)</td><td>e)</td></tr>
<tr><td colspan="5"></td></tr>
</table>

sleep in the living room. Consequently, the furniture in the living room should be changed and adjusted to the new additional function (bedroom for parents).

Table 7 The third adaptability of space

Floor plan of the apartment	Description of adaptability
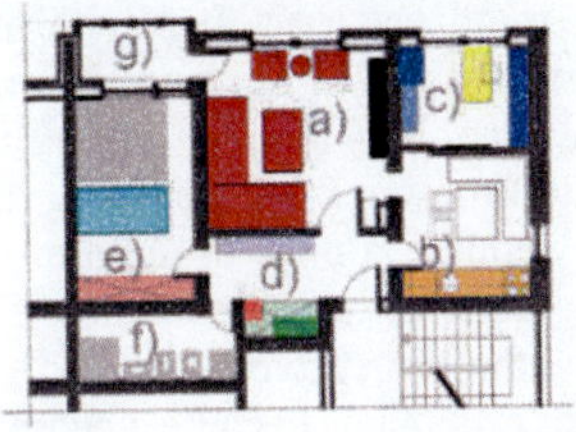 Legend label new furniture furniture that will be reused in another apartment used furniture repaired furniture	New furniture has been installed in the dining room that was adapted as an extended children's room. In addition, the children's room has been shifted into a space for work and play The bedroom was non-functional and used exclusively for the sleep of all family members in one room

a)	b)	c)	d)	e)

Review of Previously Analyzed Spatial Adaptability

We have concluded that the kitchen and bathroom have stayed the same in this apartment for five years, while all the other rooms were adaptable and constantly transformed. Analyzing all the changes in furniture use and the fact that the same furniture was constantly recycled and reused with new functions, in this paper, it has been proven that thanks to minimal intervention, it is possible to achieve the different adaptabilities of the space.

Furthermore, furniture such as closets, cabinets, shelves, and tables can be used multiple times. Therefore, it is possible to refresh them using specific self-adhesive foils (that are trendy today) or different paints for wood and plywood. According to the previously made analysis, we recommended using modular furniture elements;

Table 8 The fourth adaptability of space

Floor plan of the apartment	Description of adaptability
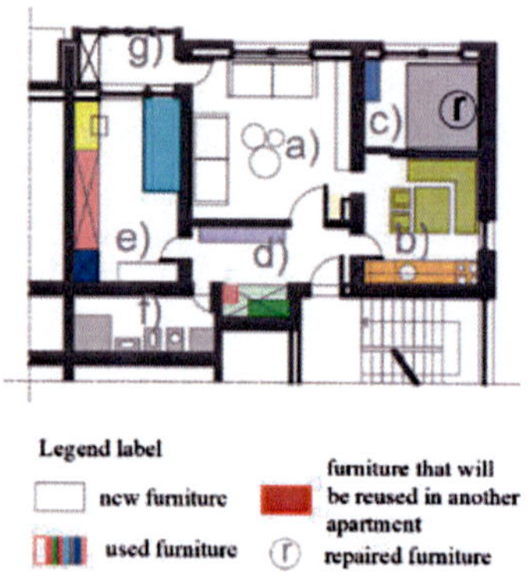	The living room was equipped with more functional furniture for the extended family. This furniture was bought in the outlet furniture stores. Old furniture from the living room was given to another family. In addition, the tiles in the kitchen were refreshed with self-adhesive foil, and the carpets were changed The children's room has become a bedroom, while the bedroom has become the children's room. All furniture in the new bedroom and children's room was previously used in the apartment, but an entirely new effect has been achieved with self-adhesive foils. In addition, the space has been further refreshed with the wallpapers placed on the wall

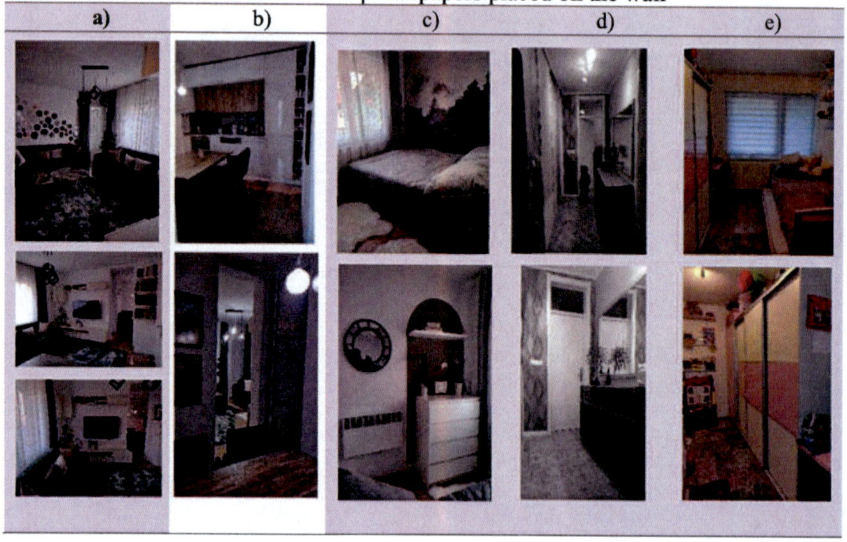

because movement throughout the apartment would be more effortless. It is important to note that disposing of and transporting old used furniture in urban areas is highly complex, and repeatedly investing in new furniture is also very expensive. Therefore, affordable homes could be created using the previously mentioned activities of recycling and reusing furniture within the minimum budget.

Figure 3 shows the space adaptability in the last five years, which has been achieved simply by providing another role to furniture elements. The following transformations are visible:

(a) adaptability 1–2 shows changes made due to the need to move the office desk from the children's room and create a new space for the children with a bed.
(b) adaptability 2–3 shows the transformation of rooms due to the need for new extra space because the family gets a new member.
(c) adaptability 3–4 shows the need for additional workspaces and more functional furniture due to a family members growth as well as the conditions of the Covid pandemic.
(d) adaptability 4–5 shows required adaptation and transformation of space because kids become more independent and needed more space in the children's room.

4 Results and Discussion

Through the case study presented in this paper, we have concluded that architects have little influence in using living space and the leading role in interior design and space adaptation play end-users. However, the flexibility of living spaces depends significantly on the architect's skills and investors' understanding of the benefits they get by applying these principles. Furthermore, the problem often occurs when architects design collective residential buildings for anonymous users. Therefore, predicting the needs and how certain space transformations will occur is complex.

Furthermore, the construction of small apartments (due to the investor's requirements) creates an additional problem for space flexibility and adaptability. Analyzing all the factors that affect the flexibility of space, we have selected three parameters of particular importance for space's flexibility and adaptability. Those parameters are apartment orientation (at least two-sided orientation), space depth, and the selection of the load-bearing structural system. Also, due to the Covid-19 pandemic, the design of residential spaces will ultimately be changed and have an entirely new dimension. Therefore, a home office space with specific characteristics and the possibility for physical separation from other spaces inside every newly designed apartment is required. Analyzing the presented case study, we have concluded that it is necessary to anticipate that the most significant changes will occur within the space of bedrooms and living rooms. Therefore, equipping the bathroom and kitchen with good quality elements is necessary because they are costly investments for young people, and these rooms have minimal changes through the years.

Furthermore, when selecting the size of the rooms in the apartment, one must consider that the purpose of the rooms in the first stage of the project has been changed three times in five years. Moreover, the positions of furniture that architects draw in apartments in the conceptual design are not used as planned because they are not designed to meet the needs of different end-users. For this reason, architects should consider the space's flexibility and focus on several solutions and possible scenarios for using the projected spaces in the conceptual design stage. According to [17], the application of the flexibility concept significantly influenced the development of other principles, such as fluid space and all-in-one space. It is important to mention that

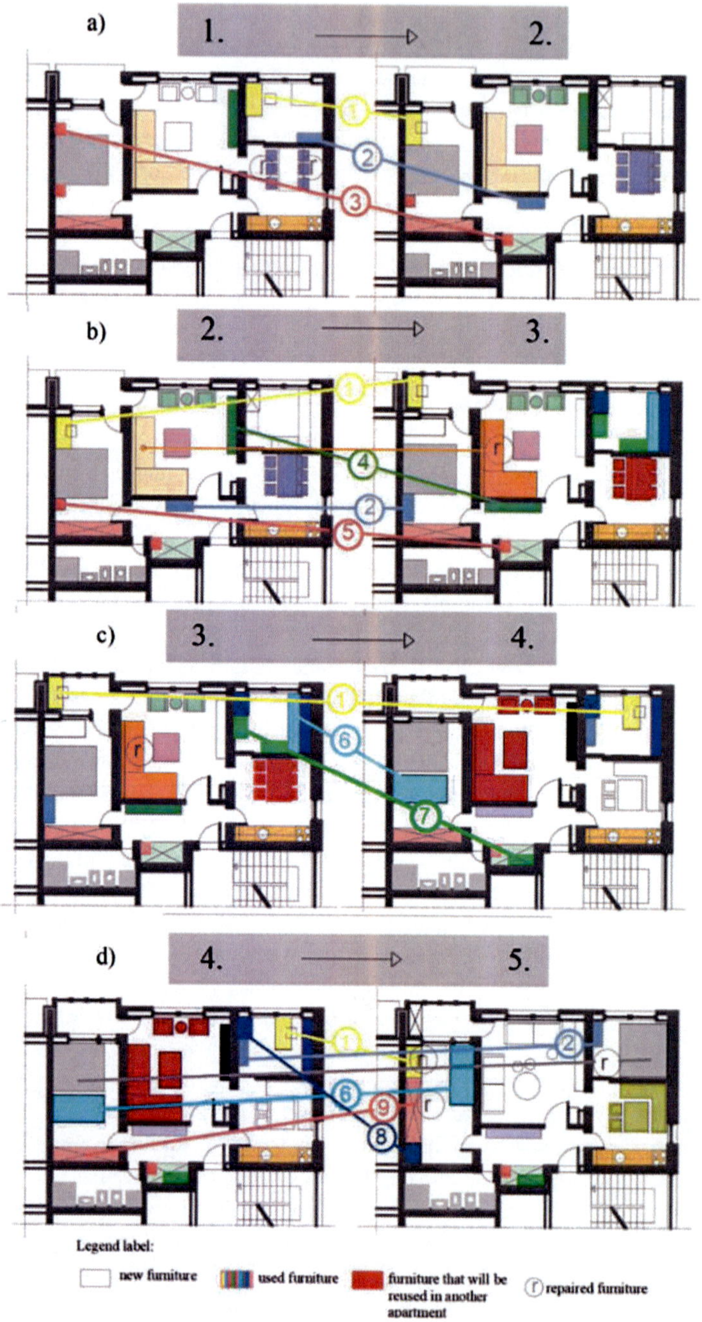

Fig. 3 Recycling and reusing existing furniture enables excellent adaptability of space

we could not find good examples of newly constructed apartments in Sarajevo that use previously mentioned principles and approach for organization of apartments.

Open-plan concepts where complete freedom is given to future users who do not have the necessary professional skills also have not given reasonable solutions in praxis. Therefore, finding the right measure to enable the transformability of living space is necessary, especially for newly constructed apartments in Sarajevo that represent rigid and inflexible structures.

In this paper, it has been shown that thanks to minimal interventions in the living space, it is possible to refresh and change its appearance by using various self-adhesive foils, remodeling old furniture, and reusing and finding a function for shabby furniture. However, the critical aspect of living space adaptability depends exclusively on the end-users, their education and level of knowledge, and their desire to change something. Since the quality of life significantly depends on the home space where citizens spend everyday life, we suggest further educating Sarajevo citizens about the possibilities of apartment adaptability and the reuse of certain pieces of furniture. All changes require a lot of effort and work, and the question is how end-users want to be innovative and spend their time enriching the space they live in and making it more beautiful. Therefore professional education of citizens who can not afford the services of architects is a crucial task.

5 Conclusion

After analyses, it can be concluded that newly designed residential apartments in Sarajevo present entirely rigid, inflexible structures due to the deep depth of the rooms, one-sided orientation, and the wrong choice of a load-bearing structural system. Nevertheless, in these rigid and inflexible apartments (for which the architect is responsible), adaptability is achievable thanks to the end-users education and desire for more appropriate conditions in the apartments that will follow their everyday needs. Unfortunately, due to the minimal budget for interior design, most citizens in Sarajevo do not hire architects for interior design consultations. Instead, they buy their apartments based on good 3D visualizations in the conceptual stage and suppose they will be the same as these rendering images. However, after constructing apartments, they realized that 3D models and photos differed from real life and their financial opportunities.

Based on the case study shown in this paper, we have concluded that end-users real life differs from the projected one and that additional education about basic interior design skills is needed for citizens who cannot afford the services of an architect. This conclusion will be presented to local authorities in Sarajevo as a project aiming to create free education for citizens on various reasonable methods that could help efficiently organize the interiors of apartments and make them adaptable for years. Therefore, this project will enable a better quality of living space that will further improve the quality of life of Sarajevo residents.

However, all these interventions are related to improving already inadequately designed apartments. Therefore, the crucial contribution in the future will be additional education of investors in the first stages of the project about the importance of apartment flexibility and adaptability to prevent the same mistakes in new buildings. Therefore, the following residential building projects in Sarajevo should respect the basic recommendations and general principles of constructing flexible housing units shown in this paper.

References

1. Nurković R (2015) Contemporary aspects of spatial and urban planning in the Canton of Sarajevo. Eur J Geogr 6(1):23–26
2. Čakarić J, Idrizbegović Zgonić A (2020) Nameless settlements of Sarajevo. IOP Conf Ser: Mater Sci Eng 960:032020
3. Federal Administration for Geodetic and Property Affairs [Online]. https://www.fgu.com.ba/en/newse-reader/real-estate-market-report-of-the-federation-of-bih-for-2021.html. Last accessed 17 April 2022
4. Bettaieb DM, Alsabban R (2021) Emerging living styles post-COVID-19: housing flexibility as a fundamental requirement for apartments in Jeddah. Archnet-IJAR: Int J Archit Res 15(1):28–50
5. Gjakun M (2015) Flexibility and comfort in limited dwelling interior. PhD thesis, Department of Architecture and Urban studies (DAStU) Politecnico di Milano, Milano
6. Yayali Yildiz B, Ek FI, Can I (2018) Transformation in a housing-design story: reading the spatial typologies of apartment project in Hatay-Izmir. ITU A|Z 15(3):123–137
7. Kim Y-J (2013) On flexibility in architecture focused on the contradiction in designing flexible space and its design proposition. Archit Res 15(4):191–200
8. De Paris SR, Lopes CN (2018) Housing flexibility problem: review of recent limitations and solutions. Front Archit Res 7:80–91
9. Živković M, Jovanović G (2012) A method for evaluating the degree of housing unit flexibility in multi-family housing. Facta Univ Ser Archit Civ Eng 10(1):17–32
10. Magdziak M (2019) Flexibility and adaptability of the living space to the changing needs of residents. IOP Conf Ser: Mater Sci Eng 471:072011
11. Schneider T, Till J (2005) Flexible housing: opportunities and limits. Archit Res Q 9(2):157–166
12. Estaji H (2017) A review of flexibility and adaptability in housing design. Int J Contemp Archit New ARCH 4(2):37–49
13. Kamara JM, Heidrich O, Tafaro VE, Maltese S, Dejaco MC, Cecconi FR (2020) Change factors and the adaptability of buildings. Sustainability 12:6585
14. Jakšić Ž, Trivunović M, Adamtsevich A (2017) Flexibility and adaptability—key elements of end-user participation in living space designing. MATEC Web Conf 106:01001
15. Sharif AA, Karvonen A (2021) Supporting, tinkering, adjusting and resisting: a typology of user translations of the built environment. Open House Int 46(2):266–280
16. Rodofsky B (1964) Architecture without architects—a short introduction to non-pedigreed architecture. Museum of Modern Art, New York
17. Alfirevic D, Simonovic Alfirevic S (2016) Open-plan in housing architecture: origin, development and design approaches for spatial integration. Arhitektura i Urbanizam 45–60

Social Aspects

Adaptive Governance of Sustainability Under Political and Security Uncertainty: A Quadruple Bottom Line Approach in Bosnia and Herzegovina

Haris Alibašić

Abstract This research examines the governance challenges and opportunities for sustainable development in Bosnia and Herzegovina, a country facing significant political uncertainties. It adopts a Quadruple Bottom Line (QBL) approach, considering sustainability's economic, social, environmental, and governance aspects. The study analyzes various reports, frameworks, and literature to identify common themes and critical elements of adaptive sustainability governance at the national level. The findings highlight the importance of adaptive governance and multi-level collaboration in achieving sustainable development goals. They emphasize the need for inclusive decision-making, transparency, and accountability in governance processes, along with promoting social inclusion, gender equality, and environmental conservation. Investment in sustainable infrastructure and renewable energy is crucial for economic growth and development. The report proposes a typology for sustainable development in unstable countries under uncertainty and risks, encompassing environmental regulation, adaptive governance, social inclusion, and sustainable economic growth. This typology is a valuable framework for assessing sustainability patterns and guiding policymakers and stakeholders in unstable contexts. While the paper focuses on Bosnia and Herzegovina, the research findings have broader implications for other countries facing similar political challenges. The study highlights the importance of stakeholder engagement, institutional arrangements, and resilience planning in achieving sustainable development goals. Integrating economic, social, governance, and environmental considerations in decision-making is crucial for promoting sustainability in diverse contexts. Overall, this research contributes to understanding sustainable development in politically uncertain environments and promotes the application of adaptive governance and the QBL approach in literature and research. The findings underscore the need for increased research, innovation, technology investment, and public participation to enhance sustainability efforts and address marginalized groups' specific challenges. The research concludes by

H. Alibašić (✉)
Public Administration Program, University of West Florida, Pensacola, FL, USA
e-mail: halibasic@uwf.edu

M. Arslanagić-Kalajdžić et al. (eds.), *Interdisciplinary Advances in Sustainable Development II*, Lecture Notes in Networks and Systems 804,
https://doi.org/10.1007/978-3-031-46692-2_19

emphasizing the importance of comprehensive and socially equitable approaches to sustainable development in Bosnia and Herzegovina.

Keywords Sustainability · Adaptive governance · Political uncertainty · Climate change · Bosnia and Herzegovina · Resilience

1 Introduction

The United Nations' sustainable development policy framework is extensive and complex, and its programs are intricately linked to environmental, social, and economic concerns and adaptive governance. Through a qualitative analysis of the existing sustainable development plans and reports in Bosnia and Herzegovina, the author suggests new empirical insight into factors that facilitate the implementation of the SDGs in countries under significant political uncertainties and offers typologies for similar experiences in the international context. The case study of Bosnia and Herzegovina illustrates the challenges in achieving and implementing UN-led sustainability-related initiatives under constantly changing political circumstances. Moreover, the author examines the facilitating elements of sustainable development and finds the existing strategies and plans suitable for implementing the resilience agenda in an environmentally-centered economy defined by resilient growth, sustainable consumption of resources, and reduced carbon emissions. The introduction section provides an overview of the political, social, economic, governance, and environmental challenges Bosnia and Herzegovina faces in advancing sustainable development. Furthermore, the section highpoints the key policies and programs, including the country's efforts to meet the United Nations SDGs and confront climate change threats. Additionally, the introduction outlines the purpose and objectives of the study, as well as the literature used to guide the research.

This research utilizes a documentary qualitative analysis to evaluate the country's efforts to meet the United Nations Sustainable Development Goals (SDGs) and confront climate change threats. Specifically, the report breaks down the analysis into social, economic, environmental, and governance sections, emphasizing adaptive governance. Bosnia and Herzegovina has been attempting to advance sustainable development while facing ongoing political uncertainty and complex social, economic, governance, and environmental aspects of sustainability. Despite significant efforts to meet the United Nations Sustainable Development Goals (SDGs), the country continues encountering challenges in implementing sustainable policies and programs. The author's research study investigates the factors contributing to the difficulties of achieving sustainability in Bosnia and Herzegovina and applies Quadruple Bottom Line and adaptatative governance as the theoretical framework to probe the various initiative.

Bosnia and Herzegovina is located in the Balkan Peninsula of Southeastern Europe with specific political arrangements of two entities and a special district. The country has been confronting numerous challenges in advancing sustainable development and has experienced significant social, economic, and environmental changes over the past few decades, including a prolonged war and post-war reconstruction [1]. These changes have significantly impacted the country's social-ecological-governance systems and have led to environmental degradation and a decline in the well-being of its citizens. Freedom House [2] reported that "the year 2021 saw the most severe political crisis in Bosnia and Herzegovina (BiH) since the end of the 1992–95 war, with several competing issues grinding the country's democratic processes to a near halt" (para 3).

The country's unique circumstances represent challenges and offer opportunities to pursue sustainable development goals. Over the past few years, the country saw a steady decline in categories defining political stability and effective governance, with the National Democratic Governance rating declining "from 2.00 to 1.75 due to the gravest secession crisis in the country's independent history and constant institutional dysfunction sowed by domestic political actors, exemplified by the ongoing failure to form a government in the Federation of Bosnia and Herzegovina (FBiH) after the 2018 elections" [2, 3]. Moreover, "Civil Society rating declined from 4.50 to 4.25 due to the gradual but marked weakening of the civic sector, which is more alienated from the general public, less responsive than ever to current events, and increasingly prone to co-optation by domestic political actors [2].

Within such political uncertainties and insecurities in the country, the analysis of sustainability plans and reports details an itineration of sustainable development goals of a balanced approach to economic, environmental, social, and governance principles. Bosnia and Herzegovina's SDG Index and Dashboards Report indicates the country ranks 59th out of 163 countries in terms of overall SDG performance, scoring 71.7 out of 100 [4]. The country performs well in SDG 1 (No Poverty) and 12 (Responsible Consumption and Production) but performs poorly in SDG 11 (Sustainable Cities and Communities). The country must improve most of the targets to achieve any SDGs by 2030, and significant efforts are needed to accelerate progress [4]. Incorporating the SDGs into all sectors' activities and action plans is critical to the effective governance of sustainability planning.

Institutionalizing governance into sustainable development planning enables accountability, answerability, and adaptive governance of sustainability activities. It ensures the indicators are measured, tracked, and reported leading to meeting objectives, goals, and targets [1, 5–8]. Moreover, Alibašić [9] defined Quadruple Bottom Line (QBL) as an "organizational capacity to embed and incorporate a set of definitive policies and programs to address economic, social, environmental, and governance aspects of sustainability, whereas governance is expressed through fiscal responsibility, community engagement for effective service delivery, transparency, accountability, and more resilient organizations and communities" (p. 11). Using the QBL approach, the author reviews and examines the current sustainable development plans through an adaptive governance lens.

Climate Change Resilience

The World Bank's [10] overview of the country's climate and climate-related risks notes that Bosnia and Herzegovina is highly vulnerable to climate change impacts, such as increased temperatures, droughts, floods, and landslides. The risks threaten the country's natural resources, agriculture, water supply, energy, infrastructure, and human health. The country's greenhouse gas emissions have increased since 1990, primarily due to energy use and transport. The current policies and measures are insufficient to achieve the country's greenhouse gas emissions reduction targets [10]. The climate models, scientific literature, and country-level data from various sources offer a detailed analysis of the country's climate and risks, including future climate scenarios and impact projections. The country's climate policies, adaptation, and mitigation measures, and implementation gaps and challenges reviews are necessary to incorporate into sustainable development reports to comprehensively analyze climate change's social and economic dimensions, such as how vulnerable populations are affected by climate risks and how climate change interacts with other drivers of poverty and inequality in the country.

Overall, the climate risks and challenges facing Bosnia and Herzegovina are significant, and urgent action to decrease greenhouse gas emissions and increase climate resilience is an increasing priority. A comprehensive and interdisciplinary approach to integrate social, governance, environmental, and economic factors into the analysis increases transparency and replicability.

Economic and Environmental Aspects of QB

A complex interaction between human and natural systems characterizes sustainability efforts and systems in Bosnia. Strambo et al. [11] explored the vulnerability of Bosnia and Herzegovina to the impacts of climate change and natural hazards, particularly regarding social equity, gender equality, and poverty reduction. The country has diverse ecosystems, including forests, rivers, wetlands, and grasslands supporting flora and fauna. However, the human population's effects on natural systems have been substantial, leading to deforestation, land degradation, water pollution, and air quality degradation [12]. Moreover, climate change has exacerbated unsustainable practices, leading to more frequent and intense natural disasters, such as floods and droughts [13].

Due to its location and reliance on agriculture and forestry sectors, Bosnia and Herzegovina is at risk of climate change impacts. Certain groups, such as children, elders, and members of minority groups, are more vulnerable due to living in hazard-prone areas and having limited resources to cope. The economic development of Bosnia has also played a significant role in shaping the country's social-ecological-governance system. The country's economy relies heavily on natural resources, including timber, minerals, and intensive agricultural land use exploitation. Unsustainable practices such as overexploitation of resources, poor waste management, and industrial pollution lead to environmental degradation and negative impacts on human health and well-being.

Small businesses have an important role in the Bosnia and Herzegovina economy. Kahriman and Tandir [14] studied the circular economy business models in Bosnia and Herzegovina, examining the country's current state of circular economy practices and analyzing the challenges and opportunities for SMEs to adopt circular business models, findings to be still in the early stages, and SMEs facing barriers related to awareness, knowledge, and financing.

Policies and initiatives to promote sustainable development and environmental protection must be considered to address various sustainability-related challenges. For example, the country has developed a National Environmental Action Plan and a National Biodiversity Strategy and Action Plan to guide its environmental policies and management practices [13]. The state of the country's biodiversity and priority areas for conservation efforts are set together with goals and targets and associated actions for the country's conservation. The plan aims to integrate biodiversity conservation into national and sectoral policies, strengthen institutional and legal frameworks, and promote public awareness and participation in biodiversity conservation. Additionally, the country has established several protected areas and nature parks to conserve its natural resources [13].

Importantly, Bosnia and Herzegovina has a significant renewable energy potential. Đurašković et al. [15] conducted a study on renewable energy in the Western Balkans, focusing on policies, developments, and perspectives. The current status of renewable energy in the region is in somewhat developmental stages with significant potential for renewable energy. Still, the sector faces policy framework, infrastructure, and investment challenges. A coordinated regional approach of governance to develop renewable energy would alleviate challenges. Moreover, Androsevic et al. [16] completed a literature review of requirements for the sustainable energy transition in Bosnia and Herzegovina in the regional context within the current energy transition state, political will, investment, and public engagement for achieving a sustainable energy transition.

Social and Governance

The social aspect of sustainable development is also relevant to the analysis. Katayanagi [17] discussed the voices of post-war reconstruction in Bosnia and Herzegovina and resilience against everyday disillusion in the context of peacebuilding theories and practices. Sustainable development in a conflict-affected society must be coupled with resilience planning and an emphasis on governance and institutional capacity building. Edgar [18] examined the rule of law in advancing and achieving the sustainable development goals of Agenda 2030. Good governance, accountability, and transparency support the implementation of the rule of law and achieving sustainable development. Furthermore, Hadžić [19] scrutinized the causes and risks of economic crime, abuse of ethnopolitical power, and systemic corruption in Bosnia and Herzegovina.

Overall, the quadruple bottom line system is complex and interconnected, and a range of factors, including human activity, natural resource use, and climate change, influences its effectiveness and applicability. Adaptive governance and policy interventions are necessary to promote sustainability and ensure the country's citizens' and natural systems' well-being. Alibašić's [9] concept of resilience and its relationship with sustainable development accentuates the strategic planning of resilience and sustainability, arguing that integrating resilience and sustainability into planning processes, communities, and organizations better adapt to changing circumstances and mitigate the risks posed by environmental, social, governance, and economic risks and uncertainties [1, 6, 7, 9]. In finality, Alibašic [9] QBL framework is an appropriate theoretical anchor to examine the ability of countries to govern sustainability. The role of policymakers is to promote positive governance and not to retreat from policymaking benefiting their constituents [20].

2 Theoretical Framework

Introduction

Countries facing severe political crises and insecurities have difficulty in promoting sustainability goals. Political instability disrupts governance systems, undermines institutions, and exacerbates social and economic inequality, impeding efforts to address environmental and social challenges [21]. Yet, despite these challenges, some countries have been able to maintain or even increase their sustainability commitments and progress toward achieving sustainability goals during periods of political crisis and increased insecurities. This paper proposes a theoretical framework for studying countries advancing sustainability under severe political crises, drawing on adaptive governance and resilience theories.

The proposed theoretical framework draws on resilience theory and adaptive governance to explore how institutions and communities in countries facing severe political crises respond to and cope with challenges while pursuing sustainability goals. Adaptive governance emphasizes the importance of formal and informal adaptability and decision-making within organizations and society more broadly [22]. Resilience theory, on the other hand, highlights the ability of social-ecological systems to adapt and transform in the face of disturbance, uncertainty, and change [9, 23].

Resilience

The research explores the underlying flexibility through adaptive governance of sustainability efforts in Bosnia. Resilience is the ability of organizations to effectively respond to and recover from crises [5, 8, 24]. Likewise, Hillman and Geunther's [25] review of organizational resilience explains the complexity of organizational dynamics, coupled with Acar and Winfrey's [26] notion of the resilient organization as a means to sustain organizational renewal and performance. Also, Clément and Rivera [27] suggested resilience was integral to adaptation to changes and transformations of the systems. Similarly, Berkes and Ross [28] understood resilience as a construct, and Klein et al. [29] as a valuable concept to explain adaptive capacity.

The resilience framework is a valuable concept to promote the exploration of adaptive governance. Alibašić's [9] view of governance included accountability as one of the critical factors in successfully implementing resilience and sustainability plans. Bosnia's ongoing political challenges and perpetual crises represent an opportunity to utilize adaptive governance design for sustainable development. Bardal et al. [30] explored the factors that facilitate the implementation of the Sustainable Development Goals (SDGs) in regional and local planning in Norway, analyzing two case studies, identifying several factors that facilitate the implementation of the SDGs, including political leadership, stakeholder engagement, and institutional arrangements. The SDG Framework for BiH suggests a roadmap for achieving sustainable development in the country through adaptive governance [4]. The framework highlights the importance of partnerships, including public–private partnerships, in achieving the SDGs. It also emphasizes the need for data-driven decision-making, monitoring, and reporting to track progress toward the goals. The framework identifies key priority areas, including poverty reduction, improving access to quality education and healthcare, promoting sustainable economic growth, and protecting the environment. The framework calls for increased investment in research, innovation, and technology, as well as greater public participation and capacity building.

Adaptive Governance

Adaptive Governance is a valuable resource for policymakers, practitioners, and researchers interested in developing more effective and sustainable governance models. The emphasis on collaboration, learning, and innovation offers a valuable framework for reviewing the sustainability challenges faced by Bosnia and Herzegovina and identifying opportunities for improving governance processes. By adopting an adaptive governance approach, Bosnia and Herzegovina can enhance its ability to address complex environmental challenges and promote more sustainable and resilient outcomes. Brunner et al. [31] offered a conceptual framework for adaptive governance and examined its applications in various environmental contexts.

Applying adaptive governance to environmental problems requires a nascent approach to collaborative efforts between scientists, policymakers, and stakeholders to develop adaptive strategies to cope with complexity, and uncertainty, demonstrating the effectiveness of adaptive governance in addressing environmental challenges. The theoretical framework is based on adaptive management principles, social-ecological systems, and collaborative governance. Chaffin et al. [32] suggested that adaptive governance emphasizes flexibility, learning, and collaboration to aid decision-makers in dealing with environmental change's uncertainty, complexity, and unpredictability. The framework involves five key elements:

1. Engaging stakeholders in decision-making;
2. Enhancing social learning;
3. Establishing flexible and adaptive institutions;
4. Using a variety of knowledge sources;
5. Promoting inter-jurisdictional collaboration.

It demonstrates its usefulness for policymakers and decision-makers in addressing complex environmental problems associated with rapid climate change.

Adaptive governance is particularly relevant for Bosnia and Herzegovina, given the country's complex political and institutional environment, multiple environmental challenges, and the need for more collaborative and participatory decision-making processes. Bosnia and Herzegovina has a record of decentralized governance and lacks effective coordination among different levels of government and stakeholders. Adaptive governance assists in overcoming the challenges of inefficacies by promoting more inclusive and collaborative approaches to involve multiple stakeholders in decision-making processes, supporting the sharing of knowledge and information, and fostering learning and innovation. Picavet et al. [33] analyzed the implication of the adaptive governance framework for advancing environmental management. Similarly, using the Quadruple Bottom Line approach, adaptive governance synthesizes and connects it to sustainable development, balancing economic, social, governance, and environmental considerations.

In Bosnia and Herzegovina, where environmental issues and climate change risks are complex and interrelated, adaptive governance is a potential tool for policymakers in addressing environmental challenges associated with sustainable development. By emphasizing collaboration, flexibility, and learning, the framework assists decision-makers with complex and interconnected sustainable development challenges. The advantage of adaptive governance is its shift from command-and-control approaches to more adaptive and collaborative governance models. Folke et al. [34] noted that adaptive governance responds to complexities, including trust and communication, the need for participatory monitoring and evaluation, and incorporating local knowledge and expertise into decision-making processes.

3 Research Design and Methodology

The author employed a qualitative case study using documentary analysis of country-level documents to address the research questions, with an in-depth analysis of a country's sustainability during a severe political crisis. Data sources include official reports of sustainability activities at the national level of government. The research aims to answer several key questions: How do institutions in countries facing severe political crises advance sustainability? How do countries' institutions maintain or adjust sustainability commitments amid political instability and uncertainty? What strategies and mechanisms do countries use to advance the quadruple bottom-line sustainability and resilience of systems? What factors facilitate or hinder sustainability efforts under political crisis? The research design and methodology are appropriate for the research question, and the author describes the methodology.

The author uses findings from the analysis of available documents to inform the typology of sustainable development under political uncertainty, identifying the potential issues that may arise and opportunities. Gerring [35] emphasized the importance of selecting appropriate cases, defining research questions, using multiple sources of evidence to ensure the validity of the research, developing theories and hypotheses based on the data considering alternative explanations and potential confounding factors [36, 37], Patton's [38], Saldana's [39], and Yin's [40] comprehensive guidance on qualitative research design and methodology described approaches to qualitative research, including case study research. Each approach is presented with a detailed explanation of its philosophical assumptions, research questions, data collection and analysis methods, and strengths and weaknesses. Documentary analysis is used to analyze existing documents to gain insights into a particular subject or phenomenon, in this case, it is national reports of sustainable development implementation. Overall, a documentary analysis proposes insights into what has been done before, what worked and what did not, making more informed decisions and increasing the chances of successfully implementing sustainable development projects.

A documentary analysis of government policies related to sustainable development involves reviewing available sustainable development documents and reports related to sustainability. Altheide et al. [41] assessed documentary analysis as a qualitative research method, describing the different types of documents that can be analyzed, the steps involved in conducting a documentary analysis, and the strengths and limitations of the method. Developing clear research questions and criteria for selecting relevant documents and the need for careful analysis and interpretation of the data are critical components of successful qualitative documentary analysis research.

In the context of researching the documentary analysis of existing sustainable development plans in Bosnia and Herzegovina, Altheide et al. [41], Creswell [37], Gerring [35], and Neufeld [42] designed a valuable setting for conducting a qualitative documentary analysis, emphasizing the need for straightforward research questions and careful analysis of the data. Selecting multiple sources of evidence is critical in analyzing complex documents such as sustainable development reports. Reflexivity and self-awareness assist the researcher in identifying potential biases and assumptions in their analysis and create a solid foundation for conducting a rigorous and insightful documentary analysis of sustainable development plans in Bosnia and Herzegovina. It enables rigorous and systematic analyses of the contents of written documents, facilitating impartial and consistent written analysis.

In the case study of sustainable development and adaptive governance, as Creswell [37] suggested, purposeful sampling is utilized for data collection and analysis and to interpret the findings. The purposeful sampling technique selects key informants or cases for the study based on their relevance to the research questions, appropriate for the study of Bosnia and Herzegovina's sustainable development. The case study research approach is applied to collect and analyze data from these cases through document analysis. The author identified patterns, themes, and connections between the data and the research questions, proposing a typology for identifying QBL elements of sustainable development in politically uncertain countries. Finally, the ethical considerations are examined with integrity and respect for the research process.

4 Results and Discussion

The Quadruple Bottom Line Framework

A review of the existing sustainable development reports, including a report on Bosnia and Herzegovina's Voluntary National Review (VNR) and its progress towards achieving the United Nations Sustainable Development Goals (SDGs), indicates the level of country's achievements, challenges, and recommendations for improving progress towards the SDGs. It underlines Bosnia and Herzegovina's efforts to address poverty, gender equality, health, education, and clean energy. It also identifies considerable economic growth, environmental sustainability, and governance trials. A comprehensive analysis of the country's progress towards achieving the SDGs, including an overview of its political, economic, and social context and its governance and institutional framework, reveals high levels of poverty and inequality, environmental degradation, and weak institutional capacity. The improvements would include strengthening governance and institutional frameworks, increasing public participation, and improving monitoring and evaluation of SDG progress to enhance policy coherence, public–private partnerships, and investment in infrastructure and education.

The Table 1 summarizes sustainable development frameworks for Bosnia and Herzegovina from various reports, arranged under the four distinct environmental, governance, social, and economic factors. Multi-level and multi-stakeholder collaboration, social cohesion, inclusive economic growth, environmental conservation and the sustainable use of natural resources, investment in sustainable infrastructure, and responsible business practices are often recommended. The SDG Framework for BiH [43] encouraged sustainable development through inclusive decision-making and collaboration, emphasized transparency and accountability in governance, addressed social inequality, and human rights, focusing on environmental sustainability, sustainable consumption, and production patterns.

The Agency for Statistics [46], UNEP [44], UNDP [45], and UNSDC BiH [47] delineated inclusive decision-making and collaboration among stakeholders, social inequality, and work opportunities. The reports noted transparency and public participation in decision-making, poverty reduction, and social inclusion, among other issues, suggesting inclusive and participatory decision-making and collaboration among stakeholders, transparency, accountability, and good governance to advance sustainable development and address social issues such as poverty reduction. While reports mention transparency, objective measures have not been discernable in the reports.

In summary, all the reports address social issues such as poverty reduction, social inclusion, and peaceful societies, focusing on environmental conservation and sustainable use of natural resources and promoting economic growth through responsible business practices, sustainable infrastructure, and decent work opportunities. The reports mentioned transparency and accountability in governance.

The Table 1 summarizes critical elements of governance, social, environmental, and economic pillars at the national level of sustainability planning and reporting in Bosnia and Herzegovina with an adaptive governance lens and Quadruple Bottom Line evaluation (Table 2).

At a national level, the SDG framework for Bosnia and Herzegovina delivers a sweeping plan for achieving sustainable development goals in the country. The framework would benefit from including adaptive governance, data-driven decision-making, and monitoring progress toward sustainability goals and objectives. However, it needs direct stakeholder engagement on a fundamental level, and local governments should be more addressed in SDG planning. Most SDG framework depends on entity-level planning and networking, making adaptive governance more relevant for national-level planning. The framework identifies key priority areas, including reducing poverty and inequality, promoting sustainable economic growth, protecting the environment, and improving access to quality education and healthcare. Sustainability planning must include increased research, innovation, and technology investment and greater public participation and capacity building.

The Quadruple Bottom Line Issues

The environmental statistics of BiH indicate the relevance of adaptive governance in achieving sustainable development. Environmental challenges persist, including air and water pollution, deforestation, and inadequate waste management systems. The country's vulnerability to climate change and emphasis and lack of consistent

Table 1 Specific reports emphasis

Report title	Governance	Social	Environmental	Economic
SDGs Framework for Bosnia and Herzegovina (United Nations, 2022)	Encouraging multi-level and multi-stakeholder collaboration and inclusive decision-making	Addressing poverty reduction, health and education, gender equality, and just societies	Emphasizing environmental conservation and the sustainable use of natural resources, including protecting biodiversity and ecosystems	Encouraging investment in sustainable infrastructure
SDG Framework for BiH [43]	Promoting inclusive decision-making and collaboration between different levels of government and stakeholders, transparency, and accountability in governance	Affecting social inequality, human rights, and the promotion of inclusive and peaceful societies	Caring about environmental sustainability, including protecting natural resources, biodiversity, ecosystems, and sustainable consumption and production patterns	Advancing economic growth and development, sound business practices, and investment in sustainable infrastructure
UN Sustainable Development Cooperation Framework 2021–2025 [58]	Creating inclusive and participatory decision-making and collaboration among stakeholders	Promoting human rights and inclusive societies, empowering women	Encouraging the promotion of sustainable consumption and production patterns	Advancing sustainable energy investments
Bosnia and Herzegovina Maps Report [44]	Advancing multi-level and multi-stakeholder collaboration and public participation	Promoting social inclusion	Protecting biodiversity, ecosystems, and the sustainable use of natural resources	Advancing sustainable economic growth and development

(continued)

Table 1 (continued)

Report title	Governance	Social	Environmental	Economic
Towards an SDG Financing Ecosystem in Bosnia and Herzegovina [45]	Calling for inclusive and participatory decision-making and stakeholder collaboration to promote sustainable development	Advancing social issues such	Environmental conservation, including protecting biodiversity, ecosystems, and the sustainable use of natural resources	Advancing economic growth and development
Bosnia and Herzegovina Agency for Statistics [46]	Reporting good governance, transparency, and accountability in decision-making to advance sustainability development			

mitigation and adaptation measures, including energy efficiency, renewable energy sources, and sustainable land use practices, are significant issues. Social development is crucial to sustainable development, achieving greater gender equality, and implementing policies, programs, and strategies to improve social protection systems, education, and health care. Challenges remain, including high unemployment rates, poverty, and income inequality. Identified gaps include the further need to strengthen social inclusion policies and measures and address the needs of marginalized groups, including Roma populations and people with disabilities. The country's economy has slowly been improving, with a positive growth rate, "with the economy of Bosnia and Herzegovina expanding by 2.6% year-on-year in the third quarter of 2022, easing from a 5.8% rise in the previous three-month period, and marking its lowest growth since the fourth quarter of 2020 [49]. However, persistent economic challenges exist, including high informality and corruption levels, limited foreign direct investment, and low labor productivity. Economic diversification and the need for private sector growth, innovation, and technology adoption are necessary to achieve sustainable economic development, improve labor market policies, promote entrepreneurship, and support small and medium-sized enterprises.

Adaptive Governance

Under governance, the analysis indicates good governance, transparency, accountability and the rule of law, strengthened public administration, answerable policies, and combating corruption is critical to advance sustainability. Effective multilevel governance and stakeholder cooperation to achieve sustainable development is

Table 2 Common themes—summary of reports

Environmental	Governance	Social	Economic
Environmental conservation and the sustainable use of natural resources, including the protection of biodiversity and ecosystems	Multi-level and multi-stakeholder collaboration to promote sustainable development through inclusive decision-making. Attention to inequality reduction, social cohesion, and inclusive economic growth	Poverty reduction, health and education, gender equality, and peaceful and just societies	Investment in sustainable infrastructure and renewables
Environmental sustainability, natural resources, biodiversity, and ecosystems. Sustainable consumption and production patterns	Sustainable development through inclusive decision-making and collaboration between different levels of government and stakeholders. Transparency and accountability in governance	Social inequality reduction, human rights, inclusive and peaceful societies. Empowerment of marginalized groups and the promotion of gender equality	Economic growth and development through the promotion of responsible business practices, investment in sustainable infrastructure, and the creation of decent work opportunities
Protection of biodiversity, ecosystems, and the sustainable use of natural resources	Transparency, accountability, and public participation in decision-making	Poverty reduction, social inclusion, and just societies	Sustainable growth and development of responsible business practices, investment in sustainable infrastructure, and creating decent work opportunities

Sources The Audit Office of the Institutions of Bosnia and Herzegovina (BiH) [48]; Agency for Statistics of Bosnia and Herzegovina [49]. Joint SDG Fund [50], United Nations [21, 51–53], Sustainable Development Report [54, 55] Sustainable Development Report [42, 56, 57], UNSDC [47], and UNSDCF [58] are all listed under references

needed to continue progress in improving governance indicators, including government effectiveness and control of corruption, but also highlight areas for improvement, such as ensuring the independence of the judiciary and strengthening the capacity of local governments. Adaptive governance suggests the prominence of monitoring and evaluation mechanisms to ensure the effective implementation of sustainable development policies. Alibašić [9] emphasized the importance of adaptive and good governance for resilience planning and suggested incorporating governance principles, such as transparency and accountability, into resilience planning frameworks can enable practical and sustainable outcomes.

Bosnia and Herzegovina's adaptive governance system must focus on the Quadruple Bottom Line (QBL), with sustainability's economic, social, environmental, and governance aspects. The QBL approach defines organizational capacity to embed and incorporate standard policies and programs to address these sustainability aspects. Furthermore, governance must be expressed through fiscal responsibility, community engagement for effective service delivery, transparency, accountability, and more resilient organizations and communities. Institutionalizing governance into the resilience planning framework enables accountability, answerability, and adaptive governance of sustainability activities. It ensures the indicators are measured, tracked, and reported leading to meeting objectives, goals, and targets.

5 Conclusion

Summary

The research yields an in-depth analysis of sustainability governance in Bosnia and Herzegovina under conditions of political uncertainty. It adopts a Quadruple Bottom Line (QBL) approach, defined by Alibašić [1, 6, 7, 9], which encompasses economic, social, environmental, and governance dimensions, to examine the country's key challenges and opportunities for sustainable development. The study identifies common themes and critical elements of sustainability governance at the national level by analyzing various reports, frameworks, and literature. The findings highlight the significance of adaptive governance and multi-level collaboration in achieving sustainable development goals. Adaptive governance, characterized by inclusive decision-making, transparency, and accountability, emerges as a crucial mechanism for resilience planning and the effective implementation of sustainable development policies. The research emphasizes the importance of stakeholder engagement and institutional arrangements, with actors from different levels of government, civil society organizations, and the private sector working together to address sustainability challenges.

The proposed typology for sustainable development under uncertainty and risks in unstable countries provides a comprehensive framework that aligns with the research findings. The typology encompasses environmental regulation, adaptive governance, social inclusion, and sustainable economic growth as critical factors for advancing sustainability goals. It emphasizes the need for setting environmental standards, promoting transparent and accountable governance, involving marginalized groups in decision-making, and fostering economic growth that is environmentally sustainable and socially inclusive.

Correspondence with Adaptive Governance and the QBL Framework

The research findings align closely with the principles of adaptive governance and the Quadruple Bottom Line (QBL) framework. Adaptive governance emphasizes the integration of science, policy, and decision-making processes to foster resilience and sustainability. The study underscores the importance of adaptive governance in addressing sustainability challenges and highlights the need for monitoring and evaluation mechanisms to ensure the effective implementation of sustainable development policies. The QBL framework, encompassing economic, social, environmental, and governance aspects, provides a lens through which to analyze and evaluate sustainability governance. The research findings demonstrate how each dimension of the QBL framework is relevant to Bosnia and Herzegovina's context. They reveal the significance of investing in sustainable infrastructure, promoting social inclusion, protecting biodiversity and ecosystems, and ensuring transparent and accountable governance.

By incorporating the principles of adaptive governance and the QBL framework, sustainability efforts in Bosnia and Herzegovina can be enhanced. The research emphasizes the importance of increased research, innovation, technology investment, and public participation to address the specific needs of marginalized groups and promote sustainability in the country. Furthermore, the findings have broader implications for other countries facing political uncertainties, as they provide valuable insights into governance mechanisms and strategies that can foster sustainable development in diverse contexts.

Typology for Sustainable Development in Unstable Countries Under Uncertainty and Risks

The author proposes a novel typology to advance sustainable development in unstable countries facing political crises based on four distinct elements: environmental, governance, social, and economic factors.

Environmental Factors: Environmental Regulation and Policies: Setting environmental standards and regulations to control pollution and protect natural resources.

Resilient Resource Management: Sustainable use and management of natural resources, water, land, and forests.

Environmental Protection and Restoration: Rehabilitating damaged or degraded ecosystems and restoring biodiversity.

Governance Factors

1. Adaptive Governance: Integrating science, policy, and decision-making to facilitate adaptive management and ensure sustainability.

2. Transparency and Accountability: Promoting transparency, participation, and accountability in decision-making to enhance public trust and legitimacy.
3. Institutional Capacity Building: Enhancing institutional and individual capacities to enable effective governance and management of resources.

Social Factors

1. Social Inclusion: Promoting social inclusion and equity in decision-making processes and ensuring that vulnerable and marginalized groups are not left behind.
2. Public Engagement: Involving local communities in decision-making and resource management processes to ensure their needs and perspectives are considered.
3. Cultural Preservation: Protecting and preserving cultural heritage and traditional knowledge.

Economic Factors

1. Sustainable Economic Growth: Promoting economic growth that is environmentally sustainable, socially inclusive, and economically beneficial.
2. Resource Efficiency: Promoting efficient use of resources, reducing waste, and increasing resource productivity.
3. Green Jobs and Resilience Innovation: Promoting green jobs and innovation in renewable energy, eco-tourism, and sustainable agriculture sectors (Table 3).

The study represents a valuable typology, identifying and assessing sustainability patterns in politically unstable countries facing crises. This framework supports policymakers and stakeholders in formulating strategies and mechanisms to enhance the resilience of quadruple bottom-line systems amidst uncertainty and risk. By addressing environmental, governance, social, and economic factors, this novel typology aids in implementation of sustainable development goals in Bosnia and Herzegovina and other similarly unstable nations. Moreover, the research provides

Table 3 Typology for sustainable development under uncertainty and risks in unstable countries

Environmental factors	Governance factors	Social factors	Economic factors
Environmental regulation and policies	Adaptive governance	Social inclusion	Sustainable economic growth
Environmental regulation and policies	Adaptive governance	Social inclusion	Sustainable economic growth
Resilient resource management	Transparency and accountability	Public engagement	Resource efficiency
Environmental Protection and restoration	Institutional Capacity building	Cultural preservation	Green jobs and resilience innovation

significant insights into the facilitators of SDG implementation in a context marked by political risks and uncertainties. It underscores the importance of stakeholder engagement and institutional arrangements in achieving sustainable development goals.

The research establishes an analytical framework that examines sustainability activities' adaptive governance and resilience under specific circumstances. Recognizing the varying financial, social, and economic circumstances of different countries, the research focuses on raising awareness and building institutional capacity, particularly among disadvantaged groups. This approach enhances their preparedness, adaptive capacity, and involvement in policymaking processes to ensure the development of policies that address their specific needs. Furthermore, the study highlights the necessity of adaptive governance approaches to policymaking and implementation, particularly in climate change and natural hazards. It emphasizes the importance of climate and disaster risk reduction policies and identifies the need to bridge the policy gap in supporting disadvantaged populations. The paper asserts that sustainable development in Bosnia and Herzegovina requires a comprehensive and socially equitable approach. By integrating adaptive governance with the resilience factors of the Quadruple Bottom Line (QBL), sustainability-led initiatives in the country can be advanced and enhanced.

The report significantly contributes to understanding sustainability governance in politically uncertain contexts, specifically in Bosnia and Herzegovina. It highlights the relevance of adaptive governance and the QBL framework in addressing sustainability challenges and advancing sustainable development goals. The recommended typology offers a comprehensive guide for policymakers and stakeholders in unstable environments. However, it is essential to acknowledge the study's limitations, such as the narrow focus on available documents from Bosnia and Herzegovina without analyzing the entities and Brčko District level documents. Consequently, the generalizability of the findings to contexts with similar political arrangements is somewhat restricted. Nonetheless, this study contributes to advancing knowledge on sustainable development in less developed countries facing political risks while promoting adaptive governance in in literature and research.

References

1. Alibašić H (2018) Sustainability and resilience planning for local governments: the quadruple bottom line strategy. Springer. https://www.springer.com/us/book/9783319725673
2. Freedom House (2022) Nations in Transit. https://freedomhouse.org/country/bosnia-and-herzegovina/nations-transit/2022
3. Freedom House (2021) Bosnia and Herzegovina. Nations in Transit 2021. https://freedomhouse.org/country/bosnia-and-herzegovina/nations-transit/2020
4. Sustainable Development Report (2022) From crisis to sustainable development: the SDGs as roadmap to 2030 and beyond. https://dashboards.sdgindex.org/static/profiles/pdfs/SDR-2022-bosnia-and-herzegovina.pdf

5. Alibašić H (2017) Measuring sustainability impact in local governments using the quadruple bottom line approach. Int J Sustain Policy Pract 13(3):37–45. https://doi.org/10.18848/2325-1166/CGP/v13i03/37-45
6. Alibašić H (2018) The nexus of sustainability and climate resilience planning: embedding climate resilience policies in local governments. Int J Clim Change: Imp Responses 10(2):21–33. https://doi.org/10.18848/1835-7156/CGP/v10i02/21-33
7. Alibašić H (2018) Paris climate accord sans the USA: the role of local governments in confronting the effects of climate change and increasing resilience. Glob Stud J 11(2):33–39. https://doi.org/10.18848/1835-4432/CGP/v11i02/33-39
8. Alibašić H (2020) The administrative and ethical considerations of climate resilience: the politics and consequences of climate change. Public Integrity. https://doi.org/10.1080/10999922.2020.1838142
9. Alibašić H (2022) Strategic resilience and sustainability planning: management strategies for sustainable and climate-resilient communities and organizations. Springer. https://link.springer.com/book/9783030911584
10. World Bank (2021) Bosnia and Herzegovina climate risk country profile. https://climateknowledgeportal.worldbank.org/sites/default/files/2021-07/15914-WB_Bosnia%20Country%20Profile-WEB%20%281%29.pdf
11. Strambo C, Jahović B, Segnestam L (2021) Climate change and natural hazards in Bosnia and Herzegovina: a gender equality, social equity and poverty reduction lens. Stockholm Environment Institute
12. Solomun MK, Ferreira C, Barger N, Tošić R, Stojanović D (2020) Policy framework on land degradation from a post conflict perspective of Bosnia and Herzegovina. Authorea
13. Ministry of Foreign Trade and Economic Relations of Bosnia and Herzegovina (2019) National biodiversity strategy and action plan of Bosnia and Herzegovina 2019–2023. https://www.cbd.int/doc/world/ba/ba-nbsap-v2-en.pdf
14. Kahriman A, Tandir N (2021) Implementation of circular economy business models by SMEs in Bosnia and Herzegovina. Stud Human Soc Sci 2(2):90–98
15. Đurašković J, Konatar M, Radović M (2021) Renewable energy in the Western Balkans: policies, developments and perspectives. Energy Rep 7:481–490. https://doi.org/10.1016/j.egyr.2021.07.104
16. Androsevic D, Camargo M, Takorabet N, Carvalho A (2022) Overview of challenges and requirements for sustainable energy transition in Western Balkans with focus on Bosnia and Herzegovina: a literature review. In: 2022 IEEE 28th international symposium on industrial electronics (ISIE), pp 464–469
17. Katayanagi M (2022) Resilience against everyday disillusion: voices for peacebuilding in Bosnia and Herzegovina. In: Micro-evidence for peacebuilding theories and practices. Springer, pp 67–89
18. Edgar AD (2021) The rule of law, peacebuilding, and agenda 2030: lessons from the Western Balkans. In: Crime prevention and justice in 2030: the un and the global alliance for a culture of peace. Springer, pp 57–72
19. Hadžić F (2020) Economic crime, abuse of ethnopolitical power, and systemic corruption in Bosnia and Herzegovina; Analysis of causes and risks of phenomena. Int J Arts Soc Sci 3(3):140–148
20. Alibašić H (2023) Hyper-engaged citizenry, negative governance and resilience: impediments to sustainable energy projects in the United States. Energy Res Soc Sci 100:103072. https://doi.org/10.1016/j.erss.2023.103072
21. United Nations (2021) Trust in public institutions: trends and implications for economic security. https://www.un.org/development/desa/dspd/2021/07/trust-public-institutions/
22. DeCaro DA, Chaffin BC, Schlager E, Garmestani AS, Ruhl JB (2017) Legal and institutional foundations of adaptive environmental governance. Ecol Soc 22(1):32. https://doi.org/10.5751/ES-09036-220132
23. Walker B, Holling CS, Carpenter SR, Kinzig A (2004) Resilience, adaptability and transformability in social–ecological systems. Ecol Soc 9(2):5. [online] http://www.ecologyandsociety.org/vol9/iss2/art5

24. Alibašić H (2020) Post-genocide recovery and governance. In: Jurkiewicz CE (ed) Global corruption and management: translating theory into action, 3rd edn. Rowman & Littlefield, Lanham, Maryland
25. Hillman J, Geunther E (2021) Organizational resilience: a valuable construct for management research? Int J Manag Rev 23:7–44. https://doi.org/10.1111/ijmr.12239
26. Acar W, Winfrey FL (1994) The resilient organization: sustaining organizational renewal and performance. J Strat Change 3:165–173
27. Clément V, Rivera J (2016) From adaptation to transformation: an extended research agenda for organizational resilience to adversity in the natural environment. Organ Environ. https://doi.org/10.1177/1086026616658333
28. Berkes F, Ross H (2013) Community resilience: toward an integrated approach. Soc Nat Resour 26(1):5–20. https://doi.org/10.1080/08941920.2012.736605
29. Klein RJ, Nicholls RJ, Thomalla F (2003) Resilience to natural hazards: how useful is this concept? Glob Environ Change Part B: Environ Hazards 5(1–2): 35–45. https://doi.org/10.1016/j.hazards.2004.02.001
30. Bardal KG, Reinar MB, Lundberg AK, Bjørkan M (2021) Factors facilitating the implementation of the sustainable development goals in regional and local planning—experiences from Norway. Sustainability 13:4282. https://doi.org/10.3390/su13084282
31. Brunner RD, Steelman TA, Coe-Juell L, Cromley CM, Edwards CM, Tucker DW (2005) Adaptive governance: integrating science, policy, and decision making. Columbia University Press
32. Chaffin BC, Gosnell H, Cosens BA (2014) A decade of adaptive governance scholarship: synthesis and future directions. Ecol Soc 19(3). https://doi.org/10.5751/es-06824-190305
33. Picavet MEB, de Macedo LSV, Bellezoni RA, Puppim de Oliveira JA (2023) How can transnational municipal networks foster local collaborative governance regimes for environmental management? Environ Manage 71(3):505–522. https://doi.org/10.1007/s00267-022-01685-w. Epub 2022 Jul 29. PMID: 35906346; PMCID: PMC9942636
34. Folke C, Hahn T, Olsson P, Norberg J (2005) Adaptive governance of social-ecological systems. Ann Rev Environ Resour 15:441–473. https://doi.org/10.1146/annurev.energy.30.050504.144511
35. Gerring J (2017) Case study research: principles and practices, 2nd edn. Cambridge University Press
36. Creswell JW (2007) Qualitative inquiry and research design: choosing among five approaches, 2nd edn. Sage, Thousand Oaks, CA
37. Creswell JW (2014) Research design: qualitative, quantitative and mixed methods approaches, 4th edn. AGE Publications, Inc.
38. Patton MQ (2002) Qualitative research and evaluation methods, 3rd edn. Sage, Thousand Oaks, CA
39. Saldana J (2021) The coding manual for qualitative researchers, 4th ed. Sage Publications, London, UK
40. Yin RK (2008) Case study research: design and methods, 4th ed. SAGE, Thousand Oaks, CA
41. Altheide DL, Coyle M, DeVriese K, Schneider CJ (2008) Emergent qualitative document analysis. In: Hesse-Biber SN, Leavy P (eds) The handbook of emergent methods. Guilford Press, pp 491–505
42. Neufeld M (1998) The restructuring of social and political theory. University of Pennsylvania Press
43. UNDP (2021) The SDGs framework in Bosnia and Herzegovina. https://www.undp.org/sites/g/files/zskgke326/files/migration/ba/SDG-Framework-for-BiH---English.pdf
44. UN Environment Program (UNEP) (2018) Bosnia and Herzegovina country data. Retrieved from https://ozone.unep.org/countries/profile/bih
45. UNDP (United Nations Development Programme) (2019) Human development report 2019: beyond income, beyond averages, beyond today: inequalities in human development in the 21st century. New York

46. Agency for Statistics of Bosnia and Herzegovina (2023) SDG portal for Bosnia and Herzegovina. Retrieved from https://bhas.gov.ba/News/Read/45?lang=en
47. UNSDC Bosnia and Herzegovina (2021) Bosnia and Herzegovina and the United Nations sustainable development cooperation framework a partnership for sustainable development 2021–2025. https://unece.org/sites/default/files/2021-05/2021-2025_UNSDC_BiH.pdf
48. The Audit Office of the Institutions of Bosnia and Herzegovina (BiH) (2019) Preparedness of the institutions of Bosnia and Herzegovina to respond to Commitments Made under the 20230 Agenda. https://www.intosai.org/fileadmin/downloads/focus_areas/SDGs_and_SAIs/SDGs_Bosnia_and_Herzegovina_EN.pdf
49. Agency for Statistics of Bosnia and Herzegovina (2023) SDG portal for Bosnia and Herzegovina. https://bhas.gov.ba/News/Read/45?lang=en
50. Joint SDG Fund (2021) Towards an SDG financing ecosystem in Bosnia and Herzegovina. https://jointsdgfund.org
51. United Nations SDG Maps Mission (2018) Agenda 2030 for Sustainable Development in Bosnia and Herzegovina. Report on Multi-Stakeholder Consultations and Technical Analysis for Advancing Sustainable Development Goals. A joint UN Report in Support of Sustainable Development Goals. Retrieved from https://wesr.unep.org/media/docs/statistics/egm/bih_maps_report_finalv1.pdf
52. United Nations. (2019) Bosnia and Herzegovina Voluntary National Review 2019. https://sdgs.un.org/sites/default/files/documents/22919BosniaHerzegovina_VNR_Key_messages_English_version.pdf
53. United Nations (2020) Voluntary national review of Bosnia and Herzegovina. https://sdgs.un.org/sites/default/files/documents/23345VNR_BiH_ENG_Final.pdf
54. Sustainable Development Report (2019) Bosnia and herzegovina. Retrieved from https://dashboards.sdgindex.org/profiles/bosnia-and-herzegovina
55. Sustainable Development Report (2020) Bosnia and Herzegovina—Indicators. Retrieved from https://www.undp.org/bosnia-herzegovina/publications/sdgs-framework-bosnia-and-herzegovina?gad_source=1&gclid=Cj0KCQiAuqKqBhDxARIsAFZELmKAe0gqqs8KsrpdIka3oRsWc0bECtCEKeIzzkzBTYoxC7Dfe7tMp0aAmQKEALw_wcB
56. Sustainable Development Report (2021) Bosnia and Herzegovina. https://dashboards.sdgindex.org/profiles/bosnia-and-herzegovina
57. Sustainable Development Report (2021) Bosnia and Herzegovina. https://dashboards.sdgindex.org/static/profiles/pdfs/SDR-2022-bosnia-and-herzegovina.pdf
58. UNSDCF (2023) A partnership for sustainable development: UNSDCF 2021–2025. https://bosniaherzegovina.un.org/en/193131-partnership-sustainable-development-unsdcf-2021-2025
59. Linnenluecke MK, Griffiths A (2012) Assessing organizational resilience to climate and weather extremes: complexities and methodological pathways. Clim Change 113:933–947
60. Munton AG, Silvester J, Stratton P, Hanks H (1999) Attributions in action: a practical approach to coding qualitative data. Wiley, Chichester
61. Saldana J (2013) The coding manual for qualitative researchers, 2nd edn. Sage, London, UK
62. Sachs et al (2022) From crisis to sustainable development: the SDGs as roadmap to 2030 and beyond. Sustainable Development Report 2022. Cambridge University Press, Cambridge
63. Sustainable Development Report (2021). Bosnia and Herzegovina—Indicators. https://dashboards.sdgindex.org/profiles/bosnia-and-herzegovina/indicators

Bibliometric Analysis of Sustainable Cities and Communities (SDG 11) Literature in South East Europe

Nikša Alfirević, Darko Rendulić, and Ivica Zdrilić

Abstract This paper analyzes the academic literature on sustainable cities and communities (related to Sustainable Development Goal 11) produced in the South East European (SEE) region in the 2017–2022 period. From the theoretical viewpoint, the SDG-11-related body of literature covers sustainable urban housing, transport, as well as urban planning and management. We identify the most productive and influential regional authors, institutions, and countries, by using the bibliometric approach, based on the Elsevier Scopus data and the Elsevier SciVal research intelligence software. In addition, we map, identify, and visualize clusters of regional literature on sustainable cities and communities.

Keywords Sustainable cities · Sustainable communities · South East Europe · Bibliometric analysis

1 Introduction

On the 70th session of the United Nations General Assembly on September 15th, 2015, the *2030 Agenda for Sustainable Development and the Sustainable Development Goals (SDG) metrics* were adopted. The SDG metrics, consisting of 17 individual goals, replaced eight *Millennium Development Goals (MDGs)* used to evaluate the sustainable development agenda throughout 2000–2015. SDGs are described as

N. Alfirević (✉)
Faculty of Economics, Business and Tourism, University of Split, Cvite Fiskovića 5, 21000 Split, Croatia
e-mail: nalf@efst.hr

D. Rendulić
Karlovac University of Applied Sciences, Trg Josipa Jurja Strossmayera 9, 47000 Karlovac, Croatia

I. Zdrilić
Department of Economics, University of Zadar, Splitska 1, 23000 Zadar, Croatia
e-mail: izdrilic@unizd.hr

M. Arslanagić-Kalajdžić et al. (eds.), *Interdisciplinary Advances in Sustainable Development II*, Lecture Notes in Networks and Systems 804,
https://doi.org/10.1007/978-3-031-46692-2_20

more ambitious and comprehensive. They also refocus the MDG attention from environmental sustainability to multiple dimensions of the sustainability construct, including economic growth and development, natural resource utilization, urban development, energy, production, and consumption, etc. [1].

Cities have a particular position in the world's sustainable development efforts since they produce 85% of the global GDP, serve as a home for approximately 55% of the world's population, and produce more than 75% of greenhouse gases [2: 174]. Their role in sustainable development is reflected by including the SDG metrics as a separate goal (SDG 11—Sustainable Cities and Communities, i.e., "Make cities and human settlements inclusive, safe, resilient, and sustainable"). This is warranted by the multiple and interconnected relationships between urban communities with different sustainability dimensions, including economic growth, transport, energy, housing, water, education, etc. The systemic nature of SDG interactions has been empirically demonstrated by Pradhan et al. [3], who identify multiple synergies and negative interactions (trade-offs) using the SDG indicator values for 227 countries. Among the synergies, the top one is identified between SDG 11 and SDG 13 (Climate change, i.e., "Take urgent action to combat climate change and its impacts").

Implementation of SDG 11-related activities and evaluation of their outcomes are tied to the national contexts, even up to significant adjustments to the originally specified challenges and metrics of sustainable urban development [4]. Therefore, there is imperative to study the regional SDG 11 literature to identify the specific characteristics of sustainable urban development.

In addition, the literature search has not identified any previous bibliometric analysis of the entire SDG 11 knowledge base. However, systematic reviews and bibliometric studies have been performed for specialized fields of sustainable urban development. Those include e.g., smart mobility [5], urban environmental governance [6], smart city development [7, 8], urban sustainability assessment [9], etc. As far as the authors are informed, there are no bibliometric or science mapping studies in the sustainable urban development field analyzing the SEE regional body of knowledge, except for the previously mentioned overview of SEE smart city literature.

These arguments support the need to develop a bibliometric and science mapping study, addressing the following research questions (RQs):

(a) *RQ1: Which SEE countries, research institutions, and researchers demonstrate the highest research productivity and impact in sustainable urban development?*
(b) *RQ2: What is the intellectual structure of sustainable urban development research in the SEE region?*

The structure of this paper is as follows: research methods and procedure are described in Sect. 2, results of the bibliometric analysis are presented in Sect. 3, while Sects. 4 and 5 (respectively) contain the discussion and implications for research, public policy, and higher education, and conclusions.

2 Methods

This paper follows the research methodology previously applied to the scientific mapping of the regional literature on sustainable production and consumption [10] and recommended by Župič and Čater [11]. The methodological steps used to perform this study are as follows:

1. Research design: determining the study aims and scope, including research questions.
2. Compilation of bibliometric data: selecting a referencing database and the supporting bibliometric analysis software solution, developing search criteria and filtering strategy.
3. Bibliometric analysis: data collection and export, data cleaning and import into the software tool of choice.
4. Interpretation of results (including visualization).

Regional literature on sustainable urban development (i.e., SDG 11-related research), mapped by the Elsevier Scopus database, has been identified using a bibliometric query developed by the Elsevier SDG 2020 mapping initiative [12]. While an identical mapping exercise has been performed annually, more recent (2021 and 2022) mapping initiatives are based on bibliometric queries, with results being refined by an additional machine learning algorithm [13]. Although the mapping results are publicly available, they are not easily reproducible. This might make the older mapping procedure more convenient, especially if another source of bibliometric data (such as Clarivate Web of Science) is also to be consulted.

We accessed the advanced Elsevier Scopus search interface and used the following bibliometric query:

```
TITLE-ABS-KEY ((city OR cities OR {human settlement} OR {human settlements} OR urban OR metropoli* OR town* OR municipal*) AND (gentrification OR congestion OR transportation OR {public transport} OR housing OR slum* OR {sendai framework} OR {Disaster Risk Reduction} OR {DRR} OR {smart city} OR {smart cities} OR {resilient building} OR {resilient buildings} OR {sustainable building} OR {sustainable buildings} OR {building design} OR {buildings design} OR urbani?ation OR {zero energy building} OR {zero energy buildings} OR {zero-energy building} OR {zero-energy buildings} OR {basic service} OR {basic services} OR {governance} OR {citizen participation} OR {collaborative planning} OR {participatory planning} OR {inclusiveness} OR {cultural heritage} OR {natural heritage} OR {UNESCO} OR {disaster} OR {ecological footprint} OR {environmental footprint} OR {waste} OR {pollution} OR {pollutant*} OR {waste water} OR {recycling} OR {circular economy} OR {air quality} OR {green space} OR {green spaces} OR {nature inclusive} OR {nature inclusive building} OR {nature inclusive buildings})) AND (LIMIT-TO (PUBYEAR, 2022) OR LIMIT-TO (PUBYEAR, 2021) OR LIMIT-TO (PUBYEAR, 2020) OR LIMIT-TO (PUBYEAR, 2019) OR LIMIT-TO (PUBYEAR, 2018) OR LIMIT-TO (PUBYEAR, 2017)) AND (LIMIT-TO (AFFILCOUNTRY, "Serbia") OR LIMIT-TO (AFFILCOUNTRY, "Croatia") OR LIMIT-TO (AFFILCOUNTRY,
```

```
"Slovenia") OR LIMIT-TO (AFFILCOUNTRY, "Bosnia and Herzegovina")
OR LIMIT-TO (AFFILCOUNTRY, "North Macedonia") OR LIMIT-TO (AFFIL-
COUNTRY, "Montenegro")) AND (LIMIT-TO (SRCTYPE, "j"))
```

We opt to include only journals into the analysis to obtain a high quality, but still comprehensive view of the regional knowledge base and benefit from a broader indexing coverage of Elsevier Scopus compared to the competitive products [14].

As previously stated, the keywords (i.e., the bibliometric query entered into the Scopus advanced search form) have been previously developed by the Elsevier BV initiative to map the Scopus-indexed research outputs with the 17 SDGs. Although not involving advanced mapping techniques introduced subsequently, such as machine learning, this approach is reliable from the bibliometric viewpoint [12].

Research outputs were filtered for the 2017–2022 period, with at least one (co)author affiliated with an institution in the SEE region (i.e., in Bosnia and Herzegovina, Croatia, Montenegro, North Macedonia, Serbia, and Slovenia).

The described advanced query resulted in 1.639 publications, available as an online appendix to this chapter [15]. All publications have been selected as relevant for SDG 11, based on the Elsevier Scopus SDG mapping initiative 2020 methodology [12]. The publication list, with the full bibliometric details, was transferred to Elsevier SciVal, which was used for bibliometric analysis. The VOS Viewer tool, produced by the Leiden University Centre for Science and Technology Studies, was used for visualization [16].

3 Results

Bibliometric Analysis of the Regional SDG 11 Research Productivity and Impact

Presentation of bibliometric analysis, related to the regional SDG 11 body of knowledge, follows the structure of the global Corporate Social Responsibility (CSR) literature review provided by Cucari et al. [17], which also uses bibliometric analysis and the Elsevier SciVal reporting tool.

Firstly, we observed the number of publications (Table 1) and the national production of the SDG 11 publications (Table 2) across the SEE region. Researchers' interest in urban sustainability topics is high. At the same time, a slight drop in 2022 scientific output can be attributed to the fact that, at the moment of writing this paper, only a part of the annual output has been processed and included in Elsevier Scopus.

Most of the regional scientific output on the analyzed topic has been produced by Serbian (646), Croatian (545), and Slovenian (407) researchers, with Italy, Spain, the UK, Germany, Portugal, Austria, the US, and Greece being the top collaborating countries.

Table 1 Regional SDG 11 publications by year

Year	Number of publications
2017	228
2018	248
2019	234
2020	294
2021	337
2022	297
2023	1

Source Elsevier SciVal (February 2023)

Table 2 National productivity and impact of SDG 11 publications (2017–2022)

Country	Number of publications	Citation count	Field-weighted citation impact
Serbia	645	6260	1.14
Croatia	545	6072	1.32
Slovenia	407	6201	1.47
Italy	120	3948	3.37
Bosnia and Herzegovina	105	1242	1.09
Spain	95	3919	3.4
United Kingdom	90	2825	2.87
North Macedonia	88	1095	1.43
Germany	77	3001	3.17
Portugal	63	2265	2.71
Austria	58	2107	3.24
United States	56	1817	2.85
Greece	54	2218	3.16
Switzerland	47	1587	2.71
Poland	45	1454	2.99
France	44	1513	3.53
Sweden	44	1211	2.38
Netherlands	43	1704	3.39
Turkey	42	1004	5.19

Source Elsevier SciVal (February 2023)

Table 2 also presents selected scientific impact indicators, including the total citation count, dependent on the size of the observed entity and the normalized Field-Weighted Citation Impact (FWCI). FWCI reflects that different scientific fields have varying citation averages and shows a relative impact by observing the number of citations received with the average expected by the scientific field and the type of study [18]. Its value is relative and can be easily interpreted (<1.0 for entities with an impact lower than the average global; 1.0 for the average global impact; >1.0 for entities with an impact higher than the average global).

While FWCI values analysis for collaborating countries may not be relevant, when analyzed for a subset of papers defined by an international collaboration with the SEE countries, such an analysis is more meaningful for specialist research in the region. FWCI values across the national SDG 11 research domains are similar, with values somewhat higher than 1.0 and the highest value for Slovenia (1.47).

Top institutions in the research of SDG 11 are the public universities located at the major regional centers, including Belgrade, Zagreb, and Ljubljana, which could be expected due to the resource availability and attraction to the researchers. Smaller and peripheral institutions with a high SDG 11 research productivity include public universities at Novi Sad (Serbia), Maribor (Slovenia), Rijeka (Croatia), Niš (Serbia), and other entities listed in Table 3. The highest impact, measured by the FWCI metric, has been achieved by the Jožef Stefan Institute (Slovenia) and the University of Split (Croatia), followed by the University of Sarajevo (Bosnia & Herzegovina) and the University of Ljubljana (Slovenia).

The top 100 regional researchers in the SDG 11-related fields, both affiliated with the regional research institutions and their collaborators, are listed in an online appendix to this chapter [19].

The entire analyzed corpus of literature received 18,170 citations, with 11.1 citations per publication. The average FWCI metric value equaled 1.23, which implies a somewhat higher impact when compared to the global field average. In the entire period, 848 papers (53%) of the corpus were published in Q1 outlets (when considering the Scopus CiteScore metric). Additional 331 papers (20.7%) were published in Q2 publication outlets, while Q3 and Q4 publications equaled 292 (18.2%) and 130 (8.1%), respectively.

The most popular publishing outlets are open-access titles, especially those published by non-traditional publishers, which will be further discussed in conclusion. Only two, out of the top five Scopus sources, chosen by the regional researchers for their SDG 11 publications belong to the traditional journals—Science of the Total Environment, published by Elsevier, and Environmental Science and Pollution Research, published by Springer (see Table 4).

Due to the rapid publication process, open-access titles attract more researchers from the SEE region. However, out of the five most influential papers from the analyzed corpus (see Table 5), only one has been published in open access in a journal from the Springer Nature portfolio (formerly Palgrave Communications, now Humanities & Social Science Communications).

In the case of SDG 11 research, international collaboration represents a significant driver of the research impact, with 17 citations per publication and the FWCI metric value of 0.74, as compared to significantly lower impact metrics for the national and institutional collaboration and the single authorship outputs (see Table 6).

Table 3 Top regional institutions in SDG 11 research productivity and impact (2017–2022)

Institution	Number of publications	Citation count	Field-weighted citation impact
University of Belgrade (RS)	306	2836	1.16
University of Zagreb (HR)	259	2885	1.29
University of Novi Sad (RS)	161	2387	1.19
University of Ljubljana (SI)	149	2345	1.55
J. Stefan Institute (SI)	71	1967	1.91
University of Maribor (SI)	59	460	1.37
University of Rijeka (HR)	54	418	1.18
University of Niš (RS)	51	355	0.98
Ruđer Boskovic Institute (HR)	49	1010	1.41
SS Cyril and Methodius University in Skopje (MKD)	48	472	1.39
University of Split (HR)	47	782	1.88
University of Montenegro (MNE)	45	641	1.28
University of Sarajevo (BIH)	33	605	1.57
Josip Juraj Strossmayer University of Osijek (HR)	31	250	1.29
University of Prishtina “Hasan Prishtina”*	30	91	0.29
University of Priština (North Mitrovica)*	28	153	0.77
Institute for medical research and occupational health (HR)	26	302	1.03
University of Kragujevac (RS)	25	234	1.32

Source Elsevier SciVal (February 2023)
Note * Elsevier SciVal does not recognize the Kosovar institutions due to the lack of international consensus in the International Organization for Standardization (ISO) and counts their output as belonging to Serbian affiliations

Table 4 Top publication outlets for regional SDG 11 research (2017–2022)

Scopus source (Publication outlet)	Scholarly output	Field-weighted citation impact	Citation count
Sustainability	109	1.03	777
Science of the total environment	56	1.88	1945
Environmental science and pollution research	34	0.57	326
International journal of environmental research and public health	23	1.52	230
Water (Switzerland)	23	1.9	279
Applied sciences (Switzerland)	22	0.61	76
Promet—Traffic—Traffico	22	0.36	83
Tehnicki Vjesnik	22	0.35	73
Journal of cleaner production	21	3.53	891
Thermal science	17	0.12	63
Energies	15	0.68	84
Sensors	14	1.92	169
Energy	13	3.19	789
Environmental pollution	13	2.52	435
Urban forestry and urban greening	13	5.55	608
Radovi Instituta za Povijest Umjetnosti	12	0.52	4
Annales-Anali za Istrske in Mediteranske Studije—Series Historia et Sociologia	11	0.66	11
Cities	11	1.18	95
Environmental monitoring and assessment	11	0.44	63
Journal of environmental management	11	1	147
Polish journal of environmental studies	10	0.19	26
Sociologija i Prostor	10	0.12	10
Spatium	10	0.99	23
Fresenius environmental bulletin	9	0.11	30
Journal of environmental protection and ecology	9	0.17	20

Source Elsevier SciVal (February 2023)

Intellectual Structure of the Regional SDG 11 Research

We performed the science mapping of the regional SDG 11-related research using the conventional keyword co-occurrence analysis, following the procedure presented in detail for the case of mapping the regional sustainable production and consumption literature [10].

We used the VOSViewer science mapping software, produced by the Leiden University CWTS research center, with default settings (minimum number of

Table 5 Most influential papers in regional SDG 11 research (2017–2022)

Publication	Citations	Field-weighted citation impact
Ranjbari, M., Shams Esfandabadi, Z., Gautam, S., et al. Waste management beyond the COVID-19 pandemic: Bibliometric and text mining analyses. Gondwana Research, 114, 124–137 (2023)	16	43.62
Ugolini, F., Massetti, L., Calaza-Martínez, P. et al. Effects of the COVID-19 pandemic on the use and perceptions of urban green space: An international exploratory study. Urban Forestry and Urban Greening, 56 (2020)	251	34.32
Kılkış, Ş., Krajačić, G., Duić, N. et al. Advances in integration of energy, water and environment systems towards climate neutrality for sustainable development. Energy Conversion and Management, 225 (2020)	34	30.18
Perc, M., Ozer, M., Hojnik, J. Social and juristic challenges of artificial intelligence. Palgrave Communications, 5 (61) (2019)	67	28.02
Formisano, A., Di Lorenzo, G., Krstevska, L. et al. Fem Model Calibration of Experimental Environmental Vibration Tests on Two Churches Hit by L'Aquila Earthquake. International Journal of Architectural Heritage, 15 (1), 113–131 (2021)	22	24.44

Source Elsevier SciVal (February 2023)

Table 6 International, national, and institutional collaboration in regional SDG 11 research (2017–2022)

Metric	% of the entire research output (%)	Number of publications	Citations per publication	Field-weighted citation impact
International collaboration	40.9	670	17	1.9
Only national collaboration	19.2	314	7.6	0.74
Only institutional collaboration	29.7	487	8	0.88
Single authorship	10.2	168	2.8	0.48

Source Elsevier SciVal (February 2023)

keyword occurrences set to five) and a minimum cluster size of 50 items. The obtained science map (see Fig. 1) consists of six clusters:

- The red cluster (487 items) focuses on the environmental and social aspects of urban sustainability across the SEE region, including the issues of municipal solid waste, energy, transportation, cultural heritage, and urban development.
- The green cluster (208 items) groups the studies on water pollution and wastewater treatment.

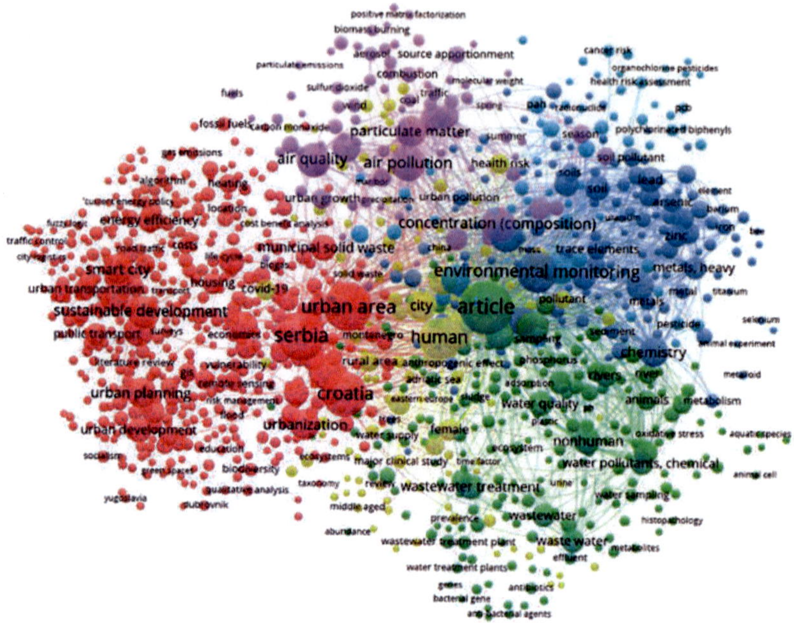

Fig. 1 Keyword co-occurrence map of the SEE literature on SDG 11 (2017–2022). *Source* Authors, based on VOSviewer visualization engine

- The blue cluster (142 items) visualizes research, focusing on soil pollution and pollutants (especially heavy metals) and environmental monitoring.
- The yellow cluster (115 items) groups research outputs focused on public health issues and the consequences of environmental pollution on the human population.
- The violet cluster (84 items) represents the research on air (atmospheric) pollution and air quality, including studies on the concentration of pollutants and their sources.
- The light blue cluster (61 items) considers a range of pollutants and their influence on soil quality and human health.

Only two sub-clusters demonstrate an interdisciplinary/transdisciplinary approach, which sustainable development research requires. Its purpose is recognized in building trust and collaboration among the multiple research fields [20] in solving 'wicked' environmental problems [21], as well as in promoting efficient pro-environmental education [22].

In the case of regional SDG 11 research, there is inter(trans)disciplinary approach in addressing the issues of comprehensive urban environmental sustainability (see Fig. 2) and environmental monitoring (see Fig. 3), considering all forms of urban pollution and limited public health consequences. Although there is limited evidence of inter(trans)disciplinarity in regional SDG 11 research, the interconnectedness of

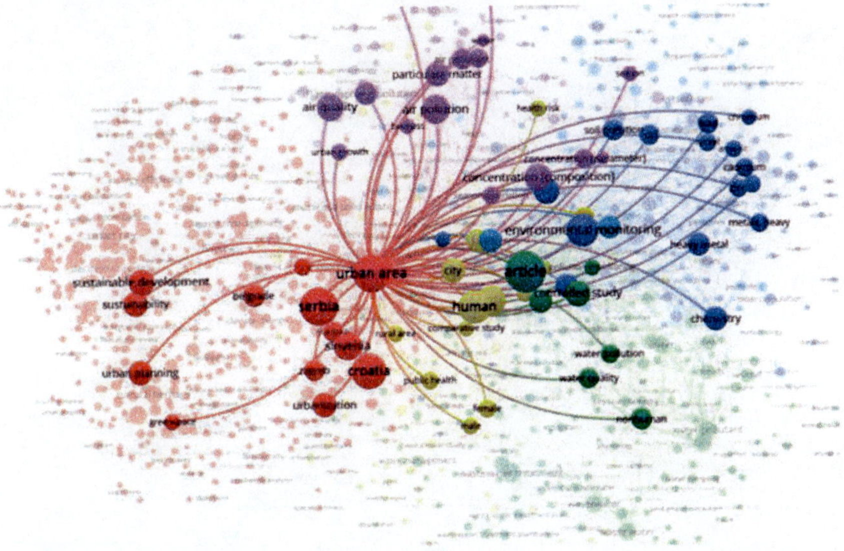

Fig. 2 Transdisciplinary links of regional research on urban environmental sustainability. *Source* Authors, based on VOSviewer visualization engine

research clusters is somewhat better than in the case of SEE sustainable production and consumption literature [10].

Limited transdisciplinarity is one of the significant issues in the analyzed regional SDG11 literature, which should be recognized and encouraged by the research and evaluation policies.

4 Discussion

In this paper, we considered regional SDG 11 research and two research questions (RQs) related to the scientific productivity and impact of the regional sustainable urban development body of knowledge and its intellectual structure.

Concerning RQ1, patterns of scientific research seem to mirror the recent findings of regional business school productivity and impact, as reported by Alfirević et al. [23]. There is a constant growth of research output, with most studies published in high-quality (Q1) publication outlets, with the interdisciplinary impact metric (FWCI) value somewhat above the global average. In addition, comparable to the regional business school study, results show that international collaboration is a significant driver of research productivity and impact. Such a finding could indicate that the local and regional scientific evaluation frameworks do not reflect the SDG priorities [23]. This could result from their orientation toward administrative and somewhat arbitrary criteria [24].

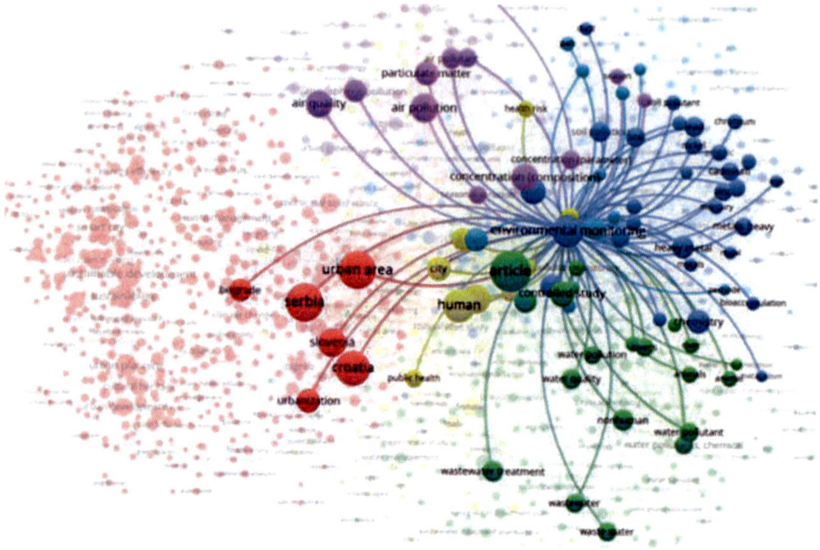

Fig. 3 Transdisciplinary links of regional research on different forms of urban pollution and environmental monitoring. *Source* Authors, based on VOSviewer visualization engine

Regarding RQ2, the observed regional literature corpus has a clear pattern of keyword grouping. Out of six research themes (clusters), we identified two interdisciplinary/transdisciplinary research topics concerning urban environmental sustainability and pollution monitoring. In terms of interdisciplinarity (transdisciplinarity), this result is somewhat better than for the case of regional SDG 12 research, characterized by the undeveloped relationships among the (isolated) research subtopics [10].

As noted previously, encouraging transdisciplinarity and ensuring the practical application of the SDG11 research remains the role of regional research and evaluation policies. In addition, future research should address the issue of the role and contribution of the regional SDG11 literature to the global body of knowledge. This study represents only a first step in such a direction since it analyzes the regional literature without an attempt to evaluate its global contribution. Although this is an obvious limitation of the current study, its objective is to obtain a realistic insight into the current state and the intellectual structure of the regional SDG11 research.

5 Conclusion

A considerable body of regional SEE research on sustainable urban development focuses on environmental and social aspects of urban sustainability, water/soil/atmospheric pollution and pollutants, and related public health issues/consequences.

SDG 11-related regional research is dominated by larger institutions in regional capitals, and driven by international collaboration, resulting in high research productivity and somewhat above-average impact. Analysis of the publication outlets shows that the open access titles, especially those from non-traditional publishers, attract most of the regional SDG 11-related research. This might result from the research productivity pressure exerted by the national research evaluation and academic promotion frameworks based on the quantitative bibliometric indicators. A recent study [25] acknowledges that the popularity and the rising market share of non-traditional, open-access publishers, at least in Central and Eastern European countries, could be attributed to the quick turnaround time. With the lengthy peer review processes [26] and publishing in traditional journals sometimes taking as long as two years from the initial submission [27], regional authors are feeling the pressure to quickly deliver research outputs indexed either in Clarivate Web of Science or Elsevier Scopus databases.

The described trends in the selection of publication outlets, often referred to as the 'SSCI syndrome' (referring to the Clarivate Web of Science Social Science Citation Index), are not unheard of, even on the global level [28]. Therefore, national policy-makers should reconsider the existing science evaluation and academic promotion frameworks, including scientific impact policies and SDG orientation(s).

References

1. Palmer E (2015) Introduction: the 2030 agenda. J Glob Ethics 11(3):262–269. https://doi.org/10.1080/17449626.2015.1119928
2. Vaidya H, Chatterji T (2020) SDG 11 sustainable cities and communities. In: Franco I, Chatterji T, Derbyshire E, Tracey J (eds) Actioning the global goals for local impact. Science for sustainable societies. Springer, Singapore. https://doi.org/10.1007/978-981-32-9927-6_12
3. Pradhan P, Costa L, Rybski D, Lucht W, Kropp JP (2017) A systematic study of sustainable development goal (SDG) interactions. Earth's Future 5(11):1169–1179. https://doi.org/10.1002/2017EF000632
4. Koch F, Krellenberg K (2018) How to contextualize SDG 11? Looking at indicators for sustainable urban development in Germany. ISPRS Int J Geo Inf 7(12):464. https://doi.org/10.3390/ijgi7120464
5. Tomaszewska EJ, Florea A (2018) Urban smart mobility in the scientific literature—bibliometric analysis. Eng Manag Prod Serv 10(2):41–56. https://doi.org/10.2478/emj-2018-0010
6. Wu F, Geng Y, Tian X, Zhong S, Wu W, Yu S, Xiao S (2018) Responding climate change: a bibliometric review on urban environmental governance. J Clean Prod 204:344–354. https://doi.org/10.1016/j.jclepro.2018.09.067
7. Zhao L, Tang ZY, Zou X (2019) Mapping the knowledge domain of smart-city research: a bibliometric and scientometric analysis. Sustainability 11(23):6648. https://doi.org/10.3390/su11236648
8. Ninčević Pašalić I, Ćukušić M, Jadrić M (2021) Smart city research advances in Southeast Europe. Int J Inf Manag 58:102127. https://doi.org/10.1016/j.ijinfomgt.2020.102127
9. Sharifi A (2021) Urban sustainability assessment: an overview and bibliometric analysis. Ecol Ind 121:107102. https://doi.org/10.1016/j.ecolind.2020.107102

10. Alfirević N, Zdrilić I, Rendulić D (2023) Mapping the sustainable production and consumption literature in South East Europe. In: Tufek-Memišević T, Arslanagić-Kalajžić M, Ademović N (eds) Interdisciplinary advances in sustainable development: international conference on sustainable development. Springer, Cham, pp 129–146. https://doi.org/10.1007/978-3-031-17767-5_10
11. Župič I, Čater T (2015) Bibliometric methods in management and organization. Organ Res Methods 18(3):429–472. https://doi.org/10.1177/1094428114562629
12. Jayabalasingham B, Boverhof R, Agnew K, Klein L (2019) Identifying research supporting the United Nations sustainable development goals (version 1). Mendeley Data. https://doi.org/10.17632/87txkw7khs.1
13. Roberge G, Kashnitsky Y, James C (2022) Elsevier 2022 Sustainable Development Goals (SDG) mapping. Mendeley Data V1. https://doi.org/10.17632/6bjy52jkm9.1
14. Norris M, Oppenheim C (2007) Comparing alternatives to the Web of Science for coverage of the social sciences' literature. J Informet 1(2):161–169. https://doi.org/10.1016/j.joi.2006.12.001
15. Alfirević N, Rendulić D, Zdrilić I, Bibliometric data on Sustainable Cities and Communities (SDG 11) literature in South East Europe. figshare. Online resource. https://doi.org/10.6084/m9.figshare.22649761
16. Van Eck N, Waltman L (2010) Software survey: VOSviewer, a computer program for bibliometric mapping. Scientometrics 84(2):523–538. https://doi.org/10.1007/s11192-009-0146-3
17. Cucari N, Tutore I, Montera R, Profita S (2022) A bibliometric performance analysis of publication productivity in the corporate social responsibility field: outcomes of SciVal analytics. Corp Soc Respons Environ Manag 30(1):1–16. https://doi.org/10.1002/csr.2346
18. Purkayastha A, Palmaro E, Falk-Krzesinski HJ, Baas J (2019) Comparison of two article level, field-independent citation metrics: Field-Weighted Citation Impact (FWCI) and Relative Citation Ratio (RCR). J Informet 13(2):635–642. https://doi.org/10.1016/j.joi.2019.03.012
19. Alfirević N, Rendulić D, Zdrilić I, Top regional authors in Sustainable Cities and Communities (SDG 11) literature in South East Europe. figshare. Online resource. https://doi.org/10.6084/m9.figshare.22649827
20. Harris F, Lyon F (2013) Transdisciplinary environmental research: building trust across professional cultures. Environ Sci Policy 31:109–119. https://doi.org/10.1016/j.envsci.2013.02.006
21. Pohl C, Truffer B, Hirsch Hadorn G (2017) Addressing wicked problems through transdisciplinary research. In: Frodeman R, Thompson Klein J, Dos Santos Pacheco RC (eds) The Oxford handbook of interdisciplinarity, 2nd edn. Oxford University Press, Oxford. pp 319–331. https://doi.org/10.1093/oxfordhb/9780198733522.013.26
22. Annan-Diab F, Molinari C (2017) Interdisciplinarity: practical approach to advancing education for sustainability and for the sustainable development goals. Int J Manag Educ 15(2):73–83. https://doi.org/10.1016/j.ijme.2017.03.006
23. Alfirević N, Pavičić J, Rendulić D (2023) A bibliometric analysis of public business school scientific productivity and impact in South-East Europe (2017–2021). South East Eur J Econ Bus 18(1):27–45. http://journal.efsa.unsa.ba/index.php/see/article/view/2084/478
24. Ciurak P, Mijač T, Wierczyński G (2021) An overview of science evaluation in Poland and Croatia. Manag: J Contemp Manag Issues 26(2):229–244. https://doi.org/10.30924/mjcmi.26.2.13
25. Csomós G, Farkas JZ (2023) Understanding the increasing market share of the academic publisher "Multidisciplinary Digital Publishing Institute" in the publication output of Central and Eastern European countries: a case study of Hungary. Scientometrics 128(1):803–824. https://doi.org/10.1007/s11192-022-04586-1
26. Nguyen VM et al (2015) How long is too long in contemporary peer review? Perspectives from authors publishing in conservation biology journals. PloS One 10(8):e0132557. https://doi.org/10.1371/journal.pone.0132557

27. Harzing AW, Adler NJ (2016) Disseminating knowledge: from potential to reality—new open-access journals collide with convention. Acad Manag Learn & Educ 15(1):140–156. https://doi.org/10.5465/amle.2013.0373
28. Chou CP, Chan CF (2017) Governance and academic culture in higher education: under the influence of the SSCI syndrome. J Int Comp Educ (JICE) 63–75. https://doi.org/10.14425/jice.2017.6.2.63

Ethical Implications of Neuromarketing: The Context of Sustainable Development Goals (SDGs)

Hamza Čengić, Maja Arslanagić-Kalajdžić, and Melika Husić-Mehmedović

Abstract In the last decade, the field of neuromarketing has undergone notable advancements owing to technological innovations, globalization, and related factors, which have resulted in its greater acceptance in both academic and business communities. Although the findings of neuromarketing have been demonstrated to have a considerable impact on marketing, the ethical implications of this field continue to be a perennial topic of debate. This study is based on desk research, which include available documents, ongoing neuromarketing projects, and previous research findings. Thus, the main goal of this paper is to provide a comprehensive overview of the ethics in neuromarketing research and provide an answer to an everlasting question of the ethical dilemma regarding the implementation of findings within the spectrum of neuromarketing methods and how the findings are being acquired and used, through the lenses of the Sustainable Development Goals, and taking into account the available information and insights from prior studies.

Keywords Neuromarketing · Ethics · Marketing ethics · Neuro-ethics

1 Introduction

Traditionally, many economic theories have posited that consumers are rational agents who make conscious and sensible decisions. However, the reality is quite different, as the majority of decisions are made at a subconscious level.

The standard methodologies employed in consumer behaviour research are frequently subjective and do not always provide the best ways to discover people's goals and wishes because most decisions we make are subconscious. One of the biggest issues today for sellers is closing the "gaps" between the subconscious of the buyer and their behaviour. These issues may be resolved with neuromarketing,

H. Čengić · M. Arslanagić-Kalajdžić · M. Husić-Mehmedović (✉)
School of Economics and Business Sarajevo, University of Sarajevo, 71000 Sarajevo, Bosnia and Herzegovina
e-mail: melika.husic@efsa.unsa.ba

M. Arslanagić-Kalajdžić et al. (eds.), *Interdisciplinary Advances in Sustainable Development II*, Lecture Notes in Networks and Systems 804,
https://doi.org/10.1007/978-3-031-46692-2_21

which replaces the subjectivism imposed by conventional research methods with objectivity [1].

Over the years, the methods employed in neuromarketing have generated much controversy. Some believe that the techniques used by scientists to access the consumer's brain pose a significant risk and are potentially harmful, while others view these methods as relatively benign. More significantly, questions have arisen as to whether companies are justified in delving into the "black box" of the mind to gain commercial advantage. With the rapid development of new technologies and the emergence of a powerful social media culture that amplifies even the smallest corporate misstep, companies are increasingly conscious of the importance of ethical practices. This is because a company's reputation is one of its most valuable assets.

Positive side of neuromarketing research is that its diligent use may be connected to several sustainability aspects, notably—specific Sustainable Development Goals (SDGs), depending upon the context and goals of the focal research. Against this background, main research questions of this study are: (1) What are the ethical implications of neuromarketing research? and (2) How can neuromarketing research be ethically applied in line with the SDGs? The intended contribution of this manuscript is twofold. First, the study will contribute to the neuromarketing research body of knowledge by providing a critical analysis of the ethical implications of neuromarketing research. In this way, we will help identify the potential ethical issues that could arise from neuromarketing research and suggest the mode of prevention of such issues. Second, the study provides a framework for ethically applying neuromarketing research in line with the SDGs. This framework consequently provides a practical guidance for businesses, policymakers, and researchers in conducting neuromarketing research while adhering to ethical standards.

2 Neuromarketing Approach and Applications

According, to Ramsøy, neuromarketing is considered to be a commercial use of neuroscience tools and insights to measure, understand, and affect consumer behaviour, whilst the consumer neuroscience, on the other hand, is regarded as the academic use of neuroscience tools and models in order to understand consumption behaviours and brain mechanisms underlying everyday choices [2].

Some of the greatest struggles nowadays, regarding the field of marketing in general, are concerned with the accuracy of the internal connection of what consumers do, i.e., buy, and what they actually want. The traditional approach to marketing research cannot always be regarded as an effective one. After a product is created, several choices are made about how to market, advertise, and position it. Academic studies have shown that neuromarketing possesses the ability to significantly improve these strategies, specially the promotional ones [3]. We are witnessing extensive use of neuroscience to gain new insights into consumers' behaviour, preferences, and decision making [4]. Neuromarketing seeks information and insights beyond that revealed by traditional techniques, with the goals of enhancing marketing theory and practice [5, 6] or improving the accuracy of predictions of consumer preferences and behaviour when combined with traditional techniques [7–9].

Because it's not always acceptable to ask a customer, "Do you like this product/design/packaging/commercial?" marketing analysts utilize neuromarketing to more accurately compare consumer preferences. With the aid of this knowledge, vendors may provide goods and services that are more useful, and marketing campaigns will have significantly more efficiency in transmitting a certain message. Some of the research areas within the field of neuromarketing, as previously mentioned, are mostly concerned with brand optimization, optimization of advertising messages' impact, whether it is through a commercial or another way of mass communication, as well as the product design and packaging especially [1].

The application of neuromarketing takes place through research conducted mainly in the laboratory setting, during which the research subjects are under a certain influence of a stimulus that affects their senses. The methods of research can be carried out by direct research of the brain or biometric research. The study of specific alterations in the human body, particularly its physiological responses, is done through biometrics. These modifications include variations in heart rate, blinking, breathing rate and many others. What is especially appropriate for gauging certain emotional reactions is exposing subjects to commercials, single ads, as well as billboard pictures and there are several techniques to record their associated responses [10]. The main goal of this type of research is to elicit the subject's reaction, and record it whilst exposed to the stimulus.

The scope of potential use of neuromarketing is extremely wide at the moment. It is expected, in this relatively early development stage of neuromarketing research, that marketers will gain deeper understanding of broad spectrum of consumer needs and be able to produce more desirable products, create better targeted promotional materials, and enhance consumers' experiences [3–5]. The standard evaluation techniques for commercial effectiveness (recognition tests, evaluating attitudes toward the product/brain, etc.) solely depend on the participants' associations, which limits them to the conscious component of mind (rational brain). The advertising message and commercial must cause the impulses to spread to the limbic system, which is in charge of emotional relationships, pleasure, and pain, as well as to the reptilian brain, which is in charge of the aim to win, of regulating quick reactions and of controlling withdrawal/waiting actions, in order to have an adequate impact on the consumer. The experiments demonstrate that there are clear distinctions between the brain's reactions to rational and emotional advertisements. The latter, in fact, provide the kind of response that every business investing in an advertising campaign is looking for, by aggressively activating the brain areas responsible for making decisions and forming social bonds [1].

3 Ethical Framework of Neuromarketing

The most common critiques of neuromarketing, which have existed since the field's foundation, focus on unethical research methods, unethical technological implementations, and consumer manipulation. Nevertheless, despite these concerns, academic research in neuromarketing and related fields has increased substantially, and more than 200 neuromarketing research and consultancy organizations have now been established globally [11].

The majority of ethical criticisms of neuromarketing focus on the potential for ethical violation and rights abuses. The relevant damages encompass both short-term consequences for specific customers and long-term consequences for society at large. The allegedly protected rights range from positive ones like the right to privacy and autonomy to negative ones like the right to not be duped or put to experiments without consent. Some of these rights are well-known and even included in generally accepted guidelines for doing research. The criticisms of neuromarketing (and our solutions) are based on what are ostensibly obvious ethical constraints that apply to neuromarketing theories rather than on any specific ethical theory [4].

One must take into consideration that some aspects of criticism on neuromarketing are generally applied to standard marketing practices. Most commonly associated fears and concerns connected to the field of neuromarketing are primarily concerned with predicting and potentially influencing consumer choices. Within the next segment, we shall go over the ethical framework of neuromarketing by providing factual aspects which will help us determine the actual current state in this field.

Marketing Ethics

One of the corporate sectors most frequently criticized is marketing. Marketing professionals frequently come under fire for being deceitful, manipulative, invasive, or taking advantage of weaker customer groups such as children, whether it be in the areas of price, advertising, marketing research, or any other [12].

The topic of marketing ethics focuses on how moral principles are applied to marketing decisions, actions, and organizations. Or, to put it another way, marketing ethics is the study of what constitutes ethically acceptable and unacceptable marketing practices. There are two methods to define more exact marketing ethics: A systematic investigation of the morality of marketing decisions, procedures, and institutions, as well as marketing ethics, might be seen as the criteria and norms that are applied when determining if a marketing activity is ethically correct or not [13].

Given that we are concerned in how marketing should be structured and carried out morally, certain norms might be considered the cornerstone of a marketing ethics. A marketing ethics is therefore a normative ethics, as it instructs marketers on how they should behave morally. However, this leaves discussions of marketing ethics that are empirical and analytical (or meta-ethical), which are essential to its normative ethics, without a home. As such, marketing ethics encompasses:

descriptive (or empirical) analyses of the moral principles, outlooks, and marketing tactics,

analytical studies of the types of reasons that may be provided for normative ethical marketing claims and the nature of ethically relevant marketing ideas.

normative analyses of the standards by which marketers ought to operate [12].

A marketer has a responsibility to uphold fundamental ethical standards, such as integrity, humility, and honesty, both internally and externally, in all marketing efforts. This is known as ethical marketing. With time, our economic system has improved in satisfying the public's demands and requirements. Due to this, the market now places a high priority on upholding moral principles while meeting client requirements, mainly because there is a stronger public perception of various services and products when an organization acts ethically. However, they must adhere to certain marketing standards if they want their efforts to be accepted by the broader public [14].

Marketing Research Ethics

Making ethical decisions is difficult in all fields, including marketing research. For the person using the research information, one providing the research information, and the chosen respondents, there are ethical concerns in marketing research. Research providers may engage in unethical general business practices, perform research below professional standards, misuse respondents, and Internet-specific problems such as privacy violations. In order to get low-cost research services, unethical customer conduct includes asking for research ideas with no intention to follow through, promising additional business that never materializes, and inflating study findings. Respondents who give false information or dishonest replies may act unethically too [15]. Today's marketing research sector is governed by internal corporate codes of ethics that many organizations have adopted. One of such is marketing research code of ethics from European Society for Opinion and Market Research (further ESOMAR), a global organization that promotes better research on markets, customers, and society [15]. ESOMAR released the first version of the Code of Marketing and Social Research Practice in 1948. Several codes created by national organizations and the International Chamber of Commerce (further ICC) came after that. A combined ICC/ESOMAR Code was issued in 1977 when the two organizations, ICC and ESOMAR, decided that having a single international code would be preferable than having two different ones. The present edition, with a slightly different title, is the fourth edition of the ICC/ESOMAR Code. It was reviewed and modified in the year 1986 and the year 1994. The key fundamentals of the code are the following [16]:

- Market researchers will conform to all relevant national and international laws.

- Market researchers will behave ethically and will not do anything that might damage the reputation of market research.
- Market researchers will take special care when carrying out research among children and other vulnerable groups of the population.
- Respondents' cooperation is voluntary and must be based on adequate, and not misleading, information about the general purpose and nature of the project when their agreement to participate is being obtained and all such statements must be honoured.
- The rights of respondents as private individuals will be respected by market researchers, and they will not be harmed or disadvantaged as the result of cooperating in a market research project.
- Market researchers will never allow personal data they collect in a market research project to be used for any purpose other than market research.
- Market researchers will ensure that projects and activities are designed, carried out, reported and documented accurately, transparently, objectively, and to appropriate quality.
- Market researchers will conform to the accepted principles of fair competition.

Hence, neuromarketing research, as a sub-branch of marketing research discipline, needs to follow closely marketing research ethical principles.

Ethics in Neuromarketing

The aim of neuromarketing is to learn more about how customers' brains are working when exposed to various market stimuli. This measuring method has the benefit that the information it gathers is unaffected by consumer prejudices and reluctance to tell the truth. Of course, there is nothing wrong with employing technologies in consumer research, but the major issue with those employed in neuromarketing is that they allow researchers to observe outside of the restrictions that research objects may establish for these studies. Because of this, the use of neuromarketing poses ethical issues that may be separated into two groups: the preservation of customers' autonomy and the protection of other parties who can be damaged or exploited by these studies [17].

Public presentation is another important issue. The lack of information about neuromarketing researches creates mistrust and rumours in the society, based mostly on inconsistent, incompetent and insufficient information. However, any new, unknown technology arises more negative rumours together with fear and mistrust than positive ones. With the lack of information, those rumours are not denied and do not disappear [13].

Since the vast majority of market research firms belong to the Neuromarketing Science and Business Association (further NMSBA), the authenticity requirements set forth in this group's ethical code serve as the primary principles for corporate behaviour. The "contribution to the creation and implementation of worldwide norms and standardization in the neuromarketing profession" is one of the NMSBA's primary duties [13].

Three significant challenges are tackled in the neuromarketing industry's code of ethics and those being: (1) to restore the confidence of the public in the legitimacy and integrity of neuromarketers, (2) to ensure neuromarketers protect the privacy of research participants and (3) to protect the buyers of neuromarketing services [18].

The underlying question in the current discussion is whether the new neuromarketing tools will offer enough understanding of how the human brain works to enable brain manipulation so that the consumer cannot detect the deception and that such manipulations result in the desired behaviour in at least some exposed individuals. With current technology, such stealth neuromarketing is not feasible, but if it were to be created, it would be a significant intrusion on personal liberty. In this research, we purposefully take into account a collection of problems that can only be solved by technological advancements that are yet in the future and may never materialize [17].

4 Methodology

This study utilizes critical assessment and thematic analysis approaches to derive conclusions and provide answers to the posed research questions. With the critical assessment approach, we aim to carefully and systematically examine the ethical issues in neuromarketing research and evaluate their strengths and weaknesses. This process involves engaging with different perspectives and viewpoints to gain a comprehensive understanding of the issue.

This assessment is complemented with the thematic analysis, with an aim to identify key themes and patterns related to the potential application of neuromarketing research to benefit the SDGs, in line with ethical principles.

The present study relies entirely on desk research and secondary data sources, which include available documents, published descriptors, and previous research findings. The analysis of these data sources aims to provide a comprehensive overview of the ethics in neuromarketing research and applications to SDGs, taking into account the available information and insights from prior studies [19, 20].

5 Results and Discussion

Neuromarketing has long been subject to canonical complaints, including unethical research methods, unethical technological applications, and consumer manipulations. These concerns date back to the field's beginnings and are still relevant today [4]. Besides the aforementioned concerns, great attention is given to general protection of rights, autonomy and privacy, as well as the protection of niche populations and potential exploitation of the same.

ICC/ESOMAR Code, primarily intended for marketing research in general, and NMSBA Code, intended specifically for neuromarketing, are currently considered to be the most credible and unbiased codes of conduct, both falling in category of ethical theory of rights and justice. Both documents were created without any relation with market research companies and currently are frequently mentioned between companies that do conduct certain neuromarketing research.

The structure of both documents is well established and for these reasons, these two codes of conduct may be highly considered for their usage and implementation within any neuromarketing laboratory. In general, most of previously mentioned concerns are well covered within both codes, especially NMSBA's Code. Participants consents, privacy and rights are covered through articles 5, 6 and 7 where it is clearly stated that one is fully protected as well as one's personal information. Within Article 6: Privacy, it is stated that: "Personal information collected shall be collected for specified Neuromarketing research purposes and not used for any other purpose", and that "Personal information may not be kept longer than is required for the purpose of the neuromarketing project" [18].

Throughout the first four articles the core principles of neuromarketing research as well as the integrity, credibility and transparency are well covered. One great aspect that is concern of many is the question of subject's security, and the same is well defined in this case within the Article 2: Integrity where it is stated that "neuromarketing researchers shall take all reasonable precautions to ensure that participants are in no way harmed or stressed as a result of their involvement in a Neuromarketing research project" [18].

The general concern regarding the niche population is relatively covered to the part where the rights, consent and the general protection of children and young people is covered within the Article 8. It is important to emphasize that an article should be created to cover every aspect of the niche population, where the elderly, people with disabilities and the like will also be taken into account. The aspect of participants free will and rights is defined within the Article 7: Participants rights, where it is clearly stated that one's rights and free will before, during, and after the research shall be respected and that one's private information will indeed be kept private and deleted per request, without any traces [18].

A worrying fact for many is that neuromarketing can potentially start to be used by tobacco companies, alcohol producers and the like which produce harmful and addictive products for the consumer. This matter is rather covered within the Article 1: Core principles rather loosely, where it is stated that "Neuromarketing Researchers shall not act in any way that could negatively impact the reputation and the integrity of the Neuromarketing research profession" [18]. For this, a distinctive article should be created which would firmly state that Neuromarketing research cannot be conducted for companies that are producing addictive or harmful products/content/substances.

The moment of potential consumer manipulation is of great importance to many worried and unaware of the current possibilities of neuromarketing. The tools used within this field are just measurement tools. The important thing to conclude is that such usage of this science is neither currently possible nor ethical for that matter,

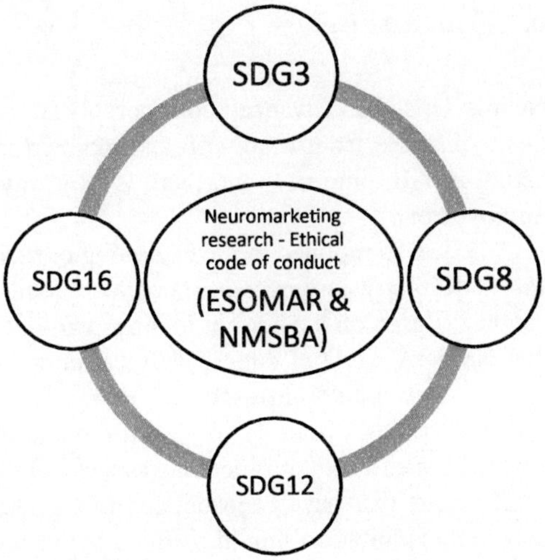

Fig. 1 Framework of the use of ethical neuromarketing research for achieving SDGs

and potentially will not be neither technically achieved in the future for the mere protection of humankind and neuroscience.

When it comes to potential positive implications of neuromarketing research and its correlation to SDGs, we have identified four dominant topics (see Fig. 1 for the illustration of the framework).

Following sustainability topics are ones that can benefit the most from neuromarketing research: (1) SDG 3: "Good Health and Well-being", neuromarketing research can contribute to the development of products and services that promote health and well-being; when it comes to the SDG 8: "Decent Work and Economic Growth", neuromarketing research can contribute to the development of more effective marketing strategies, in particular for internal marketing aspects.

Furthermore, for the SDG 12: "Responsible Consumption and Production", neuromarketing research can help to promote responsible consumption in many ways. In example, by understanding how consumers respond to messages about sustainability or the environmental impact of products, businesses can create campaigns that promote more sustainable consumption. This is applicable to other environmental and social issues, for example: by understanding how consumers perceive and value waste, businesses can develop strategies to encourage waste reduction. Finally, in the context of SDG 16: "Peace, Justice and Strong Institutions", neuromarketing research can contribute to the development of stronger consumer protection policies and regulations. These topics are not exclusive since all SDGs assume ethical considerations and have ethical implications, but we have identified fields that could have the most benefits. However, all these implications assume ethical code of conduct in the neuromarketing research. The proposed framework that provides answers to both research questions of this study is presented in Fig. 1.

6 Conclusion

Neuroscience has advanced considerably in the last two decades, and the achievements possible from it are still an unknown fact to many. The field of neuromarketing is still quite new and there is as always with anything, potential room for improvement.

As it was mentioned in the beginning of this paper, one of the goals of this study was to develop a certain code of conduct based on the current most relevant codes of conduct. Based on the current findings within this field of science, we can conclude that NMSBA Code of ethics, with a support of ICC/ESOMAR Code of ethics, is the one that should currently be followed and implemented by most companies and laboratories, primarily for the aforementioned credibility and unbiased detailed coverage of currently problematic aspects within this field.

The gaps that have been noticed in the aspect of the niche population and its potential exploitation should be filled by a separate article within the code, as well as the potential use of neuromarketing research by companies that produce addictive and harmful products.

Based on the current findings, it can be safely concluded that the main reason for the taboo situation within the field of neuromarketing is the lack of communication between neuromarketing companies and the public. Currently, companies operate more transparent than several years ago, which is a significant improvement in this field. In the future, it is important to convey to consumers the very concept of neuromarketing and that scientists in this field do not operate with mind-controlling devices or magic wands, but with techniques that simply measure person's arousal while being exposed to a certain stimulus and record things that are truisms.

In terms of the implications of the SDG-related framework, in the context of the SDG 3: "Good Health and Well-being", for example, by studying the brain's response to certain stimuli, such as food or advertising, researchers can identify factors that influence healthy lifestyle and behaviour and develop interventions to promote healthier choices. Further, in the SDG 8 "Decent Work and Economic Growth" one avenue can be applying neuromarketing research on workers in order to learn about their job-related traits—to enable employees the main premises of decent work and work-life balance. Furthermore, by developing more effective marketing strategies that can increase sales and revenue for businesses, firms can generate economic growth and hence the creation of new jobs in a long run.

In the SDG 12 "Responsible Consumption and Production": neuromarketing research can help to promote responsible consumption by identifying the factors that influence consumer behaviour and developing strategies to encourage more sustainable choices. One of the directions is further exploring the concept of Consumer Social Responsibility (CnSR). Finally, in terms of the SDG 16 "Peace, Justice and Strong Institutions" policymakers can benefit from the findings of the neuromarketing research.

One of the key limitations of our study is the exclusive use of desk research in generating analysis and discussion. However, while the exclusive use of this data may have limitations, it enables the study to leverage existing knowledge and data in the fields of neuromarketing research, ethics and sustainability. Future research could complement this study by collecting the primary data (qualitative and/or quantitative) and verifying the proposed applicability of neuromarketing research on SDGs.

References

1. Dragolea L, Cotîrlea D (2011) Neuromarketing-between influence and manipulation. Pol J Manage Stud 3(1):79–89
2. Ramsøy T (2020) Neurons Inc. https://youtu.be/FpCkKORmrDA. Last accessed 20 June 2022
3. Reimann M, Zaichowsky J, Neuhaus C, Bender T, Weber B (2010) Aesthetic package design: a behavioral, neural, and psychological investigation. J Consum Psychol 20:431–441
4. Stanton SJ, Sinnott-Armstrong W, Huettel SA (2016) Neuromarketing: ethical implications of its use and potential misuse, Springer
5. Plassmann H, Venkatraman V, Huettel SA, Yoon C (2015) Consumer neuroscience: applications, challenges, and possible solutions. J Mark Res 52(4):427–435
6. Yoon C, Gonzalez R, Bechara A, Berns GS, Dagher AA, Dube L, Huettel SA, Kable JW, Liberzon I, Plassmann H, Smidts A, Spence C (2012) Decision neuroscience and consumer decision making. Mark Lett 23:473–485
7. Boksem MAS, Smidts A (2015) Brain responses to movie trailers predict individual preferences for movies and their population-wide commercial success. J Mark Res 52(4):482–492. https://doi.org/10.1509/jmr.13.0572
8. Smidts A, Hsu M, Sanfey AG, Boksem MAS, Ebstein RB, Huettel SA et al (2014) Advancing consumer neuroscience. Mark Lett 25:257–267
9. Venkatraman V, Dimoka A, Pavlou PA, Vo K, Hampton W, Bollinger B (2015) Predicting advertising success beyond traditional measures: new insights from neurophysiological methods and market response modeling. J Mark Res 52(4):436–452. https://doi.org/10.1509/jmr
10. Zurawicki L (2010) Neuromarketing—exploring the brain of the consumer. Springer, Boston
11. Plassmann H, Ramsoy T, Milosavljevic M (2012) Branding the brain: a critical review and outlook. J Consum Psychol 22:18–36
12. Jamnik A (2011) The question of ethical decision in marketing and ethics. Revista Cult Econ 28(80):41–53
13. Arlauskaitė E, Sferle A (2013) Ethical issues in neuromarketing. Lund University, School of Economics and Management, Lund
14. Shivani (2021) Marketing ethics: meaning, importance, and examples. https://www.invitereferrals.com/blog/marketing-ethics/. Last accessed 23 February 2023
15. Joseph J, Hair F, Celsi MW, Ortinau DJ, Bush RP (2017) Essentials of marketing research, 4th edn. McGraw-Hill Education, New York
16. ICC/ESOMAR (1994) ICC/ESOMAR international code on market and social research (1994). https://iccwbo.org/publication/iccesomar-international-code-on-market-and-social-research/. Last accessed 15 February 2023
17. Murphy ER, Illes J, Reiner PB (2008) Neuroethics of neuromarketing. J Consum Behav 7:293–302
18. NMSBA.: NMSBA Code of Ethics (2013) https://nmsba.com/neuromarketing-companies/code-of-ethics. Last accessed 18 February 2023
19. Krajina A, Husić-Mehmedović M, Koštrebić K (2021) Can you see how it smells? what eye tracking can tell us about the shelf management of luxury perfumes. South East Eur J Econ Bus 16(1):93–106
20. Kolar T, Batagelj Z, Omeragić I, Husić-Mehmedović M (2021) How moment-to-moment EEG measures enhance ad effectiveness evaluation. J Advert Res 61(4). https://doi.org/10.2501/JAR-2021-014

Enhancing Organizational Resilience Through Corporate Social Responsibility: The Case of Firms in Bosnia and Hercegovina During COVID-19 Pandemic

Sabina Đonlagić Alibegović, Vedrana Ajanović, and Lejla Dajdžić

Abstract In December 2019, the WHO declared COVID-19 a global pandemic, which greatly affected the global economy, since it created a crisis that had not been seen before, affecting not only peoples' lives, but also the survival of organizations. National economies and businesses around the world have had to suspend or reduce their operations, reduce the number of employees, reschedule financial obligations, intermitcompletely or slow down the movement of goods and people. However, the crisis caused by COVID-19 can also be assumed as an opportunity for companies to see how resilient and well prepared they are for managing risks and adapting to new circumstances. In times of crisis, the public is much more sensitized to the behavior of companies towards society, which was especially evident when many employers unscrupulously laid off workers, closed plants, increased the prices of products or services, asked for state aid, while at the same time paying large managerial bonuses or dividends to shareholders, which the public strongly condemned. Therefore, in this paper we present the results of research related to the investigation of organizational resilience observed through corporate social responsibility in the period of the COVID-19 pandemic. The aim of this research is to determine the level of social responsibility of companies towards employees in a period of crisis, i.e. during the duration of the global COVID-19 pandemic, and to show what is important for achieving greater resilience of companies in periods of crisis. The research was designed as a survey research that included 241 respondents. The research results indicate that Bosnian companies demonstrated a high level of social responsibility

S. Đonlagić Alibegović
Faculty of Economics, University of Tuzla, Urfeta Vejzagića 8, 75 000 Tuzla, Bosnia and Herzegovina
e-mail: sabina.djonlagic@unitz.ba

V. Ajanović (✉)
BIT Centar Tuzla, Trg Slobode 16, 75 000 Tuzla, Bosnia and Herzegovina
e-mail: vedrana@bit.ba

L. Dajdžić
JU "Centar Za Kulturu" Tuzla, DragišeTrifkovića 2, 75 000 Tuzla, Bosnia and Herzegovina

M. Arslanagić-Kalajdžić et al. (eds.), *Interdisciplinary Advances in Sustainable Development II*, Lecture Notes in Networks and Systems 804,
https://doi.org/10.1007/978-3-031-46692-2_22

towards their employees, understanding the importance of preserving their health and life, as well as the resilience and sustainability of their organization.

Keywords Organizational resilience · Social responsibility · The COVID-19 pandemic

1 Introduction

Economic development increases output of goods and services but it also pollutes the environment and contributes to extreme environmental damage which may even exceed the benefits of these increased outputs. These conditions have put the focus on sustainable development. Climate changes, natural disasters etc. are all consequences of irresponsible business practices and mans' doing. Therefore, in recent years crisis and disaster management has become a key topic for both practitioners and academics. Natural disasters, pandemic diseases, terrorist attacks, economic recession, equipment failure and human error can represent a potentially unpredictable and serious threat to the continuity of an organization's work [1–3]. Therefore, resilience isn't only important to organizations, in fact it matters also for economic development, as well. Resilience of businesses is important for the economic resilience of municipalities or countries. Businesses will need to adapt to the new environment full of challenges and changes and fulfill the needs of their stakeholders in order to have sustainable business futures [4]. Businesses have their own organizational routines, rules and procedures that are predictable and often neglected in stable times. However, companies faced with high uncertainty, such as in times of crisis, are unable to follow and comply with rules, routines and procedures in such moments. Organizational resilience is distinct from adaptability, agility, flexibility, improvisation, recovery, redundancy and robustness. Resilience involves the response of an organization that abolishes at a given moment, emphasizes the ability to recover and develops in a state of uncertainty, discontinuity and urgency. Organizational resilience refers precisely to the company's capability to adapt to the environment in unexpected shocks, crises and uncertainty arising from them [5]. When the COVID-19 pandemic was declared in December 2019, a crisis of the global economy arose that the world has not encountered in modern times and that affected not only people's lives, but also the survival of organizations. All over the world, overnight, companies had to suspend or reduce their operations, reduce the number of employees, reschedule financial obligations, which led to the slowdown of national economies. Seen from the other side, the crisis caused by the COVID-19 pandemic can also be understood as an opportunity for companies to show how resilient and well prepared they are for managing risks and adapting to new circumstances. National Governments around the world, in accordance with the instructions of the World Health Organization, took epidemiological measures that affected many economic sectors and activities, which led to the collapse of many companies and the loss of a large number of jobs. Companies that had the capacity to financially overcome the crisis had to make significant adjustments in their operations, and the issue of social responsibility of those companies came to the fore during this period. Social responsibility in this period was not only related to "doing good", but referred to the overall relationship that companies have towards employees, clients and the environment and represented the way in which companies communicate their principles and values to employees and the public. In times of crisis, as a rule, the public is much more sensitized to the behavior of companies towards society and the community, which was especially evident during the COVID-19 crisis when many employers unscrupulously fired workers, closed plants, increased the prices of products or services, asked for help from the state, and at the same time paid out large managerial bonuses or dividends to shareholders, which was condemned by the public. According to McKinsey (2020) the COVID-19 pandemic has demonstrated the need for all organizations to be able to weather major, unforeseen disruption as some research indicated that 30% of businesses in the UK only had cash for a three month period or that 24% even paused trading.

In order to survive a crisis, organizations need to build resilience that allows them to effectively face uncertainty, overcome the crisis, and stabilize the organization [6]. There is still no consensus among authors about which elements contribute to building that resilience, some authors believe that these are organizational strategies, processes, but also people [7]. While other authors, especially in the context of the COVD—19 pandemic, point out that social responsibility is an important predictor of organizational resilience [8].

2 Organizational Resilience and Social Responsibility

Today's world is characterized by increasing economic, social and environmental challenges, turbulences and uncertainties and organizations struggle to adapt [2]. Organizations which are capable of surviving challenges are referred to as "resilient" and they are essential due to their contribution to the on-going viability of the economy and the wider community [8]. Organizational resilience can be defined as the ability of an organization to anticipate, prepare and respond to change and sudden disruptions in its environment in order to survive and continue its business operations. Organizational resilience has to do with the capability of an organization to bounce back from negative experiences by adapting to the business environment in a flexible manner using its available resources [10]. According to McKinsey resilient organizations don't just bounce back from misfortune, they actually bounce forward and absorb shocks and turn them into opportunities to grow further sustainably. Thus, [11] expressed that resilience is the ability of business enterprise to recover from challenges to move beyond survival to organizational prosperity and success in the face of challenging conditions, in a way that allows the business firm to be more flexible in adapting to future challenges. The sudden outbreak of COVID-19 sparked widespread concern about organizational resilience. [8] During COVID-19 many organizations were forced to revamp their strategies and operations, and employee value propositions due to supply shortages, customers' new needs and employees' changing expectations for the working environment in this time. For those reasons organizational resilience is also related to the capability of the organization to meet the needs of its stakeholders [12]. This is known as corporate social responsibility.

Corporate social responsibility is the obligation of organizations to satisfy the interest of its shareholders and to meet the needs of the society [13]. Corporate social responsibility is therefore linked to the organization's mission, which should answer the question of the company's attitude towards its stakeholders [14]. In order to successfully respond to the challenges, companies should base their actions "on ethical principles and principles of business, ethical attitudes and beliefs. The organizations that succeed in harmonizing and establishing a balance, as well as a kind of symbiosis between the principles of profitability and ethics, will be those organizations that will have a long-term perspective of growth and development." [15].

Organizational resilience is defined in the research literature as a set of capabilities that enable the organization to cope with stressful situations, maintain its competitive position, and even take advantage of potential adverse conditions [16]. Organizational resilience can be developed and managed through "a set of specific organizational capabilities, routines, practices, and processes by which an enterprise conceptually orients itself, acts to move forward, and creates an environment of diversity and adaptive integration" [17]. Although not much time has passed since the onset of the COVID-19 spread, numerous studies have been conducted [18, 19] on the impact of this pandemic on global businesses. At a time when the spread of the disease was still in its early stages, [19] predicted in an article titled "COVID-19 and finance: Agendas for future research" that the financial world will see a major impact of the disease on financial health in the near future. In their research, [20] deeply connected CSR and organizational resilience. In their observations, Simionescu and Dumitrescu deeply connected socially responsible business and organizational resilience [21].

Companies are increasingly aware of the fact that corporate social responsibility is becoming an indispensable part of business strategy, but also a significant investment in safe and sustainable business [22]. In addition, the public is becoming increasingly sensitive to the decisions of business systems and their impact on the quality of life, working conditions and the environment. In fact, companies today have not only an economic function, but equally a social one, which ultimately contributes to the sustainable development of the company itself [23].

In previous research on socially responsible business, questions were most often asked about how a company can achieve sustainable development in the long term, how to improve its image on the market, how to improve business, etc. However, in unknown and crisis situations like the one caused by the COVID-19 pandemic, companies once again found themselves faced with a new challenge that needed to be solved and which will surely be the subject of future research for a long time.

Numerous studies predict that it is precisely those companies that invest in corporate social responsibility, will respond much more easily to new challenges and emerge from the crisis modeled by the pandemic. The results of the research show that the companies that operate as socially responsible, primarily by strengthening ties with stakeholders, such are customers, suppliers and other interested parties, will make easier survival in the future period of crisis [24]. It is interesting that the COVID-19 pandemic had a significant impact on the price movement of shares of companies operating in the same market. [25] states in his research that, for example, the average American manufacturing company recorded a 29% drop in stock prices during the first quarter of 2020. The authors of the research found that socially responsible business increased the resilience of companies to the crisis caused by the COVID-19 virus. Companies with better performances in the field of corporate social responsibility before the pandemic had better stock price results than similar companies in the same industry and market.

Continuous development of relations with interest groups is also important for the successful improvement of corporate social responsibility, especially in times of crisis such is the coronavirus pandemic. Understanding and building trust is crucial for successfully overcoming the negative consequences that have affected the entire world. Also, according to [25] companies can strengthen their ties with interest groups through social responsibility activities, such as creating safe workplaces, using ethical business practices, providing lasting, reliable services to customers and investing in the local environment and community.

This research leads us to conclude that a crisis (as the one during the global COVID-19 pandemic) provides an excellent opportunity for companies to return to the original principles of socially responsible business, in order to respond as successfully, as possible to global social and environmental challenges. Many companies displayed a high level of corporate social responsibility during the pandemic by providing protective equipment and means, some companies donated disinfectant and protective products instead of selling them, some supermarkets set special working hours for the elderly who represented the most vulnerable category, by reorganzing work and prviding work from home and other remote work opportunities etc. [25].

The COVID-19 pandemic had an economic impact on Bosnia and Herzegovina, with the most severely affected service industries, representing industries in which a large number of micro-enterprises and trades operate. The assumption, that the COVID-19 pandemic will cause serious economic consequences, has come true, so many companies' primary goal was to maintain business stability. It is evident that some organizations have been more resilient than others during the pandemic, and some companies were able to develop new products, as a result of the pandemic and thus improve profits and business.

With its emerge, the COVID-19 pandemic created a completely new environment and thus changed not only living conditions, but also businesses. As a result, companies were also exposed to changes, to which they had to successfully respond, in order to continue operations on the market. As a logical response to the emerging crisis, socially responsible business, that develops and changes in accordance with the environment and needs, stands out.

The pandemic greatly affected all aspects of life and society, in the economic, financial, health, ecological and sociological sense, which brought the social responsibility of companies into focus. Companies were forced to adapt to the new situation, which meant a new business strategy and activities aimed at meeting the needs of the company and its stakeholders, but at the same time surviving on the market and achieving business results.

In crisis situations, the expectations of the environment towards companies are higher and in focus when it comes to applying the principles of socially responsible business.

3 Strengthening Organizational Resilience Through Social Responsibility Towards Employees

In crises affecting the society and the economy, one of the ways to react, and a logical answer is corporate social responsibility (CSR) in such a way, that companies show concern for their employees and, therefore, the local community. At the moment when the global pandemic was declared and when the national crisis headquarters adopted measures to fight the virus in order to preserve health, many companies were directly affected by the measures. Taking into account the prescribed curfew and movement restriction measures, some companies have enabled their employees to work from home or have adjusted working hours and shifts. Large and small companies that have a long-standing CSR practices remained faithful to the principles and strategies of CSR even during the coronavirus pandemic. Some decided to apply the CSR concept only with the emerge of the pandemic, but many have lost the battle on the market, as shown by data on layoffs, increased unemployment rates, mostly in developing countries such as Bosnia and Herzegovina. In the paper, Economy in the Post-Panemic Era, [27] states that according to the data of the Labor and Employment Agency of Bosnia and Herzegovina, there were 398,270 registered unemployment in Bosnia and Herzegovina in March 2020, it increased to 419,589 persons in April 2020, and ending with 421,747 unemployed persons in Bosnia and Herzegovina, as of May 31st 2020, according to the data published by the Labour Agency.

The application of socially responsible business is specific for each company separately. There is no universal method of applying socially responsible business because companies are characterized by different circumstances and business conditions. Therefore, every company should devise its own strategy for implementing socially responsible business in accordance with the company's fundamental values, business culture, main business activities, and the environment in which it operates. Socially responsible business is mostly applied in the following areas: market, working environment and employees, local community and environment [28].

The working environment and employees, as an area of social responsibility of the company, includes the adoption of business policies and practices that include the care of employees, the preservation of their health, safety at the workplace and the like, all with the aim of improvement and motivation. It is precisely the company coprorate social resposnbility towards its employees that is the best indicator of the extent to which the company operates in a socially responsible manner.

In order for the company to be responsible towards society, it is necessary to pay attention to the activities carried out within the company. Because in this way the company lays the foundation for responsible operations and making positive impact on the quality of life in the environment.

4 Research Methodology

The focus of this research is to investigate organizational resilience observed through corporate social responsibility for the period of the COVID-19 pandemic and to investigate the behavior of Bosnian companies in the context of corporate social responsibility during the duration of the COVID-19 pandemic. Taking into account the latest trends in the field of perception of the corporate social responsibility of companies towards different categories of stakeholders and the aforementioned, especially during the crisis caused by the COVID-19 pandemic, several research question arise: to what extent did Bosnian companies act socially responsibly and have they displayed a high level of corporate social responsibility and solidarity towards their employees during the COVID-19 pandemic?

Primary research (field research) was conducted using the method of examination, i.e. written examination technique. In doing so, a survey questionnaire was used as an instrument for data collection. The survey questionnaire was created in accordance with the substantive elements of the research subject. The questions in the questionnaire are grouped as follows: (1) basic information about the respondent, (2) social responsibility towards employees, (3) treatment of employees during the COVID-19 pandemic. The survey questionnaire includes questions with a nominal scale, as well as questions with an interval Likert scale, with five degrees of agreement. A total of 350 survey questionnaires were sent via email, out of which, a total of 241 respondents replied, which implies a return rate of approximately 68%. Processing of data collected by survey questionnaires was carried out with the help of appropriate software support, where descriptive statistical analysis, reliability analysis of measurement scales using the Cronbach alpha model, connection analysis using the Pearson correlation coefficient, and simultaneous multiple regression analysis were applied.

5 Research Results

Descriptive Statistical Analysis

The first part of the questionnaire refers to the sociodemographic characteristics of the respondents (employees). According to the results most of the employees are from the healthcare and pharmaceutical sector, 14.5% from the public administration,

13.7% from trade sector, 12.9% from the education section, 9.1% from the banking sector, 7.9% from.

Production sector, 6.6% and the rest of the respondents, in lower percentages, belong to other service industries (telecommunications, tourism and hotel industry, etc.). According to the obtained data, 45.2% of respondents are employed in large companies with over 250 employees, 19.9% in medium-sized companies, and the rest in micro and small companies.

14.5% of respondents belonged to top management, 19.9% to middle management, and 25.3% to operational management, while the remaining employees do not belong to managerial structures in their companies. When we look at the age of the respondents, 40.2% are between 30 and 40 years old, 22.4% are under 30 years old, 28.6% are between 40 and 50 years old, and the remaining 8.8% are respondents over 50 years old.

Analyzing the structure of respondents according to length of service, we can state that the majority of respondents or 40.7% have work experience of less than 5 years, 19.9% experience of up to 10 years, 19.5% experience of 10 to 15 years, 8.7% experience of 15 to 20 years, and 11.2% with more than 20 years of experience.

Given that the functioning of pre-school and school institutions was hampered during the COVID-19 pandemic, respondents were asked whether they have children of that age in the household. Of all the respondents, 25.3% of the respondents have no children, 30.3% have children under 7 years old (kindergarten age), 17% have children from 7 to 15 years old, 12% have children between 15 and 18 years old, and 15.4% children older than 18 years. It is interesting for the research to see how BiH companies behaved towards employees with children during the pandemic period who could not go to preschool or school institutions.

Companies were able to reorganize work in the company and enable work from home, which many companies used both in BiH and around the world. According to the research results, 39.0% of respondents confirmed that they worked from home during the COVID-19 pandemic.

Reliability Analysis of the Corporate Social Responsibility Scale

Cronbach's alpha coefficient was used for the reliability of the corporate social responsibility scale. Cronbach's alpha coefficient for the variable "corporate social responsibility" for all 6 statements is 0.903, which shows good reliability and internal consistency for this sample. This is supported by the value of the "Standardized Cronbach'salpha coefficient" which is 0.903, which additionally confirms the reliability of the scale—socially responsible business (Table 1).

The following table shows the independent variable "corporate social responsibility" with all the elements that this variable includes.

Table 1 Cronbach's alpha coefficient for socially responsible business

Cronbach's alpha coefficient	Standardized Cronbach's alpha coefficient	Total number of statements
0.903	0.903	6

Source Author's calculations based on data from the sources

Table 2 shows the average ratings of all indicators of the independent variable "corporate social responsibility" and statistical parameters: median, mode and standard deviation. It can be seen that the arithmetic mean ranges from 3.01 to 3.85, which implies the conclusion that socially responsible business is recognized and valued. This is supported by the fact that the most common rating is 5 (mod). The value of the median is 4. The table also lists the values of the standard deviation, which measures the dispersion of the sample data. It is evident from the table that the values of standard deviations range between 1.193 and 1.470. Less dispersion of the data always means more representativeness of the mean.

Reliability Analysis of the Scale—Socially Responsible Behavior of Managers Towards Employees

When it comes to the reliability of the scale of the independent variable related to the socially responsible behavior of managers towards employees, it should be noted that this scale is also reliable, considering that the value of the Cronbach's coefficient is 0.755. This is supported by the value of the "Standardized Cronbach's alpha coefficient" which is 0.770, which additionally confirms the reliability of the mentioned scale (Table 3).

The following table shows the variable "socially responsible behavior of managers towards employees" with all the elements that this variable includes.

Table 4 shows the average ratings of all indicators of the independent variable "socially responsible behavior of managers towards employees". Other statistical parameters are also shown: median, mode and standard deviation. The arithmetic mean ranges from 2.12 to 4.04 (on a scale of 1 to 5, 1—do not agree at all, 5—completely agree). The most common rating is 5 (mode). The median value is generally 3.00 or 4.00. The table also lists the values of the standard deviation, which measures the dispersion of the sample data. It is evident from the table that the values of standard deviations range between 1.258 and 1.576. Less dispersion of the data always means more representativeness of the mean.

Table 2 Reliability analysis—descriptive statistics for the corporate social responsibility variable

Indicators	Arithmetic mean	Median	Mod	Standard deviation
The realization that the needs of the public have changed, there are high social expectations and companies benefit from a better local community is characteristic of your company	3.71	4.00	5	1.193
For our company, responsibility towards employees is the most important area of social responsibility	3.59	4.00	5	1.357
The company has flexible policies that enable employees to better coordinate work and private life (e.g. days off, daycare within the company, work from home, flexible working hours, etc.)	3.01	3.00	1	1.470
In my company, all employees feel like members of one big family	3.13	3.00	3	1.374
In my company, they encourage teamwork and cooperation among employees	3.55	4.00	5	1.303
My company regularly communicates with its employees before and during the COVID-19 pandemic	3.85	4.00	5	1.329

Source Author's calculations based on data from the sources

Table 3 Cronbach's coefficient alpha for the socially responsible behavior of managers towards employees

Cronbach'salpha coefficient	Standardized Cronbachalpha coefficient	Total number of statements
0.755	0.770	9

Source Author's calculations based on data from the sources

Reliability Analysis of the Scale of the Level of Corporate Social Responsibility Towards Employees in Bosnian Enterprises During the COVID-19 Pandemic

Cronbach's alpha coefficient was used for the reliability of the scale of the level of corporate social responsibility towards employees in Bosnian enterprises during the COVID-19 pandemic. Cronbach's alpha coefficient for the variable "level of

Table 4 Reliability analysis—descriptive statistics for the socially responsible behavior of managers towards employees variable

Indicators	Arithmetic mean	Median	Mod	Standard deviation
Employees are involved in the decision-making process and are consulted on important issues	2.90	3.00	1	1.459
The company provides employees with training and education, as well as ways to be promoted within the company	3.13	3.00	3	1.419
The company has a clear commitment against discrimination, encourages a multicultural work environment and promotion of women, employment of people with disabilities and the like	3.48	4.00	5	1.426
Employees in my company are accepted, their superiors understand and respect them	3.32	3.00	5	1.363
In my company, employees could choose whether to work from home	2.12	1.00	1	1.411
I can talk to my supervisor to help me if there is a problem at work	3.40	4.00	5	1.576
If I have difficulties while doing my work, I can turn to my work colleagues for help	4.04	5.00	5	1.258
I have the ability to choose how I complete tasks at work	3.71	4.00	5	1.271
I don't feel a sense of belonging to the collective where I work because my hard work and dedication is not recognized and adequately rewarded	2.61	2.00	1	1.516

Source Author's calculations based on data from the sources

corporate social responsibility towards employees in Bosnian enterprises during the COVID-19 pandemic" for all 12 statements is 0.886, which shows good reliability and internal consistency for this sample. This is supported by the value of the "Standardized Cronbach coefficient alpha" which is 0.888, which further confirms the reliability of the scale (Table 5).

The following table shows the dependent variable "level of corporate social responsibility towards employees in Bosnian enterprises during the COVID-19 pandemic" with all the elements that this variable includes.

Table 5 Cronbach'salpha coefficient for the level of corporate social responsibility towards employees in BiH companies during the COVID-19 pandemic

Cronbach's alpha coefficient	Standardized Cronbach alpha coefficient	Total number of statements
0.886	0.888	12

Source Author's calculations based on data from the sources

Table 6 shows the average ratings of all indicators of the dependent variable "level of corporate social responsibility towards employees in Bosnian enterprises during the COVID-19 pandemic". In addition, other statistical parameters are shown: median, mode and standard deviation. It can be seen that the arithmetic mean ranges from 1.69 to 4.08 (on a scale of 1 to 5, 1—I do not agree at all, 5—I completely agree), which implies the conclusion that most companies from the sample reacted and prepared for the COVID-19 pandemic in a timely manner. This is supported by the fact that the most common rating is 5 (mod). The value of the median is 5. The table also lists the values of the standard deviation, which measures the dispersion of the sample data. It is evident from the table that the values of standard deviations range between 1.163 and 1.647. Less dispersion of the data always means more representativeness of the mean.

Analysis of the Relationship Between the Corporate Social Responsibility Dimension and the Level of Corporate Social Responsibility Towards Employees Dimension in BiH Enterprises During the COVID-19 Pandemic

In this part, we will present the results of the analysis of the relation of corporate social responsibility, which includes the dimensions: socially responsible business and socially responsible behavior of managers towards employees, as an independent variable on the one hand, and the dimension of the level of corporate social responsibility towards employees in Bosnian enterprises during the COVID-19 pandemic, as a dependent variable on the other hand (Table 7).

Based on the data in the previous table, we can state the following:

- The research found that the socially responsible business dimension correlates highly at a significance level of 1% with the dimension of corporate social responsibility towards employees in Bosnian enterprises during the COVID-19 pandemic ($r = 0.803$; $p < 0.01$).
- A high correlation at the level of significance of 1% is present between the socially responsible behavior of managers towards employees and the dimension of corporate social responsibility towards employees in Bosnian enterprises during the COVID-19 pandemic ($r = 0.809$; $p < 0.01$).

Table 6 Reliability analysis—descriptive statistics for the variable level of corporate social responsibility towards employees in Bosnian enterprises during the COVID-19 pandemic

Indicators	Arithmetic mean	Median	Mod	Standard deviation
My company takes care of the working conditions of the employees, the equipment of the workplaces, the airiness of the rooms, light, heat and other essential conditions for the performance of work	3.57	4.00	5	1.439
My company responded in a timely manner and adapted all procedures to the COVID-19 pandemic	4.08	5.00	5	1.163
In my company, they were particularly sensitive to sensitive categories (older employees, employees with illnesses, women with children, etc.)	3.65	4.00	5	1.379
The way of performing work and tasks is fully adapted to epidemiological and health requirements (shift work, number of people in the office, use of common rooms, etc.)	3.84	4.00	5	1.269
Employees in my company have at their disposal disinfectants, hygiene products, masks, visors, gloves, suits, etc	4.00	5.00	5	1.302
In my company, everyone is forced to work from home	2.00	1.00	1	1.383
During the pandemic, work from home was mostly represented in the company	2.53	2.00	1	1.547
The job content and job description is adapted for working from home	2.44	2.00	1	1.494
In my company, employees are paid the same as they are for working at the workplace in the company	3.78	5.00	5	1.537
Employees who work from home through different platforms (Skype, Zoom, Hangouts or company platforms) continuously communicate with other colleagues and superiors	3.18	3.00	5	1.612
Flexible working hours have been introduced in my company due to the COVID-19 pandemic	3.00	3.00	1	1.647
My company introduced the possibility of counseling (psychological support) due to increased stress, uncertainty or anxiety	1.69	1.00	1	1.224

Source Author's calculations based on data from the sources

Table 7 The connection between the dimension of CSR and the dimension of the level of CSR towards employees in BiH companies during the COVID-19 pandemic

Level of CSR towards employees in BiH companies during the COVID-19 pandemic		
Corporate social responsibility	Pearson Correlation	0.803*
	Sig. (2-tailed)	0
	N	241
Socially responsible behavior of managers towards employees	Pearson Correlation	0.809*
	Sig. (2-tailed)	0
	N	241

* Correlation is significant at the 0.01 level (2-tailed)
Source Author's calculations based on data from the sources

So, generally speaking, we can observe that the analyzed variables are significantly linearly positively related to each other, that is, through research, we have established statistically significant correlations at the significance level of 1%.

Analysis of the Impact of the CSR Dimension and the CSR Level Towards Employees Dimension in Bosnian Enterprises During the COVID-19 Pandemic

In simultaneous multiple regression analysis, the predictive power of each independent variable is evaluated, i.e. it is measured at what degree it would improve the model consisting of a series of independent variables. Multiple regression actually shows how much of the variance of the dependent variable is explained by the variance of the independent variables. The analysis presented below was aiming to provide an answer to the research question whether *Bosnian companies demonstrated a high level of social responsibility towards their employees during the COVID-19 pandemic.*

The independent variable, which is corporate social responsibility, included the following dimensions: socially responsible business and socially responsible behavior of managers towards employees, while the dependent variable is defined by the dimension of the level of corporate social responsibility towards employees in Bosnian enterprises during the COVID-19 pandemic.

The following table shows the parameters of the simultaneous multiple regression analysis model, where it can be seen that the coefficient of multiple linear correlation is 0.833, which points to the conclusion that there is an extremely strong positive linear relationship between the observed variables (Table 8).

Table 8 The model of simultaneous multiple regression analysis for the dependent variable, the level of CSR towards employees in BiH companies during the COVID-19 pandemic

Model	Multiple linear correlation coefficient	Coefficient of determination	Corrected coefficient of determination	Standard error of estimate
1	0.833[a]	0.693	0.691	6.33353

Source Author's calculations based on data from the sources
[a] Independent variable: corporate social responsibility and socially responsible behavior of managers towards employees

A specific indicator of the representativeness of simultaneous multiple regression is the coefficient of multiple determination. The coefficient of multiple determination shows the percentage of variation in the dependent variable that is explained by the joint influence of the independent variables included in the model. It can take values in the interval [0.1]. The model is more representative if the coefficient is closer to unity. [29]. Based on the parameters in the model, we can conclude that the coefficient of determination is R2 = 0.693, which means that the selected model of simultaneous multiple regression analysis explained 69.3% of all deviations. In fact, this coefficient shows how much of the variance of the dependent variable (in our case, the level of CSR towards employees in Bosnian enterprises during the COVID-19 pandemic) is explained by the model that includes the variables: corporate social responsibility and socially responsible behavior of managers towards employees. Since the coefficient of determination is 0.693, and in accordance with Chaddock scale, we can state that there is a strong connection regarding the direction and intensity of the correlation between the observed variables. The table also shows the size of the corrected coefficient of determination. It is calculated using the coefficient of determination, and in its calculation the number of degrees of freedom, i.e. the sample size, is taken into account. This value is given because the coefficient of determination is an overly optimistic estimate of the true value of the coefficient of determination in the population when calculated on a smaller sample. The corrected coefficient of determination corrects that value and gives a better estimate of the true value of the coefficient of determination.

Analysis of variance determined that the presented model of simultaneous multiple regression analysis is statistically significant, since the value of the empirical level of significance is less than 0.05 ($p \approx 0$); F (2.238) = 269.198. Simply put, the model is suitable for further data processing (Table 9).

In the following presentation, we will try to determine how much each variable in the model contributed to the prediction of the dependent variable based on standardized beta coefficients. Standardized beta coefficients mean that the values of the variables are converted to the same scale so that they can be compared, while unstandardized coefficients are used to set up a multiple regression equation. In our case, we want to compare the contributions of all independent variables, so we used beta coefficients (Table 10).

Table 9 ANOVA[a]—analysis of variance of the model of simultaneous multiple regression analysis

Model	Sum of squares	Number of degrees of freedom	Means of squares	F ratio	Empirical level of significance
Interpreted by the model	21,597.013	2	10,798.507	269.198	0.000[b]
Not interpreted by the model	9,547.028	238	40.114		
Total	31,144.041	240			

Source Author's calculations based on data from the sources
[a] Dependent variable: level of corporate social responsibility towards employees in BiH companies during the COVID-19 pandemic
[b] Independent variable: corporate social responsibility and socially responsible behavior of managers toward employees

Table 10 Results of simultaneous multiple regression analysis - evaluation of independent variables

Model	Non-standardized coefficient		Standardizedcoefficient	t	Sig.	VIF
	B	St. error	Beta			
(Constant)	3.066	1.648		1.860	0.625	
Corporate social responsibility	0.703	0.127	0.407	5.538	0.000	4.190
Socially responsible behavior of managers towards employees	0.698	0.113	0.454	6.176	0.000	4.190

Source Author's calculations based on data from the sources

a. Dependent variable: level of corporate social responsibility towards employees during covid-19.

Based on the previous table, we can determine whether the dimensions of socially responsible business and socially responsible behavior of managers towards employees have a positive impact on the level of CSR towards employees in Bosnian enterprises during the COVID-19 pandemic. Also, in the table we can see how much each variable in the multiple regression model contributed to the prediction of the dependent variable. Therefore, we will use beta coefficients to compare the contributions of all independent variables, that is, to establish their influence on the dependent variable. Based on the presented beta coefficients, we can state that the highest beta coefficient is 0.454, which is actually the value for the variable "socially responsible behavior of managers towards employees". This means that this variable individually contributes the most to explaining the dependent variable "level of CSR towards employees in Bosnian enterprises during the COVID-19 pandemic", and the variable "corporate social responsibility" also has a significant contribution of 0.407. The same conclusion can be made on the basis of significance. Namely, if the value of sig. is less than 0.05, we can state that the independent variable has a significant influence, i.e. if the empirical significance level is greater than 0.05, then the variable has no significant influence. The values of the standard errors are very small for both dimensions and range from 0.113 to 0.127, which implies the conclusion that all the results of the conducted simultaneous multiple regression analysis can be accepted.

In addition to the interpretation of the results of standard beta coefficients and the empirical level of significance, it is necessary to examine whether there is a problem of multicollinearity between the predictor variables? The value of this coefficient in the model is 4.190, which is considered adequate values (VIF<10).

6 Discussion and Conclusion

The focus of this research is to investigate organizational resilience observed through corporate social responsibility during the period of the COVID-19 pandemic and to investigate the behavior of Bosnian and Herzegovinian companies in the context of social responsibility during the COVID-19 pandemic. Based on the presented results, we can state that the dimensions of socially responsible business and socially responsible behavior of managers towards employees have a positive impact on the level of CSR towards employees in Bosnian enterprises during the COVID-19 pandemic. The results of the research idnicate that BH companies showed a high level of social responsibility towards their employees during the COVID-19 pandemic. Considering that Bosnia and Herzegovina is a developing country, with a complex political system and an unfavorable socio-economic climate, the fact that companies showed a high level of CSR towards employees during the pandemic, despite numerous restrictions, is a very positive indicator and can only be a motivation for companies to improve their business in the coming period as well. The necessary improvements mostly relate to the introduction of more flexible policies that would enable employees to better coordinate their business and private lives, such as, for example, days off, daycare within the company, work from home, flexible working hours. In addition, the results indicate that the majority of employees do not participate in the decision-making process, such as, among other things, the decision about working from home during the pandemic. Namely, 60% of respondents had to come to work and were forced to expose themselves to the risk of contracting the virus.

The research presented in this paper led us to conclude that we fulfilled the goal of the research, according to which companies showed a high level of corporate social responsibility towards employees during the COVID-19 pandemic. Based on this fact, we can also conclude that the companies they are employed in have been more resilient in times of the COVID-19 pandemic as they were able to continue with business operations.

Furthermore, based on the experiences during the COVID-19 pandemic companies should pay more attention to Corporate social responsibility towards employees and human resource management and introduce more flexible policies that would enable employees to better coordinate their private and business life, by having days off, daycare within the company and more flexible working hours- the overall well-being of their employees. In addition, the authors are of the opinion that employees need to be more involved in the decision-making process and provide them with greater psychological support, especially in times of crisis such as the COVID-19 pandemic.

One of the important limiting factors of more socially responsible behavior of Bosnian companies and the reduction of organizational resilience in times of crisis, such as the COVID-19 pandemic, is the unfavorable and complex socio-economic and political climate, as well as the absence of legislation and standards relevant to the implementation of the concept of socially responsible business. In developed countries, paying taxes, properly registering employees, regularly paying salaries, as well as cultivating good relations with customers and suppliers are self-evident. In Bosnia and Herzegovina, first of all, it is necessary to affirm the concept of socially responsible business through more intensive and better information to the public in order to increase its "awareness" and better interest in issues from this field of science, especially in the time of the COVID-19 pandemic, because according to researchers' announcements, it is not the last global pandemic of a deadly virus.

References

1. Gebhardt C, Piplas H, Blagojevic M (2020) From science parks to science cities. Organizational and planning perspectives of the new paradigm. Triple Helix: J Univ-Ind-GovMent Innov Entrep
2. Bhamra R, Dani S, Burnard K (2011) Resilience: the concept, a literature review and future direction. Int J Prod Res 49(18):5375–5393
3. Zolli A, Healy AM (2012) Resilience: why things bounce back. Simon and Schuster
4. Burnhard, K, Bhamra R Organisational resilience: development of a conceptual framework for organizational responses. Int J Prod Res, 49(8), pp 5581–5599
5. Xiao L, Cao H (2017) Organizational resilience: The theoretical model and research implication. In: ITM Web of Conferences 12
6. Huang W, Chen S, Nhuyen LT (2017) Corporate social responsibility and organizational resilience to COVID-19 Crisis: An empirical study of Chinese Firms. Sustain 2020, 12(21)
7. Demmer WA, Shawnee KV, Roger C (2011) Engendering resilience in small- and medium-sized enterprises (SMEs): A case study of Demmer Corporation. Int J Prod Res 49:5395–5413
8. Ates A, Umit B (2011) Change process: A key enabler for building resilient SMEs. Int J Prod Res 49:5601–5618
9. Quendler E Organisational Resilience: Building business value in a changing world. J Int Bus Entrep Dev, 10(2):101, https://doi.org/10.1504/JIBED.2017.10005151
10. Grote G (2009) Management of uncertainty: theory and application in the design of systems and organizations, decision engineering. Springer, London
11. Denhardt J, Denhardt R (2010) Building organizational resilience and adaptive management. The Guilford Press, New York
12. Yang H, Tag M, Li P (2021) Organisational resilience in the COVID-a9:A case study from China. J Manag & Organ, 1–19

13. Oga KC, Evwierhumhoma ED (2020) Enhancing organizational resilience through corporate social responsinility. Int J Sci Res Publ, 10 (12)
14. Pearce J, Robinson R (2011) Strategic management, 12th edn. McGraw-Hill, New York
15. Alibegović Đonlagić S, Delić A (2018) Uvod u strateški menadžment, OFF-SET, Tuzla
16. Kotler P, Lee N (2009) Društveno odgovorno poslovanje—Suvremena teorija i najbolja praksa, M.E.P. d.o.o., Zagreb
17. Kantur D, Isery-Say A (2015) Measuring organizational resilience: a scale development. J Bus, Econ Financ 4:456–472
18. Bouaziz F, Hachica ZS (2018) Strategic human resource management practices and organizational resilience, J Manag Dev 37 (10), 2018
19. Goodell JW (2021) Effect of COVID-19 Pandemic on the performance of small and medium business enterprises in Abuja-FCT, Nigeria, Open J Bus Manag 9(5)
20. Rizwan MS, Ahmad G, Ashraf D (2020) Systemic risk: The impact of COVID-19, Financ Res Lett, 36(C)
21. Emamverdy P (2022) Organizational resilience during the corona virus pandemic: The influence of corporate social responsibility strategy and moderating role of organizational citizenship and lnnovative behavior. Int J Recent Res Thesis Diss 3(2):19–22
22. Bahtijarević-Šiber F, Sikavica P, Pološki Vokić N (2008) Suvremeni menadžment, Vještine, sustavi, izazovi, SK Školska knjiga, Zagreb, str. 561
23. Kurtić A (2009) Osnove menadžmenta, OFF-SET štamparija, Tuzla
24. Ding W, Levine R, Lin C, Xie W (2021) Corporate immunity to the COVID -19 pandemic. J Financ Econ 141:802–830
25. Ezgeta M (2021) Istraživanje: društvenoodgovornekompanijelakšećepreživjetiovukrizu, Društvenoodgovornoposlovanje u Hrvatskoj, https://dop.hr/istrazivanje-drustveno-odgovorne-kompanije-lakse-ce-prezivjeti-ovu-krizu/, (pristupljeno: 08.03.2021)
26. He H, Harris L (2020) The impact of Covid-19 pandemic on corporate social responsibility and marketing philosophy. J Bus Res, 116(C), 176–182
27. Kahrović A (2020) Ekonomija u postpandemijskom vremenu, Prilike za razvoj lokalnog poduzetništva
28. Glumac M (2018) Društvena odgovornost kao važan aspekt uspješnog poslovanja, Paragraf 2(1)
29. Fazlović S (2013) Primijenjena statistika, OFF-SET, Tuzla